W9-COZ-408

The IQ Controversy

The IQ Controversy

Critical Readings

Edited by
N. J. Block & Gerald Dworkin

Pantheon Books
A Division of Random House
New York

Library of Congress Cataloging in Publication Data

Main entry under title:
The IQ Controversy
Includes index.
1. Intellect. 2. Intelligence tests. 3. Nature and nurture. I. Block, Ned Joel, 1942– II. Dworkin, Gerald, 1937–
BF431.I2 1976 153.9 75-38113
ISBN 0-394-49056-8
ISBN 0-394-73087-9 pbk.

Since this copyright page cannot accommodate all acknowledgments, they are to be found on the following pages.

Manufactured in the United States of America

Grateful acknowledgment is made to the following for permission to reprint previously published material:

American Association for the Advancement of Science and Sandra Scarr-Salapatek: "Unknowns in the IQ Equation," by Sandra Scarr-Salapatek, from *Science*, vol. 174, pp. 1223–28, December 17, 1971. Copyright © 1971 by the American Association for the Advancement of Science.

American Psychological Association and David C. McClelland: "Testing for Competence Rather Than for 'Intelligence,'" by David C. McClelland, from *American Psychologist*, vol. 29, p. 107, January 1973. Copyright © 1973 by the American Psychological Association.

Brunner/Mazel, Inc.: "Genetics and Educability: Educational Implications of the National Debate," by Carl Bereiter, from *Disadvantaged Child*, vol. 3, edited by J. Hellmuth, 1970.

Educational Theory and Clarence Karier: "Testing for Order and Control in the Corporate Liberal State," by Clarence Karier, from *Educational Theory*, vol. 22, no. 2, pp. 154–80, Spring 1972.

Lawrence Erlbaum Associates, Inc., Publishers, and Leon Kamin: "Heredity, Intelligence, Politics, and Psychology, I and II," by Leon Kamin, to be included in the forthcoming book *Tutorial Essays in Psychology*, edited by N. S. Sutherland.

Arthur S. Goldberger: "Mysteries of the Meritocracy," by Arthur S. Goldberger.

Grune & Stratton, Inc. and Jerry Hirsch: "Behavior-Genetic Analysis and Its Biosocial Consequence," by Jerry Hirsch, from *Seminars in Psychiatry*, no. 2, pp. 89–105, 1970.

Harper's Magazine: "Five Myths About Your IQ," by Christopher Jencks and Mary Jo Bane, from *Harper's*, February 1973. Copyright © 1973 by Harper's Magazine.

Journal of Biosocial Science: "Limitations To Genetic Comparison of Populations," by J. M. Thoday, from Supplement 1 to the *Journal of Biosocial Science*, January 1969. Copyright © 1969 by the Galton Foundation.

Little, Brown and Company, in association with The Atlantic Monthly Press: excerpt from *I.Q. in the Meritocracy*, by Richard J. Herrnstein. Copyright © 1971, 1973 by Richard J. Herrnstein.

Mouton Publishers: "The Fallacy of Richard Herrnstein's I.Q.," by Noam Chomsky, appeared as a part of "Psychology and Ideology," from *Cognition*, vol. 1, no. 1, 1973. "Comments on Herrnstein's Response" by Noam Chomsky from *Cognition*, vol. 1, no. 3, 1972. "Science or Superstition? A Physical Scientist Looks at the I.Q. Controversy," by David Layzer, from *Cognition*, vol. 1, no. 2, 1973. "Whatever Happened to Vaudeville? A Reply to Professor Chomsky," by Richard Herrnstein, from *Cognition*, vol. 1, no. 1, 1973.

Nature: "Educability and Group Differences," by J. M. Thoday, from *Nature*, vol. 245, October 26, 1973.

Princeton University Press: "I.Q., Heritability, and Inequality," by N. J. Block and Gerald Dworkin, from *Philosophy and Public Affairs*, vol. 3, no. 4 and vol. 4, no. 1, pp. 331–409 (part 1) and pages 40–99 (part 2). Copyright © 1974 by Princeton University Press.

Science and Public Affairs: "Race and Intelligence," by Richard Lewontin, from *Bulletin of the Atomic Scientists,* pp. 2–8, March 1970. "Further Remarks on Race and Intelligence," by Richard Lewontin, from *Bulletin of the Atomic Scientists,* pp. 23–25, May 1970. Both articles Copyright © 1970 by the Educational Foundation for Nuclear Science. "Race and Gentics of Intelligence: A Reply to Lewontin," by Arthur R. Jensen, from *Bulletin of the Atomic Scientists,* May 1970. Copyright © 1970 by the Educational Foundation for Nuclear Science.

The University of Chicago Press and Richard Lewontin: "The Analysis of Variance and the Analysis of Cause," by Richard Lewontin, from *American Journal of Human Genetics,* vol. 26, pp. 400–411, 1974. Copyright © 1974 by The University of Chicago Press.

Contents

Introduction

THERE IS PERHAPS NO ISSUE IN THE HISTORY OF SCIENCE THAT PRESENTS such a complex mingling of conceptual, methodological, psychological, ethical, political, and sociological questions as the controversy over whether intelligence has a substantial genetic component. The discussion and debate, which may be dated from the publication of Galton's *Hereditary Genius* in 1869, was brought most prominently to public attention by the publication one hundred years later of Arthur Jensen's article "How Much Can We Boost IQ and Scholastic Achievement?" in the *Harvard Educational Review*. By 1972, Jensen estimated that over 120 articles had been "provoked" by his article alone. Among the articles that have called attention to social and political implications of the controversy, the most prominent is Richard Herrnstein's "IQ," published in the *Atlantic Monthly* in the fall of 1971. Since then, Herrnstein has expanded his article into a book, *IQ in the Meritocracy*, and Jensen has published two books, *Genetics and Education* and *Educability and Group Differences*.

The title of this reader is accurate. We do not attempt to present "all sides of the issue." We have brought together the best of the critical literature. These include articles directed specifically at the Herrnstein and Jensen pieces as well as readings dealing with the issues raised by those articles from a more general—but critical—historical, sociological, or biological perspective. The reader who wishes to pursue the controversy further may consult the notes and reference lists that accompany many of the selections.

The problems dealt with in these readings are so rich in complexity and interest that it would theoretically be possible to build an entire education around them. In crudest summary form, the following questions would arise: What is the nature of science? What is measurement? What is an explanation? What are theory, hypothesis, inference? What is an experiment? How do we distinguish causes from correlations? What is the role of mathematics in science? What do we know about human nature? What features of a person can be changed? What are intelligence, ability, learning, motivation? How can we best educate people? How should we take into account differences between people in organizing a society? How have the findings of the sciences been used in the past to affect policy? What is the relationship of expertise to politics? What are the ethics of research, of publishing, of journalism? The great diversity in these issues is reflected in the disciplines of our contributors—psychology, journalism, genetics, astronomy, linguistics, economics, history, education, sociology, and philosophy.

The division of the readings into sections is, as is usually the case with such divisions, somewhat arbitrary. Some of the readings could with equal justice appear in other categories. However, the headings are sufficiently suggestive and coherent that they serve their organizational purpose. We begin, in Part I, with a retrospective glance at the debate over IQ testing in the 1920s and conclude with a recent article on the same subject. In Part II, we consider the controversy concerning the evidence for a genetic component in IQ differences. In Part III, we consider the implications of this possible genetic component for various issues of social and educational policy. Finally, in Part IV, the editors exercise their prerogative of having the last word—in this case, many.

None of the readings are technical in the sense of requiring mathematical techniques or information not contained in the articles themselves. A few technical terms are explained in the Appendix. We hope that persons other than those already in the relevant disciplines will take the opportunity to become acquainted with the controversy dealt with in this reader. For this controversy is a concrete example of an issue that is becoming more acute with the rapid advance of science and technology—the use and abuse of science in the formation of public opinion and the promotion of public policy. A crucial task facing the educational system is that of teaching people to understand and assess the findings of scientists and their policy implications. A healthy and democratic society must bear in mind that although not all can make social policy, all may judge it.

PART I

IQ

Introduction

The first major American IQ test, the Stanford-Binet, was introduced by Lewis M. Terman in 1916. An adaptation of it was put together, in a period of a few weeks, for the army in World War I; after the war, the use of such tests spread rapidly. An enormous number of books and articles appeared, based largely on Yerkes' report on the army test data. It was claimed that the average American has a mental age of 14,[1] and thus that democracy cannot work,[2] and even that efforts to improve standards of living, health, and education are folly.[3] A number of books[4] discussed the fact that recent Polish, Russian, Jewish, and Italian immigrants scored well below persons who had immigrated earlier from England and Western Europe, arguing that the former groups were genetically inferior. The alternative hypothesis, that the latter groups were more assimilated into the dominant culture, was rejected or ignored. Such claims, in turn, were used to bolster arguments that immigration from Eastern and Southern Europe should be restricted. In the period 1922–1923 an enormous controversy[5]—of which Lippmann's essays are an important part—flared up over these claims. But the controversy died down as quickly as it flared up; the day of the pre-eminence of IQ tests was at hand. Today, IQ tests are not essentially different, and neither are many of the objections of their critics.

Both Walter Lippmann's (1922) and David C. McClelland's (1973) essays were provoked by popular expositions of claims about innate intelligence based on IQ test data. Both argue that there is no scientific basis for the claim that IQ tests measure intelligence: that IQ tests consist of arbitrary stunts; that emotional responses to the tests influence scores; that correlations with school success are nonvalidatory because school success itself is not a reliable index of intelligence; and that even if the tests measure ability, they probably measure a rather narrow kind of scholastic ability. Neither is opposed to the use of the tests per se, but rather to abuses of the tests that involve treating them as if they were measures of fixed innate intelligence. Finally, both suggest that the practical problems of selecting people for tasks or programs would be better accomplished by tests of competence in the area involved, rather than by attempting to measure "intelligence."

"The Great Conspiracy" is Lewis Terman's reply to a few of Lippmann's points and "The Great Confusion" is Lippmann's rejoinder. We also include a letter from Terman to *The New Republic,* along with Lippmann's answer.

Notes

1. R. M. Yerkes, *Psychological Examining in the United States Army,* vol. 15, *Memoirs of the National Academy of Sciences* (Washington, D.C., 1921).
2. P. Popenoe, "Measuring Human Intelligence," *Journal of Heredity* 12 (May 1921):231–236; G. B. Cutten, "The Reconstruction of Democracy," *School and Society* 16 (October 1922):477–489.
3. A. Wiggam, "The New Decalogue of Science," *Century Magazine* 103 (March 1922):643–650.
4. C. W. Gould, *America, A Family Matter* (New York: Charles Scribner's Sons, 1922); M. Grant, *The Passing of the Great Race* (New York: Charles Scribner's Sons, 1922); T. L. Stoddard, *The Revolt Against Civilization* (New York: Macmillan Co., 1922); C. C. Brigham, *A Study of American Intelligence* (Princeton, N.J.: Princeton University Press, 1923).
5. *See* L. J. Cronbach, "Five Decades of Public Controversy over Mental Testing," *American Psychologist* 30, no. 1 (January, 1975): 1–14, for an account of this dispute.

The Lippmann-Terman Debate

The Mental Age of Americans

Walter Lippmann

A STARTLING BIT OF NEWS HAS RECENTLY BEEN UNEARTHED AND IS NOW being retailed by the credulous to the gullible. "The *average* mental age of Americans," says Mr. Lothrop Stoddard in *The Revolt Against Civilization,* "is only about fourteen."

Mr. Stoddard did not invent this astonishing conclusion. He found it ready-made in the writings of a number of other writers. They in their turn got the conclusion by misreading the data collected in the army intelligence tests. For the data themselves lead to no such conclusion. It is impossible that they should. It is quite impossible for honest statistics to show that the average adult intelligence of a representative sample of the nation is that of an immature child in that same nation. The average adult intelligence cannot be less than the average adult intelligence, and to anyone who knows what the words "mental age" mean, Mr. Stoddard's remark is precisely as silly as if he had written that the average mile was three-quarters of a mile long.

The trouble is that Mr. Stoddard uses the words "mental age" without explaining either to himself or to his readers how the conception of "mental age" is derived. He was in such an enormous hurry to predict the downfall of civilization that he could not pause long enough to straighten out a few simple ideas. The result is that he snatches at a few scarifying statistics and uses them as a base upon which to erect a glittering tower of generalities. For the statement that the average mental age of Americans is only about fourteen is not inaccurate. It is not incorrect. It is nonsense.

Mental age is a yardstick invented by a school of psychologists to measure "intelligence." It is not easy, however, to make a measure of intelligence and the psychologists have never agreed on a definition. This quandary presented itself to Alfred Binet. For years he had tried to reach a definition of intelligence and always he had failed. Finally he gave up the attempt, and started on another tack. He then turned his attention to the practical problem of distinguishing the "backward" child from the "normal" child in the Paris schools. To do this he had to know what was a normal child. Difficult as this promised to be, it was a good deal easier than the attempt to define intelligence. For Binet concluded, quite logically, that the standard of a normal child of any particular age was something or other which an arbitrary percentage of children of that age could do. Binet therefore decided to consider "normal" those abilities which were common to between 65 and 75 percent of the children of a particular age. In deciding on these percentages, he thus decided to consider at least 25 percent of the children as backward. He might just as easily have fixed a percentage which would have classified 10 percent of the children as backward, or 50 percent.

Having fixed a percentage which he would henceforth regard as "normal," he devoted himself to collecting questions, stunts, and puzzles of various sorts, hard ones and easy ones. At the end he settled upon fifty-four tests, each of which he guessed and hoped would test some element of intelligence; all of which together would test intelligence as a whole. Binet then gave these tests in Paris to 200 school children who ranged from three to fifteen years of age. Whenever he found a test that about 65 percent of the children of the same age could pass he called that a Binet test of intelligence for that age. Thus a mental age of seven years was the ability to do all the tests which 65 to 75 percent of a small group of seven-year-old Paris schoolchildren had shown themselves able to do.

This was a promising method, but of course the actual tests rested on a very weak foundation indeed. Binet himself died before he could carry his idea much further, and the task of revision and improvement was then transferred to Stanford University. The Binet scale worked badly in California. The same puzzles did not give the same results in California as in Paris. So about 1910, Professor L. M. Terman undertook to revise them. He followed Binet's method. Like Binet he would guess at a stunt which might indicate intelligence, and then try it out on about 2,300 people of various ages, including 1,700 children "in a community of average social status." By editing, rearranging, and supplementing the original Binet tests he finally worked out a series of tests for each age which the average child of that age in about one hundred Californian children could pass.

The puzzles which this average child among one hundred Californian children of the same age about the year 1913 could answer are the yardstick by which mental age is measured in what is known as the Stanford Revision of the Binet-Simon scale. Each correct answer gives a credit of two months' mental age. So if a child of seven can answer all tests up to the seven-year-old tests perfectly, and cannot answer any of the eight-year-old tests, his total score is seven years. He is said to test "at age," and his "intelligence quotient" or "IQ" is unity or 100 percent. Anybody's IQ can be figured, therefore, by dividing his mental age by his actual age. A child of five who tests at four years' mental age has an IQ of 80 (4/5 = .80). A child of five who tests at six years' mental age has an IQ of 120 (6/5 = 1.20).

The aspect of all this which matters is that mental age is simply the average performance with certain rather arbitrary problems. The thing to keep in mind is that all the talk about "a mental age of fourteen" goes back to the performance of eighty-two California school children in 1913–1914. Their success and failures on the days they happened to be tested have become embalmed and consecrated as the measure of human intelligence. By means of that measure writers like Mr. Stoddard fix the relative values of all the peoples of the earth and of all social classes within the nations. They don't know they are doing this, however, because Mr. Stoddard at least is quite plainly taking everything at second hand.

However, I am willing for just a moment to grant that Mr. Terman in California has worked out a test for the different ages of a growing child. But I insist that anyone who uses the words "mental age" should remember that Mr. Terman reached his test by seeing what the average child of an age group could do. If his group is too small or is untypical, his test is in the same measure inaccurate.

Remembering this, we come to the army tests. Here we are dealing at once with men all of whom are over the age of the mental scale. For the Stanford-Binet scale ends at "sixteen years." It assumes that intelligence stops developing at sixteen, and everybody sixteen and over is therefore treated as "adult" or as "superior adult." Now the adult Stanford-Binet tests were "standardized chiefly on the basis of results from 400 adults [Terman, p. 13] of moderate success and of very limited educational advantages" and also thirty-two high school pupils from sixteen to twenty years of age. Among these adults, those who tested close together have the honor of being considered the standard of average adult intelligence.

Before the army tests came along, when anyone talked about the average adult he was talking about a few hundred Californians. The army tested about 1,700,000 adult men. But it did not use the Binet system of scoring by mental ages. It scored by a system of points which we need not stop to describe. Naturally enough, everyone interested in mental testing wanted to know whether the army tests agreed in any way with the Stanford-Binet mental-age standard. So by another process, which need also not be described, the results of the army tests were translated into Binet terms. The result of this translation is the table which has so badly misled poor Mr. Stoddard. This table showed that the average of the army did not agree at all with the average of Mr. Terman's Californians. There were then two things to do. One was to say that the average intelligence of 1,700,000 men was a more representative average than that of 400 men. The other was to pin your faith to the 400 men and insist they gave the true average.

Mr. Stoddard chose the average of 400 rather than the average of 1,700,000 because he was in such haste to write his own book that he never reached page 785 of *Psychological Examining in the United States Army*, the volume of the data edited by Major Yerkes.* He would have found there a clear warning against the

* "For norms of adult intelligence, the results of the Army examinations are undoubtedly the most representative. It is customary to say that the mental age of the average adult is about sixteen years. This figure is based, however, upon examinations of only 62 persons. . . . This group is too small to give very reliable results and is furthermore probably not typical." *Psychological Examining in the United States Army*, p. 785.

The reader will note that Major Yerkes and his colleagues assert that the Stanford standard of adult intelligence is based on only sixty-two cases. This is a reference to page 49 of Mr. Terman's book on the Stanford Revision of the blunder he was about to commit, the blunder of treating the average of a small number of instances as more valid than the average of a large number.

But instead of pausing to realize that the army tests had knocked the Stanford-Binet measure of adult intelligence into a cocked hat, he wrote his book in the belief that the Stanford measure is as good as it ever was. This is not intelligent. It leads one to suspect that Mr. Stoddard is a propagandist with a tendency to put truth not in the first place but in the second. It leads one to suspect, after such a beginning, that the real promise and value of the investigation which Binet started is in danger of gross perversion by muddle-headed and prejudiced men.

The Mystery of the "A" Men

Walter Lippmann

BECAUSE THE RESULTS ARE EXPRESSED IN NUMBERS, IT IS EASY TO MAKE the mistake of thinking that the intelligence test is a measure like a foot rule or a pair of scales. It is, of course, a quite different sort of measure. For length and weight are qualities which men have learned how to isolate no matter whether they are found in an army of soldiers, a heap of bricks, or a collection of chlorine mole-

---

Binet-Simon Scale. But page 13 of the same book speaks of 400 adults being the basis on which the adult tests were standardized. I have used this larger figure because it is more favorable to the Stanford-Binet scale.

It should also be remarked that the army figures are not the absolute figures but the results of a "sample of the white draft" consisting of nearly 100,000 recruits. In strictest accuracy, we ought to say then that the disagreement between army and Stanford-Binet results derives from conclusions drawn from 100,000 cases as against 400.

If these 100,000 recruits are not a fair sample of the nation, as they probably are not, then in addition to saying that the army tests contradict the Stanford-Binet Scale, we ought to add that the army tests are themselves no reliable basis for measuring the average American mentality.

cules. Provided the foot rule and the scales agree with the arbitrarily accepted standard foot and standard pound in the Bureau of Standards at Washington they can be used with confidence. But intelligence is not an abstraction like length and weight; it is an exceedingly complicated notion which nobody has as yet succeeded in defining.

When we measure the weight of a schoolchild we mean a very definite thing. We mean that if you put the child on one side of an evenly balanced scale, you will have to put a certain number of standard pounds in the other scale in order to cancel the pull of the child's body toward the center of the earth. But when you come to measure intelligence you have nothing like this to guide you. You know in a general way that intelligence is the capacity to deal successfully with the problems that confront human beings, but if you try to say what those problems are, or what you mean by "dealing" with them, or by "success," you will soon lose yourself in a fog of controversy. This fundamental difficulty confronts the intelligence tester at all times. The way in which he deals with it is the most important thing to understand about the intelligence test, for otherwise you are certain to misinterpret the results.

The intelligence tester starts with no clear idea of what intelligence means. He then proceeds by drawing upon his common sense and experience to imagine the different kinds of problems men face which might in a general way be said to call for the exercise of intelligence. But these problems are much too complicated and too vague to be reproduced in the classroom. The intelligence tester cannot confront each child with the thousand and one situations arising in a home, a workshop, a farm, an office or in politics, that call for the exercise of those capacities which in a summary fashion we call intelligence. He proceeds, therefore, to guess at the more abstract mental abilities which come into play again and again. By this rough process the intelligence tester gradually makes up his mind that situations in real life call for memory, definition, ingenuity, and so on. He then invents puzzles which can be employed quickly and with little apparatus, that will, according to his best guess, test memory, ingenuity, definition, and the rest. He gives these puzzles to a mixed group of children and sees how children of different ages answer them. Whenever he finds a puzzle that, say, 60 percent of twelve-year-old children can do, and 20 percent of the eleven-year-olds, he adopts that test for the twelve-year-olds. By a great deal of fitting, he gradually works out a series of problems for each age group which 60 percent of his children can pass, 20 percent cannot pass and, say, 20 percent of the children one year younger can also pass. By this method he has arrived, under the Stanford-Binet system, at a conclusion of this sort: Sixty percent of children twelve years old should be able to define three out of the five words: pity, revenge, charity, envy, justice. According to Professor Terman's instructions, a child passes this test if he says that "pity" is "to be sorry for someone"; the child fails if he says "to help" or "mercy." A correct definition of "justice" is as follows: "It's what you get when you go to court"; an incorrect definition is "to be honest."

A mental test, then, is established in this way: The tester himself guesses at a large number of tests which he hopes and believes are tests of intelligence. Among these tests, those finally are adopted by him which 60 percent of the children under his observation can pass. The children whom the tester is studying select his tests.

There are, consequently, two uncertain elements. The first is whether the tests really test intelligence. The second is whether the children under observation are a large enough group to be typical. The answer to the first question—whether the tests are tests of intelligence—can be determined only by seeing whether the results agree with other tests of intelligence, whatever they may be. The answer to the second question can be had only by making a very much larger number of observations than have yet been made. We know that the largest test made, the army examinations, showed enormous error in the Stanford test of adult intelligence. These elements of doubt are, I think, radical enough to prohibit anyone from using the results of these tests for large generalization about the quality of human beings. For when people generalize about the quality of human beings, they assume an objective criterion of quality, and for testing intelligence there is no such criterion. These puzzles may test intelligence, and they may not. They may test an aspect of intelligence. Nobody knows.

What then do the tests accomplish? I think we can answer this question best by starting with an illustration. Suppose you wished to judge all the pebbles in a large pile of gravel for the purpose of separating them into three piles, the first to contain the extraordinary pebbles, the second the normal pebbles, the third the insignificant pebbles. You have no scales. You first separate from the pile a much smaller pile and pick out one pebble which you guess is the average. You hold it in your left hand and pick up another pebble in your right hand. The right pebble feels heavier. You pick up another pebble. It feels lighter. You pick up a third. It feels still lighter. A fourth feels heavier than the first. By this method, you can arrange all the pebbles from the smaller pile in a series running from the lightest to the heaviest. You thereupon call the middle pebble the standard pebble, and with it as a measure you determine whether any pebble in the larger pile is a subnormal, a normal, or a supernormal pebble.

This is just about what the intelligence test does. It does not weigh or measure intelligence by any objective standard. It simply arranges a group of people in a series from best to worst by balancing their capacity to do certain arbitrarily selected puzzles, against the capacity of all the others. The intelligence test, in other words, is fundamentally an instrument for classifying a group of people. It may also be an instrument for measuring their intelligence, but of that we cannot be at all sure unless we believe that M. Binet and Mr. Terman and a few other psychologists have guessed correctly when they invented their tests. They may have guessed correctly but, as we shall see later, the proof is not yet at hand.

The intelligence test, then, is an instrument for classifying a group of people, rather than "a measure of intelligence." People are classified within a group according to their success in solving problems which may or may not be tests of intelligence. They are classified, according to the performance of some Californians in the years 1910 to about 1916, with Mr. Terman's notion of the problems that reveal intelligence. They are not classified according to their ability in dealing with the problems of real life that call for intelligence.

With this in mind, let us look at the army results as they are dished up by writers like Mr. Lothrop Stoddard and Professor McDougall of Harvard. The following table is given:

4½%	of the army were	A	men
9 %	of the army were	B	men
16½%	of the army were	C+	men
25 %	of the army were	C	men
20 %	of the army were	C−	men
15 %	of the army were	D	men
10 %	of the army were	D−	men

But how, you ask, did the army determine the qualities of an "A" man? For an "A" man is supposed to have "very superior intelligence," and of course, mankind has wondered for at least two thousand years what were the earmarks of very superior intelligence. McDougall and Stoddard are quite content to take the army's word for it, or at least they never stop to explain, before they exploit the figures, what the army meant by "very superior intelligence." The army, of course, had no intention whatever of committing itself to a definition of very superior intelligence. The army was interested in classifying recruits. It therefore asked a committee of psychologists to assemble from all the different systems, Binet and otherwise, a series of tests. The committee took this series and tried it out in a few camps. They timed the tests. "The number of items and the time limits were so fixed that five percent or less in any average group would be able to finish the entire series of items in the time allowed."* It is not surprising, therefore, that five percent or less (4½ percent actually) of the army made a top score. It is not surprising that tests devised to pass 5 percent or less "A" men should have passed 4½ percent "A" men.

The army was quite justified in doing this because it was in a hurry and was looking for about 5 percent of the recruits to put into officers' training camps. I quarrel only with the Stoddards and McDougalls who solemnly talk about the 4½ percent of "A" men in the American nation without understanding how these 4½ percent were picked. They do not seem to realize that if the army had wanted half the number of officers, it could by shortening the time have made the scarcity of "A" men seem even more alarming. If the army had wanted to double the "A" men, it could have done that by lengthening the time. Somewhere, of course, in the whole group would have been found men who could not have answered all the questions correctly in any length of time. But we do not know how many men of the kind there were because the tests were never made that way.†

The army was interested in discovering officers and in eliminating the feeble-minded. It had no time to waste, and so it adopted a rough test which would give a quick classification. In that, it suc-

* Yoakum and Yerkes, *Army Mental Tests,* p. 3.

† *Psychological Examining in the United States Army,* p. 419. "The high frequencies of persons gaining at the upper levels (often 100%) indicate for the people making high scores on single time the 'speed' element is predominant."

ceeded on the whole very well. But the army did not measure the intelligence of the American nation, and only very loose-minded writers imagine that it did. When men write as Mr. Stoddard does that "only four and a half millions [of the whole population] can be considered 'talented,'" the only possible comment is that the statement has no foundation whatsoever. We do not know how many talented people there are: first, because we have no measure of talent, and second, because we have never made the attempt to devise one or apply one. But when we see how men like Stoddard and McDougall have exploited the army tests, we realize how necessary, but how unheeded, is the warning of Messrs. Yoakum and Yerkes that "the ease with which the army group test can be given and scored makes it a dangerous method in the hands of the inexpert. It was not prepared for civilian use, and is applicable only within certain limits to other uses than that for which it was prepared."

The Reliability of Intelligence Tests

Walter Lippmann

SUPPOSE, FOR EXAMPLE, THAT OUR AIM WAS TO TEST ATHLETIC RATHER than intellectual ability. We appoint a committee consisting of Walter Camp, Percy Haughton, Tex Rickard, and Bernard Darwin; and we tell them to work out tests which will take no longer than an hour and can be given to large numbers of men at once. These tests are to measure the true athletic capacity of all men anywhere for the whole of their athletic careers. The order would be a large one, but it would certainly be no larger than the pretensions of many well known intelligence testers.

Our committee of athletic testers scratch their heads. What shall be the hour's test, they wonder, which will "measure" the athletic "capacity" of Dempsey, Tilden, Sweetser, Siki, Suzanne Lenglen, and Babe Ruth; of all sprinters, Marathon runners, broad jumpers, high divers, wrestlers, billiard players, marksmen, cricketers, and pogo bouncers? The committee has courage. After much guessing and some experimenting, the committee works out a sort of condensed Olympic games which can be held in any empty lot. These games consist of a short sprint, one or two jumps, throwing a ball at a bull's eye, hitting a punching machine, tackling a dummy, and a short game of clock golf. They try out these tests on a mixed assortment of champions and duffers and find that on the whole the champions do all the tests better than the duffers. They score the result and compute statistically what is the average score for all the tests. This average score then constitutes normal athletic ability.

Now it is clear that such tests might really give some clue to athletic ability. But the fact that in any large group of people 60 percent made an average score would be no proof that you had actually tested their athletic ability. To prove that, you would have to show that success in the athletic tests correlated closely with success in athletics. The same conclusion applies to the intelligence tests. Their statistical uniformity is one thing; their reliability another. The tests might be a fair guess at intelligence, but the statistical result does not show whether they are or not. You could get a statistical curve very much like the curve of "intelligence" distribution if, instead of giving each child from ten to thirty problems to do, you had flipped a coin the same number of times for each child and had credited him with the heads. I do not mean, of course, that the results are as chancy as all that. They are not, as we shall soon see. But I do mean that there is no evidence for the reliability of the tests as tests of intelligence in the claim, made by Terman,* that the distribution of intelligence quotients corresponds closely to "the theoretical normal curve of distribution (the Gaussian curve)." He would, in a large enough number of cases, get an even more perfect curve if these tests were tests not of intelligence but of the flip of a coin.

Such a statistical check has its uses of course. It tends to show, for example, that in a large group, the bias and errors of the tester have been canceled out. It tends to show that the gross result is reached in the mass by statistically impartial methods, however wrong the judgment about any particular child may be. But the fairness in giving the tests and the reliability of the tests themselves must not

* *Stanford Revision of Binet-Simon Scale*, p. 42.

be confused. The tests may be quite fair applied in the mass, and yet be poor tests of individual intelligence.

We come then to the question of the reliability of the tests. There are many different systems of intelligence testing and, therefore, it is important to find out how the results agree if the same group of people take a number of different tests. The figures given by Yoakum and Yerkes* indicate that people who do well or badly in one are likely to do more or less equally well or badly in the other tests. Thus the army test for English-speaking literates, known as Alpha, correlates with Beta, the test for non-English speakers or illiterates at .80. Alpha with a composite test of Alpha, Beta, and Stanford-Binet gives .94. Alpha, with Trabue B and C completion-tests, combined gives .72. On the other hand, as we noted in the first article of this series, the Stanford-Binet system of calculating "mental ages" is in violent disagreement with the results obtained by the army tests.

Nevertheless, in a rough way the evidence shows that the various tests in the mass are testing the same capacities. Whether these capacities can fairly be called intelligence, however, is not yet proved. The tests are all a good deal alike. They all derive from a common stock, and it is entirely possible that they measure only a certain kind of ability. The type of mind which is very apt in solving Sunday newspaper puzzles, or even in playing chess, may be specially favored by these tests. The fact that the same people always do well with puzzles would in itself be no evidence that the solving of puzzles was a general test of intelligence. We must remember, too, that the emotional setting plays a large role in any examination. To some temperaments the atmosphere of the examination room is highly stimulating. Such people "outdo themselves" when they feel they are being tested; other people "cannot do themselves justice" under the same conditions. Now in a large group these differences of temperament may neutralize each other in the statistical result. But they do not neutralize each other in the individual case.

The correlation between the various systems enables us to say only that the tests are not mere chance, and that they do seem to seize upon a certain kind of ability. But whether this ability is a sign of general intelligence or not, we have no means of knowing

* *Army Mental Tests,* p. 20.

from such evidence alone. The same conclusion holds true of the fact that when the tests are repeated at intervals on the same group of people they give much the same results. Data of this sort are as yet meager, for intelligence testing has not been practiced long enough to give results over long periods of time. Yet the fact that the same child makes much the same score year after year is significant. It permits us to believe that some genuine capacity is being tested. But whether this is the capacity to pass tests or the capacity to deal with life, which we call intelligence, we do not know.

This is the crucial question, and in the nature of things there can as yet be little evidence one way or another. The Stanford-Binet tests were set in order about the year 1914. The oldest children of the group tested at that time were 142 children ranging from fourteen to sixteen years of age. Those children are now between twenty-two and twenty-four. The returns are not in. The main question of whether the children who ranked high in the Stanford-Binet tests will rank high in real life is now unanswerable, and will remain unanswered for a generation. We are thrown back, therefore, for a test of the tests on the success of these children in school. We ask whether the results of the intelligence test correspond with the quality of school work, with school grades, and with school progress.

The crude figures, at first glance, show a poor correspondence. In Terman's studies,* the intelligence quotient correlated with school work, as judged by teachers, only .45 and with intelligence as judged by teachers, only .48. But that in itself proves nothing against the reliability of the intelligent tests. For after all the test of school marks, of promotion or the teacher's judgments, is not necessarily more reliable. There is no reason certainly for thinking that the way public school teachers classify children is any final criterion of intelligence. The teachers may be mistaken. In a definite number of cases Terman has shown that they are mistaken, especially when they judge a child's intelligence by his grade in school and not by his age. A retarded child may be doing excellent work, an advanced child poorer work. Terman has shown also that teachers make their largest mistakes in judging children who are above or below the average. The teachers become confused by the fact that the school system is graded according to age.

* *Stanford Revision of Binet-Simon Scale,* chap. 6.

A fair reading of the evidence will, I think, convince anyone that as a *system of grading* the intelligence tests may prove superior in the end to the system now prevailing in the public schools. The intelligence test, as we noted in an earlier article, is an instrument of classification. When it comes into competition with the method of classifying that prevails in school, it exhibits many signs of superiority. If you have to classify children for the convenience of school administration, you are likely to get a more coherent classification with the tests than without them. I should like to emphasize this point especially, because it is important that in denying the larger pretensions and misunderstandings we should not lose sight of the positive value of the tests. We say, then, that none of the evidence thus far considered shows whether they are reliable tests of the capacity to deal intelligently with the problems of real life. But as gauges of the capacity to deal intelligently with the problems of the classroom, the evidence justifies us in thinking that the tests will grade the pupils more accurately than do the traditional school examinations.

If school success were a reliable index of human capacity, we should be able to go a step further and say that the intelligence test is a general measure of human capacity. But of course no such claim can be made for school success, for that would be to say that the purpose of the schools is to measure capacity. It is impossible to admit this. The child's success with school work cannot be a measure of the child's success in life. On the contrary, his success in life must be a significant measure of the school's success in developing the capacities of the child. If a child fails in school and then fails in life, the school cannot sit back and say: You see how accurately I predicted this. Unless we are to admit that education is essentially impotent, we have to throw back the child's failure at the school, and describe it as a failure not by the child but by the school.

For this reason, the fact that the intelligence test may turn out to be an excellent administrative device for grading children in school cannot be accepted as evidence that it is a reliable test of intelligence. We shall see in the succeeding articles that the whole claim of the intelligence testers to have found a reliable measure of human capacity rests on an assumption, imported into the argument, that education is essentially impotent because intelligence is hereditary and unchangeable. This belief is the ultimate foundation of the claim that the tests are not merely an instrument of classification but a true measure of intelligence. It is this belief which has been seized upon eagerly by writers like Stoddard and McDougall. It is a belief which is, I am convinced, wholly unproved, and it is this belief which is obstructing and perverting the practical development of the tests.

The Abuse of the Tests

Walter Lippmann

We have found reason for thinking that the intelligence test may prove to be a considerable help in sorting out children into school classes. If it is true, as Professor Terman says,* that between a third and a half of the school-children fail to progress through the grades at the expected rate, then there is clearly something wrong with the present system of examinations and promotions. No one doubts that there is something wrong, and that in consequence both the retarded and the advanced child suffer.

The intelligence test promises to be more successful in grading the children. This means that the tendency of the tests in the average is to give a fairly correct sample of the child's capacity to do school work. In a wholesale system of education, such as we have in our public schools, the intelligence test is likely to become a useful device for fitting the child into the school. This is, of course, better than not fitting the child into the school, and under a more correct system of grading, such as the intelligence test promises to furnish, it should become possible even where education is conducted in large classrooms to specialize the teaching, because the classes will be composed of pupils whose capacity for school work is fairly homogeneous.

* *The Measurement of Intelligence*, p. 3.

Excellent as this seems, it is of the first importance that school authorities and parents realize exactly what this administrative improvement signifies. For great mischief will follow if there is confusion about the spiritual meaning of this reform. If, for example, the impression takes root that these tests really measure intelligence, that they constitute a sort of last judgment on the child's capacity, that they reveal "scientifically" his predestined ability, then it would be a thousand times better if all the intelligence testers and all their questionnaires were sunk without warning in the Sargasso Sea. One has only to read around in the literature of the subject, but more especially in the work of popularizers like McDougall and Stoddard, to see how easily the intelligence test can be turned into an engine of cruelty, how easily in the hands of blundering or prejudiced men it could turn into a method of stamping a permanent sense of inferiority upon the seal of a child.

It is not possible, I think, to imagine a more contemptible proceeding than to confront a child with a set of puzzles, and after an hour's monkeying with them, proclaim to the child, or to his parents, that here is a C− individual. It would not only be a contemptible thing to do. It would be a crazy thing to do, because there is nothing in these tests to warrant a judgment of this kind. All that can be claimed for the tests is that they can be used to classify into a homogeneous group the children whose capacities for school work are at a particular moment fairly similar. The intelligence test shows nothing as to why those capacities at any moment are what they are, and nothing as to the individual treatment which a temporarily retarded child may require.

I do not mean to say that the intelligence test is certain to be abused. I do mean to say it lends itself so easily to abuse that the temptation will be enormous. Suppose you have a school in which there are fifty ten-year-old children in the seventh grade and fifty eleven-year-olds in the eighth. In each class you find children who would jump ahead if they could and others who lag behind. You then regrade them according to mental age. Some of the ten-year-olds go into the eighth grade, some of the elevens into the seventh grade. That is an improvement. But if you are satisfied to leave the matter there, you are doing a grave injustice to the retarded children and ultimately to the community in which they are going to live. You cannot, in other words, be satisfied to put retarded eleven-year-olds and average ten-year-olds together. The retarded eleven-year-olds need something besides proper classification according to mental age. They need special analysis and special training to overcome their retardation. The leading intelligence testers recognize this, of course. But the danger of the intelligence tests is that in a wholesale system of education, the less sophisticated or the more prejudiced will stop when they have classified and forget that their duty is to educate. They will grade the retarded child instead of fighting the causes of his backwardness. For the whole drift of the propaganda based on intelligence testing is to treat people with low intelligence quotients as congenitally and hopelessly inferior.

Readers who have not examined the literature of mental testing may wonder why there is reason to fear such an abuse of an invention which has many practical uses. The answer, I think, is that most of the more prominent testers have committed themselves to a dogma which must lead to just such abuse. They claim not only that they are really measuring intelligence, but that intelligence is innate, hereditary, and predetermined. They believe that they are measuring the capacity of a human being for all time and that this capacity is fatally fixed by the child's heredity. Intelligence testing in the hands of men who hold this dogma could not but lead to an intellectual caste system in which the task of education had given way to the doctrine of predestination and infant damnation. If the intelligence test really measured the unchangeable hereditary capacity of human beings, as so many assert, it would inevitably evolve from an administrative convenience into a basis for hereditary caste.

In the next article we shall examine the evidence for the claim that the intelligence tests reveal the fixed hereditary endowment.

Tests of Hereditary Intelligence

Walter Lippmann

THE FIRST ARGUMENT IN FAVOR OF THE VIEW THAT THE CAPACITY FOR intelligence is hereditary is an argument by analogy. There is a good deal of evidence that idiocy and certain forms of degeneracy are transmitted from parents to offspring. There are, for example, a number of notorious families—the Kallikaks, the Jukes, the Hill Folk, the Nams, the Zeros, and the Ishmaelites—who have a long and persistent record of degeneracy. Whether these bad family histories are the result of a bad social start or of defective germplasm is not entirely clear, but the weight of evidence is in favor of the view that there is a taint in the blood. Yet even in these sensational cases, in fact just because they are so sensational and exceptional, it is important to remember that the proof is not conclusive.

There is, for example, some doubt as to the Kallikaks. It will be recalled that during the Revolutionary War, a young soldier, known under the pseudonym of Martin Kallikak, had an illegitimate feeble-minded son by a feeble-minded girl. The descendants of this union have been criminals and degenerates. But after the war was over, Martin married respectably. The descendants of this union have been successful people. This is a powerful evidence, but it would, as Professor Cattell* points out, be more powerful, and more interesting scientifically, if the wife of the respectable marriage had been feeble-minded, and the girl in the tavern had been a healthy, normal person. Then only would it have been possible to say with complete confidence that this was a pure case of biological rather than of social heredity.

Assuming, however, that the inheritance of degeneracy is established, we may turn to the other end of the scale. Here we find

* *Popular Science Monthly,* May 1915.

studies of the persistence of talent in superior families. Sir Francis Galton, for example, found "that the son of a distinguished judge had about one chance in four of becoming himself distinguished, while the son of a man picked out at random from the general population had only about one chance in four thousand of becoming similarly distinguished."* Professor Cattell, in a study of the families of 1,000 leading American scientists, remarks in this connection: "Galton finds in the judges of England a notable proof of hereditary genius. It would be found to be much less in the judges of the United States. It could probably be shown by the same methods to be even stronger in the families conducting the leading publishing and banking houses of England and Germany." And in another place he remarks that "my data show that a boy born in Massachusetts or Connecticut has been fifty times as likely to become a scientific man as a boy born along the Southeastern seaboard from Georgia to Louisiana."

It is not necessary for our purpose to come to any conclusion as to the inheritance of capacity. The evidence is altogether insufficient for any conclusion, and the only possible attitude is an open mind. We are, moreover, not concerned with the question of whether intelligence is hereditary. We are concerned only with the claim of the intelligence tester that *he reveals and measures* hereditary intelligence. These are quite separate propositions, but they are constantly confused by the testers. For these gentlemen seem to think that if Galton's conclusion about judges and the tale of the Kallikaks are accepted, then two things follow: first, that by analogy† all the graduations of intelligence are fixed in heredity, and second, that the tests measure these different grades of heredity intelligence. Neither conclusion follows necessarily. The facts of heredity cannot be proved by analogy; the facts of heredity are what they are. The question of whether the intelligence test measures heredity is a wholly different matter. It is the only question which concerns us here.

We may start then with the admitted fact that children of favored classes test higher on the whole than other children. Binet tests made in Paris, Berlin, Brussels, Breslau, Rome, Petrograd, Moscow, in England, and in America agree on this point. In Cali-

* Galton, *Hereditary Genius* (1869), cited by Stoddard, *Revolt Against Civilization*, p. 49.

† Cf. McDougall, p. 40.

fornia, Professor Terman* divided 492 children into five social classes and obtained the following correlation between the median intelligence quotient and social status:

Social Group	*Median IQ*
Very Inferior	85
Inferior	93
Average	99.5
Superior	107
Very Superior	106

On the face of it, this table would seem to indicate, if it indicates anything, a considerable connection between intelligence and environment. Mr. Terman denies this, and argues that "if home environment really has any considerable effect upon the IQ we should expect this effect to become more marked, the longer the influence has continued. That is, the correlation of IQ with social status should increase with age." But since his data show that at three age levels (5–8 years) and (9–11 years) and (12–15 years) the coefficient of correlation with social status declines (it is .43, .41, and .29 respectively), Mr. Terman concludes that "in the main, native qualities of intellect and character, rather than chance [*sic*] determine the social class to which a family belongs." He even pleads with us to accept this conclusion: "After all does not common observation teach us that etc. etc." and "from what is already known about heredity should we not naturally expect" and so forth and so forth.

Now I propose to put aside entirely all that Mr. Terman's common observation and natural expectations teach him. I should like only to examine his argument that if home environment counted much, its effect ought to become more and more marked as the child grew older.

It is difficult to see why Mr. Terman should expect this to happen. To the infant, the home environment is the whole environment. When the child goes to school, the influences of the home are merged in the larger environment of school and playground. Gradually, the child's environment expands until it takes in a city, and the larger invisible environment of books and talk and movies and newspapers. Surely Mr. Terman is making a very strange as-

* *Stanford Revision of Binet-Simon Scale*, p. 89.

sumption when he argues that as the child spends less and less time at home the influence of home environment ought to become more and more marked. His figures, showing that the correlation between social status and intelligence declines from .43 before eight years of age to .29 at twelve years of age, are hardly an argument for hereditary differences in the endowment of social classes. They are a rather strong argument, on the contrary, for the traditional American theory that the public school is an agency for equalizing the opportunities of the privileged and the unprivileged.

But Mr. Terman could by a shrewder use of his own data have made a better case. It was not necessary for him to use an argument which comes down to saying that the less contact the child has with the home the more influential the home ought to be. That is simply the gross logical fallacy of expecting increasing effects from a diminishing cause. Mr. Terman would have made a more interesting point if he had asked why the influence of social status on intelligence persists so long after the parents and the home have usually ceased to play a significant part in the child's intellectual development. Instead of being surprised that the correlation has declined from .43 at eight to .29 at twelve, he should have asked why there is any correlation left at twelve. That would have posed a question which the traditional eulogist of the little red schoolhouse could not answer offhand. If the question had been put that way, no one could dogmatically have denied that differences of heredity in social classes may be a contributing factor. But curiously, it is the mental tester himself who incidentally furnishes the most powerful defense of the orthodox belief that in the mass differences of ability are the result of education rather than of heredity.

The intelligence tester has found that the rate of mental growth declines as the child matures. It is faster in infancy than in adolescence, and the adult intelligence is supposed to be fully developed somewhere between sixteen and nineteen years of age. The growth of intelligence slows up gradually until it stops entirely. I do not know whether this is true or not, but the intelligence testers believe it. From this belief it follows that there is "a decreasing significance of a given amount of retardation in the upper years."* Binet, in fact, suggested the rough rule that under ten years of age a retardation of two years usually means feeble-mindedness, while

* *Ibid.*, p. 51.

for older children feeble-mindedness is not indicated unless there is a retardation of at least three years.

This being the case, the earlier the influence the more potent it would be, the later the influence the less significant. The influences which bore upon the child when his intelligence was making its greatest growth would leave a profounder impression than those which bore upon him when his growth was more nearly completed. Now in early childhood you have both the period of the greatest growth and the most inclusive and direct influence of the home environment. Is it surprising that the effects of superior and inferior environments persist, though in diminishing degree, as the child emerges from the home?

It is possible, of course, to deny that the early environment has any important influence on the growth of intelligence. Men like Stoddard and McDougall do deny it, and so does Mr. Terman. But on the basis of the mental tests they have no right to an opinion. Mr. Terman's observations begin at four years of age. He publishes no data on infancy and he is, therefore, generalizing about the hereditary factor after four years of immensely significant development have already taken place. On his own showing as to the high importance of the earlier years, he is hardly justified in ignoring them. He cannot simply lump together the net result of natural endowment and infantile education and ascribe it to the germplasm.

In doing just that, he is obeying the will to believe, not the methods of science. How far he is carried may be judged from this instance which Mr. Terman cites* as showing the negligible influence of environment. He tested twenty children in an orphanage and found only three who were fully normal. "The orphanage in question," he then remarks, "is a reasonably good one and affords an environment which is about as stimulating to normal mental development as average home life among the middle classes." Think of it. Mr. Terman first discovers what a "normal mental development" is by testing children who have grown up in an adult environment of parents, aunts, and uncles. He then applies this foot rule to children who are growing up in the abnormal environment of an institution and finds that they are not normal. He then puts the blame for abnormality on the germplasm of the orphans.

* *Ibid.*, p. 99.

A Future for the Tests

Walter Lippmann

HOW DOES IT HAPPEN THAT MEN OF SCIENCE CAN PRESUME TO DOGmatize about the mental qualities of the germplasm when their own observations begin at four years of age? Yet this is what the chief intelligence testers, led by Professor Terman, are doing. Without offering any data on all that occurs between conception and the age of kindergarten, they announce on the basis of what they have got out of a few thousand questionnaires that they are measuring the hereditary mental endowment of human beings. Obviously, this is not a conclusion obtained by research. It is a conclusion planted by the will to believe. It is, I think, for the most part unconsciously planted. The scoring of the tests itself favors an uncritical belief that intelligence is a fixed quantity in the germplasm and that, no matter what the environment, only a predetermined increment of intelligence can develop from year to year. For the result of a test is not stated in terms of intelligence, but as a percentage of the average for that age level. These percentages remain more or less constant. Therefore, if a child shows an IQ of 102, it is easy to argue that he was born with an IQ of 102.

There is here, I am convinced, a purely statistical illusion, which breaks down when we remember what IQ means. A child's IQ is his percentage of passes in the test which the average child of a large group of his own age has passed. The IQ measures his place in respect to the average at any year. But it does not show the rate of his growth from year to year. In fact, it tends rather to conceal the fact that the creative opportunities in education are greatest in early childhood. It conceals the fact, which is of such far-reaching importance, that because the capacity to form intellectual habits decreases as the child matures, the earliest education has a cumulative effect on the child's future. All this the static percentages of the

IQ iron out. They are meant to iron it out. It is the boast of the inventors of the IQ that "the distribution of intelligence maintains a certain constancy from five to thirteen or fourteen years of age, *when the degree of intelligence is expressed in terms of the intelligence quotient.*"* The intention is to eliminate the factor of uneven and cumulative growth, so that there shall be always a constant measure by which to classify children in classrooms.

This, as I have pointed out, may be useful in school administration, but it can turn out to be very misleading for an unwary theorist. If instead of saying that Johnny gained thirty pounds one year, twenty-five the next, and twenty the third, you said that measured by the average gain for children of his age, Johnny's weight quotients were 101, 102, 101, you might, unless you were careful, begin to think that Johnny's germplasm weighed as much as he does today. And if you dodged that mistake, you might nevertheless come to think that since Johnny classified year after year in the same position, Johnny's diet had no influence on his weight.

The effect of the intelligence quotient on a tester's mind may be to make it *seem* as if intelligence were constant, whereas it is only the statistical position in large groups which is constant. This illusion of constancy has, I believe, helped seriously to prevent men like Terman from appreciating the variability of early childhood. Because in the mass the percentages remain fixed, they tend to forget how in each individual case there were offered creative opportunities which the parents and nurse girls improved or missed or bungled. The whole more or less blind drama of childhood, where the habits of intelligence are formed, is concealed in the mental test. The testers themselves become callous to it. What their foot rule does not measure soon ceases to exist for them, and so they discuss heredity in schoolchildren before they have studied the education of infants.

But of course, no student of human motives will believe that this revival of predestination is due to a purely statistical illusion. He will say with Nietzsche that "every impulse is imperious, and, as *such,* attempts to philosophize." And so behind the will to believe he will expect to find some manifestation of the will to power. He will not have to read far in the literature of mental testing to discover it. He will soon see that the intelligence test is being sold to the public on the basis of the claim that it is a device which will

* *Stanford Revision of Binet-Simon Scale,* p. 50.

measure pure intelligence, whatever that may be, as distinguished from knowledge and acquired skill.

This advertisement is impressive. If it were true, the emotional and the worldly satisfactions in store for the intelligence tester would be very great. If he were really measuring intelligence, and if intelligence were a fixed hereditary quantity, it would be for him to say not only where to place each child in school, but also which children should go to high school, which to college, which into the professions, which into the manual trades and common labor. If the tester could make good his claim, he would soon occupy a position of power which no intellectual has held since the collapse of theocracy. The vista is enchanting, and even a little of the vista is intoxicating enough. If only it could be proved, or at least believed, that intelligence is fixed by heredity, and that the tester can measure it, what a future to dream about! The unconscious temptation is too strong for the ordinary critical defenses of the scientific methods. With the help of a subtle statistical illusion, intricate logical fallacies and a few smuggled obiter dicta, self-deception as the preliminary to public deception is almost automatic.

The claim that we have learned how to *measure hereditary intelligence* has no scientific foundation. We cannot measure intelligence when we have never defined it, and we cannot speak of its hereditary basis after it has been indistinguishably fused with a thousand educational and environmental influences from the time of conception to the school age. The claim that Mr. Terman or anyone else is measuring hereditary intelligence has no more scientific foundation than a hundred other fads—vitamins and glands and amateur psychoanalysis and correspondence courses in will power—and it will pass with them into that limbo where phrenology and palmistry and characterology and the other Babu sciences are to be found. In all of these, there was some admixture of primitive truth which the conscientious scientist retains long after the wave of popular credulity has spent itself.

So, I believe, it will be with mental testing. Gradually, under the impact of criticism, the claim will be abandoned that a device has been invented for measuring native intelligence. Suddenly it will dawn upon the testers that this is just another form of examination, differing in degree rather than in kind from Mr. Edison's questionnaire or a college entrance examination. It may be a better form of examination than these, but it is the same sort of thing. It tests, as they do, an unanalyzed mixture of native capacity, acquired habits and stored-up knowledge, and no tester knows at any moment which factor he is testing. He is testing the complex result of a long and unknown history, and the assumption that his questions and his puzzles can in fifty minutes isolate abstract intelligence is, therefore, vanity. The ability of a twelve-year-old child to define pity or justice and to say what lesson the story of the fox and crow "teaches" may be a measure of his total education, but it is no measure of the value or capacity of his germplasm.

Once the pretensions of this new science are thoroughly defeated by the realization that these are not "intelligence tests" at all nor "measurements of intelligence," but simply a somewhat more abstract kind of examination, their real usefulness can be established and developed. As examinations they can be adapted to the purposes in view, whether it be to indicate the feeble-minded for segregation, or to classify children in school, or to select recruits from the army for officers' training camps, or to pick bank clerks. Once the notion is abandoned that the tests reveal pure intelligence, specific tests for specific purposes can be worked out.

A general measure of intelligence valid for all people everywhere at all times may be an interesting toy for the psychologist in his laboratory. But just because the tests are so general, just because they are made so abstract in the vain effort to discount training and knowledge, the tests are by that much less useful for the practical needs of school administration and industry. Instead, therefore, of trying to find a test which will with equal success discover artillery officers, Methodist ministers, and branch managers for the rubber business, the psychologists would far better work out special and specific examinations for artillery officers, divinity school candidates, and branch managers in the rubber business. On that line they may ultimately make a serious contribution to a civilization which is constantly searching for more successful ways of classifying people for specialized jobs. And in the meantime the psychologists will save themselves from the reproach of having opened up a new chance for quackery in a field where quacks breed like rabbits, and they will save themselves from the humiliation of having furnished doped evidence to the exponents of the New Snobbery.

The Great Conspiracy

The Impulse Imperious of Intelligence Testers, Psychoanalyzed and Exposed by Mr. Lippmann

Lewis M. Terman

AFTER MR. BRYAN HAD CONFOUNDED THE EVOLUTIONISTS, AND VOLIVA the astronomers, it was only fitting that some equally fearless knight should stride forth in righteous wrath and annihilate that other group of pseudoscientists known as "intelligence testers." Mr. Walter Lippmann, alone and unaided, has performed just this service. That it took six rambling articles to do the job is unimportant. It is done. The world is deeply in debt to Mr. Lippmann. So are the psychologists, if they only knew it, for henceforth, they should know better than to waste their lives monkeying with those silly little "puzzles" or juggling IQs and mental ages.

What have intelligence testers done that they should merit such a fate? Well, what have they not done? They have enunciated, *ex cathedra*, in the guise of fact, law, and eternal verity, such highly revolutionary and absurd doctrines as the following; to wit:

1 that the strictly average representative of the genus homo is not a particularly intellectual animal;

2 that some members of the species are much stupider than others;

3 that school prodigies are usually brighter than school laggards;

4 that college professors are more intelligent than janitors, architects than hod carriers, railroad presidents than switch tenders, and (most heinous of all),

5 that the offspring of socially, economically, and professionally successful parents have better mental endowment, on the average, than the offspring of said janitors, hod carriers, and switch tenders.

These are indeed dangerous doctrines, subversive of American democracy. The crime of the intelligence testers is made worse by the fact that they have attempted to gain credence for their nefarious theories by resort to cunningly devised statistical formulae which common people do not understand. It is true that some of these doctrines had been voiced before, but as long as they were expressed in ordinary language, they passed as mere opinion and did little harm. But to talk about mental differences in terms of IQs, or to reckon mental inheritance in terms of a ".50 coefficient of resemblance between parent and offspring," is a far more serious matter. In the interest of freedom of opinion, there ought to be a law passed forbidding the encroachment of quantitative methods upon those fields which from time immemorial have been reserved for the play of sentiment and opinion. For example, why should not one be allowed to take his political or social theory as he takes his religion, without having it all mixed up with IQs, probable errors and coefficients of correlation?

At any rate, it will not do to let the idea get abroad that human beings differ in any such vital trait as ability to think, comprehend, reason; or, if such differences really exist, that there is the remotest possibility of anyone ever being able to measure them. If the psychologists should succeed in getting the intelligentsia to swallow this vanity-satisfying doctrine, who knows that they would not next succeed in putting over a system of plural voting based upon intelligence indices (to be determined by these self-same psychologists)? Absurd? By no means. Suppose, for example, they should somehow manage to give a test to the members of Congress (it might be done without their knowing it) and should then shrewdly award to each and every one a flatteringly high IQ. Sheer instinct on the part of the recipients could be depended upon to do the rest.

Let there be no misapprehension: the principle of democracy is at stake. The essential thing about a democracy is not equality of opportunity, as some foolish persons think, but equality of mental endowment. Where would our American democracy be if it should turn out that people differ in intelligence as they do in height; especially if the psychologist could make it appear that he had discovered a method of triangulating everybody's intellectual altitude? The argument of the psychologists that they would use their method in the discovery and conservation of talent, among rich and poor alike, is brazen camouflage. They don't care a twirl-o'-your-thumb about the conservation of talent. Their real purpose is to set up a neo-aristocracy, more snobbish, more tyrannical, and on every count more hateful than any that has yet burdened the earth. Inasmuch as the psychologists know their little puzzle stunts better than anyone else can hope to know them, they are doubtless entertaining ambitious visions of themselves forming the capstone of this new political and social structure. As Mr. Lippmann well says, "if the tester could make good his claim that his tests test intelligence he would soon occupy a position of power which no intellectual has held since the collapse of theocracy." In short, the whole thing is motivated by the Nietzschean Impulse Imperious.

It is high time that we were penetrating the wiles of this crafty cult. We have been entirely too unsuspecting. For example, the innocent-minded Germans are being shamefully taken in at this very moment. Hardly had the old government of Germany crashed, when the educational authorities of the newly established republic allowed the psychologists to launch a wild orgy of intelligence testing in the schools. The orgy continues unabated. The ostensible purpose is to sift the schools for superior talent in order to give it a chance to make the most of itself, in whatever stratum of society it may be found. The psychologists pretend that they are trying to break up the old Prussian caste system. They are not. It is the Impulse Imperious. If the German people don't wake up, they will soon find themselves in the grip of a super-Junker caste that will out-Junker anything Prussia ever turned loose. England and the other European countries are in similar danger. The conspiracy has even spread to Australia, South Africa, and Japan. It is world-wide.

Now it is evident that Mr. Lippmann has been seeing red; also, that seeing red is not very conducive to seeing clearly. The impassioned tone of these six articles gives their case away. Clearly, something has hit the bull's-eye of one of Mr. Lippmann's emotional complexes. From the concentration of attack upon me, one would infer that I had caused all the trouble, even to the point of seducing such an eminent psychologist as William McDougall. If such is the case, my responsibility is very great, for a majority of the psychologists of America, England, and Germany are now enrolled in the ranks of the intelligence testers, and all but a handful of the rest use their results.

The six articles are introduced by the editors of *The New Republic* as a critical "analysis and estimate of intelligence tests." As it turns out, the estimate is considerably more in evidence than the analysis. The former rings out clearly in every paragraph; the latter, when it is not downright loco, is vague and misleading. One gathers, however, that Mr. Lippmann thinks he has a mission to perform and that the end justifies the means. This is not an accusation; it is a charitable way of explaining his misuse of facts and quotations.

The validity of intelligence tests is hardly a question the psychologist would care to debate with Mr. Lippmann; nor is there any reason to engage in so profitless a venture. It is only necessary to examine casually a few samples of his allegations in order to show what weight they should carry.*

Sample No. 1. *The belief that our draftees in the war had an average mental age of only fourteen years rests entirely upon the mental age standards embodied in the Stanford Revision of the Binet tests. These standards were based upon a mere handful of samplings and are entirely overthrown by the army results. The army tests have "knocked the Stanford Revision into a cocked hat."*

As a matter of fact, the belief in question does not rest at all upon the correctness of the Stanford mental age norms. Independent age norms have several times been derived for the army tests by applying them to large groups of unselected school children. I have presented some of these norms in the very report from which Mr. Lippmann quotes a few of the facts he is unable to interpret.† Such independently derived norms for the Alpha test, for the Beta test, and for the Yerkes-Bridges Point Scale (all used in the army) agree with the Stanford-Binet in the verdict as to average mental age of our drafted soldiers. On every kind of test that was employed, even the most nonverbal, the average score earned by draftees was less than that earned by average fourteen-year-old schoolchildren. Psychologists are not entirely agreed as to how this

* His allegations are here stated in highly condensed form, as the text is much too verbose for literal quotation.

† R. M. Yerkes, *Psychological Examining in the United States Army*, vol. 15, *Memoirs of the National Academy of Science* (Washington, D.C., 1921), pp. 536 ff.

fact should be interpreted, but that is beside the point. Those who accept the army data at their face value think that the "Fourteen-Year" tests of the Stanford-Binet should be renamed "Average Adult" tests. The possible desirability of such renaming has no bearing whatever on the average mental age of soldiers or, for that matter, on the validity of the Stanford tests as a measure of intelligence.

Sample No. 2. *The intelligence rating earned by a soldier was determined chiefly by the time limits used in giving the tests.*

The effects of increased time limits were thoroughly investigated by the Division of Psychology, Surgeon General's Office. The results of the experiment, which was carried out under my direction by Dr. Mark A. May, are stated in the following words: "In general, then, we have no reason to assume that an extension of time limits would have improved the test or have given an opportunity to many individuals materially to alter their ratings."* In fact, scores earned by 510 men on regular time correlated with scores earned by the same group on double time to the extent of .965. This means, of course, that the top 5 percent by one method included almost exactly the same men as were in the top 5 percent by the other method, and similarly for a cross section in any range of the score distribution. These facts are to be found just three pages from a statement which Mr. Lippmann takes out of its setting and quotes in a manner certain to mislead.

Sample No. 3. *The symmetrical distribution of IQs resulting from application of the Stanford-Binet to unselected children is no proof whatever of the validity of the test.*

Perfectly true and perfectly irrelevant. I have never made such a claim, although Mr. Lippmann tries to give the impression that I have. It is true, as he asserts, that coin-tossing gives an even more symmetrical curve of distribution. Mr. Lippmann uses this illustration in order to suggest that intelligence score distributions, like those for coin-tossing, are mainly a product of chance. (He does admit they are "not quite as chancy as that.") What are the facts? Over and over again, the experiment has been made of testing a large group of children twice, with an interval of several days, or months, or even years between the tests. Each pupil's original score

* *Psychological Examining in the U. S. Army*, p. 416.

is then paired with his later score, and a correlation coefficient is computed for the two series of tests. If the scores were due to chance, the resulting correlations would of course be .00. Actually, they are nearly always above .80, and occasionally above .90. If Mr. Lippmann will make two 1,000 series of coin tosses and then correlate the results of the two series by pairing first toss with first toss, second toss with second toss, etc., he will get, not .80 or .90, but .00, plus or minus a small probable error.

Sample No. 4. *The tests are doubtless useful in classifying schoolchildren, but this is no evidence that they test intelligence.*

Possibly, it is not; or possibly, it depends upon one's definition of intelligence. Most of us have uncritically taken it for granted that children who attend school eight or ten years without passing the fourth grade or surmounting long division, are probably stupider than children who lead their classes into high school at twelve years and into college at sixteen. Mr. Lippmann contends that we can't tell anything about how intelligent either one of these children is until he has lived out his life. Therefore, for a lifetime, at least, Mr. Lippmann considers his position impregnable!

Sample No. 5. *Although intelligence tests are capable of rendering valuable service in classifying schoolchildren, they are in great danger of becoming an "engine of cruelty" by being turned into "a method of stamping a permanent sense of inferiority upon the soul of the child." Nothing could be more contemptible than to—etc., etc.*

Mr. Lippmann does not charge that the tests have been thus abused, but that they easily could be. Very true; but they simply aren't. That is one of the recognized rules of the game. Isn't it funny what horrible possibilities an excited brain can conjure up? I recall a patient who had worked himself into a wretched stew from thinking how terrible it would be if butchers, by concerted action all over the country, should suddenly take it into their heads to slaughter their unsuspecting customers. He was actually determined to get a law passed that would deprive these potential murderers of their edged and pointed tools.

Sample No. 6. *There is no proof that mental traits are inherited. Goddard thought he had proved it for mental deficiency, but Cattell questions his evidence. Galton thought he had proved it for genius, but Cattell does not seem to think much of that proof either.*

Note how cleverly Mr. Lippmann strives for effect by playing off one psychologist against another. He resorts to this frequently. The trick is very simple; all you do is take an isolated statement out of its original setting and quote it in a setting made to order. In that way, you can have all the expert opinion on your side. Mr. Bryan is said to use this method with telling effect against the evolutionists. Not that psychologists don't sometimes disagree, even as doctors do. It would be a sorry outlook for their young science if they did not. But when the outsider comes along and tries to make capital out of such differences, it is well to be on one's guard. In ninety-nine cases out of a hundred, it means that an unfair advantage is being taken both of the reader and of the author quoted. Think, for example, of Mr. Lippmann's quoting Cattell in support of his tirade against intelligence testing—Cattell, the pupil of Galton, the father of mentality testing in America, the inventor of new methods for the study of individual differences, the author of important studies (in progress) on the inheritance of genius!

Sample No. 7. (Main Allegation, asserted at least three times in every paragraph, always with signs of greatly increased blood pressure.) *The intelligence tests do not test pure intelligence. Any appearance to the contrary is due to "a subtle statistical illusion." The psychologist's assumption ". . . that his questions and his puzzles can in fifty minutes isolate abstract intelligence is vanity." It is worse than vanity; it is an attempt to restore the "doctrine of predestination and infant damnation" in favor of an "intellectual caste system."*

It is evident that Jack has prepared an imposing giant for the slaughter. No matter that it is stuffed with straw or that it is set up in a fashion to make it the easy victim of a few vigorous puffs of superheated atmosphere. As a matter of fact, all the intelligence testers will readily agree with Mr. Lippmann that their tests do not measure simon-pure intelligence, but always native ability plus other things, with no final verdict yet as to exactly how much the other things affect the score. However, nearly all the psychologists believe that native ability counts very heavily. Mr. Lippmann does not. He prefers to believe that more probably an individual's IQ is determined by what happens to him in the nursery before the age of four years, in connection with the "creative opportunities which the parents and nurse girls improved or missed or bungled"! After all, if our experiences in the nursery gave us our emotional complexes, as the Freudians say, why shouldn't they have determined our IQs at the same time?

One wonders why Mr. Lippmann, holding this belief, did not suggest that we let up on higher education and pour our millions into kindergartens and nurseries. For, really and truly, high IQs are not to be sneezed at. The difference between 150 IQ and 50 IQ is the difference between an individual who will be able, if he half tries, to graduate from college with Phi Beta Kappa honors at twenty, and an individual who at that age can hardly do long division, make change for four cents out of twenty-five, or name the months of the year. Even the difference between 100 IQ and 75 IQ is the difference between the ability to graduate from high school or possibly from college, and inability to do even first year of high school work satisfactorily.

And just to think that we have been allowing all sorts of mysterious, uncontrolled, chance influences in the nursery to mold children's IQs, this way and that way, right before our eyes. It is high time that we were investigating the IQ effects of different kinds of baby talk, different versions of Mother Goose, and different makes of pacifiers and safety pins. If there is any possibility of identifying, weighing, and bringing under control these IQ stimulants and depressors, we can well afford to throw up every other kind of scientific research until the job is accomplished. That problem once solved, the rest of the mysteries of the universe would fall easy prey before our made-to-order IQs of 180 or 200.

Does not Mr. Lippmann owe it to the world to abandon his role of critic and to enter this enchanting field of research? He may safely be assured that if he unravels the secret of turning low IQs into high ones, or even into moderately higher ones, his fame and fortune are made. If he could guarantee to raise certified 100s to certified 140s, or even certified 80s to certified 100s, nothing but premature death or the discovery and publication of his secret would keep him out of the Rockefeller-Ford class if he cared to achieve it. I know of a certain modern Croesus who alone would probably be willing to start him off with 10 or 20 million if he could only raise one particular little girl from about 60 to 70 to a paltry 100 or so. Of course, if this man had only understood the secrets of "creative opportunity" in the nursery, he might have had all this and more for nothing. Who knows but if the matter were put up to him in the right way, he would be willing to endow for Mr. Lippmann a Bureau of Nursery Research for the Enhancement of the IQ?

If Mr. Lippmann gets this Bureau started, there are several questions I shall want to submit to it for solution. Some of these have been bothering me for a long time. One is, why both high and low IQs are so often found in children of the same family and of the same nursery. To be sure, parental habits change more or less as children come and grow up; nurse girls arrive and depart; toys wear out. The problem admittedly is complex, but by successive experiments in which one factor after another was kept constant while the others were varied, the evil and beneficent influences might gradually be sorted out.

Next, I should want to propose a minute comparative study of the influences operative in our California Japanese nurseries and those of our California Portuguese. Here is mystery enough to challenge any group of scientists Mr. Lippmann can get together, for notwithstanding the apparent similarity of nursery environment in the two cases, the IQ results are markedly different. Our average Portuguese child carries through school and into life an IQ of about 80; the average Japanese child soon develops an IQ not far below that of the average California white child of Nordic descent. In this case, the nurse-girl factor is eliminated; one might almost say, the nursery itself. But of course, there are the toys, which are more or less different. It is also conceivable that the more liquid Latin tongue exerts a sedative effect on infants' minds as compared with the harsher Japanese language, which may be stimulating in comparison.

Another problem would relate to the IQ resemblance of identical twins as compared with that of fraternal twins. The latest and most extensive investigation of this problem* indicates a considerably greater IQ resemblance for the former than for the latter. This is a real poser, which I leave to Mr. Lippmann without attempting an explanation.

* By a Stanford student and not yet published at the time.

The Great Confusion

A Reply to Mr. Terman

Walter Lippmann

IN LAST WEEK'S ISSUE, PROFESSOR LEWIS M. TERMAN STATED THAT IT would be "profitless" to debate the validity of the intelligence tests with me. Nevertheless, it seems that he felt impelled to pause in those labors, which he modestly compares with Darwin's, in order to write eight columns of print about the contemptible creature who had challenged his dogmas. I cannot imagine who Mr. Terman thinks is going to be fooled by this pretense.

Certainly I feel entitled to assume that Mr. Terman's reply contains the most damaging criticism of my articles which he is able to make. I assume that he has not spared me in his contempt. Therefore, I shall take no space to restate any of the conclusions which Mr. Terman has not challenged.

The first three columns of Professor Terman's reply come down to the charge that I criticized him because I think his arguments are subversive of the old naïve theory of democracy. All that I can say to that is that I have written a book of over four hundred pages devoted to arguing the fallacy of the naïve democratic theory. He must not think, therefore, that a criticism of the Terman dogmas constitutes a defense of this old theory. It often happens that an error is attacked in an erroneous and misleading way. Mr. Bryan's theory of creation, for example, is no doubt wrong, but it does not follow that everyone who answers Mr. Bryan is a sound biologist. In the same way, it does not follow that Mr. Terman is a sound critic of democracy, because the democratic theory is open to criticism.

Nor does it necessarily follow, as Mr. Terman seems to imply, that because intelligence testing has spread all over the wide world,

Mr. Terman's conclusions about these tests are valid. Popular success is not scientific evidence, and even the practical usefulness of the tests in schools is no evidence whatever that the theoretical assumptions are true or that the social generalizations are correct.

At last after three columns of ridicule, Mr. Terman stoops to argument. He denies that the talk about the average mental age of the army being fourteen rests upon the Stanford standardization of mental age. He refers me to pages 535 ff. of the army report where he has presented what he calls "independent" age norms derived from large groups of unselected schoolchildren. Mr. Terman has missed the point of the matter entirely. An age norm is *internally standardized* according to the average performance of an age group. The army tests were given to the largest group ever collected. Therefore, the mental age of the army cannot be measured by norms, however "independent," which are derived from smaller and less representative groups. If the army tests were tests, you have to admit that they created their own norms. Otherwise, you arrive at such logical absurdities as arguing that the average mental age of adults is less than the average mental age of adults.

The Stanford norms gave sixteen as adult. The army norms give a little less than fourteen on the Stanford scale. Either the army tests are unreliable, or the Stanford norm is incorrect. Mr. Terman more or less realizes this, for he says that "Psychologists are not entirely agreed as to how this fact [this disparity between army 'adults' and Stanford 'adults'] should be interpreted, but that is beside the point." I beg his pardon. It is the whole point when you are talking about "the average mental age" of Americans.

The second time Mr. Terman stoops to argument is on the factor of speed in the army tests. He refers me to an experiment conducted under his direction by Doctor Mark A. May as proof that time played no part in the army results. The only trouble with Dr. May's experiment is that it does not prove Mr. Terman's point. What is more, the editors of the army report stated quite definitely that Dr. May's experiment did not prove the point.

Dr. May tested about a thousand men with double time, and found that "the order of abilities of various persons" was not materially altered. That is to say, the relative positions of the men remained about the same. The better men still graded ahead of the poorer. But in actual performance, time *was* a very great factor, as Dr. May's own figures amply prove. Added time did not help the very poorest men much. Of those who scored zero on single time in each of seven tests, an average of even 17 percent made a better showing when the time was doubled. But of those who scored ten points in each of the seven tests, over 80 percent gained by doubling the time. Naturally, then, the editors of the army report, who are men of quite different scientific temper from Mr. Terman, took good care not to let Dr. May's experiment be used for just such careless generalization as Mr. Terman has let himself in for. Therefore, after presenting Dr. May's data, they hasten to state that "the high frequencies of persons gaining at the upper levels (often 100 percent) indicate that for the people making high scores on single time the 'speed' element is predominant."

Mr. Terman's difficulty is a very simple one. Finding that the best men are still the best with double time, he notes triumphantly that the top 5 percent are almost exactly the same men under double as under single time. The top 5 percent are still the top 5 percent. I never denied it. What he fails to note is that although the relative positions are the same, the absolute performances are often radically different. "The top 5 percent" has no relevance to the argument. The relevant fact is that a very large percentage of men made much better scores when the time was doubled. And since the classification of any man as an "A," a "B," or a "C" was determined by his score, not by his relative position, there could have been more "A men" in the American nation if the army had thought it convenient to double the time.

Mr. Terman ought at the very least to interpret correctly work done by his own pupils under his own direction.

These are the only two instances where Mr. Terman ventures to debate. Throughout the rest of his reply, he is consumed with laughter. It strikes him as enormously funny that data based on a few years out of the lives of a rather small number of schoolchildren should not be deemed proof that Mr. Terman has measured "native intelligence." It strikes him as enormously funny that I should argue that his tests do not provide convincing canons by which to distinguish hereditary from acquired ability. He jumps to the conclusion that I deny the hereditary factor. I did not deny it. I denied Mr. Terman's unproved claim that he had isolated the hereditary factor. Mr. Terman's logical abilities are so primitive that he finds this point impossible to grasp. But nothing seems to him so ludicrous as my argument that no testing which begins at four years of age can possibly claim that its results are uninfluenced by infantile education. The notion that the first four years of experience are highly formative arouses all his gifts of satire.

I said in my article that Mr. Terman was insensitive to childhood. I repeat the statement and add that a psychologist who sneers at the significance of the earliest impressions and habits is too shallow to write about education.

Finally, a word about Mr. Terman's notion that I have an "emotional complex" about this business. Well, I have. I admit it. I hate the impudence of a claim that in fifty minutes you can judge and classify a human being's predestined fitness in life. I hate the pretentiousness of that claim. I hate the abuse of scientific method which it involves. I hate the sense of superiority which it creates, and the sense of inferiority which it imposes.

And so, while I honestly think that there is a considerable future for mental testing, if it is approached with something like the caution employed by the editors of the army report, I believe also that the whole field is destined to be the happy hunting ground of quacks and snobs if loose-minded men are allowed to occupy positions of leadership much longer.

Letter to The New Republic, *January 17, 1923*

SIR:

Mr. Lippmann evidently does not know that I myself edited a large part of the official report of army mental testing, including the section he quotes against me.

Mr. Lippmann's insistence that the fourteen-year mental age of soldiers rests on the Stanford-Binet norms is entirely incorrect. Mental age standards for the army tests have been established independently of any other test.

Mr. Lippmann says the letter classification was not by relative position but by score, which was affected by the time limits. Actually the one purpose of the classification was to get the relative position. It is irrelevant what time limits were used so long as that purpose was accomplished.

LEWIS M. TERMAN

Palo Alto, California

Letter to The New Republic, *January 17, 1923*

SIR:

(1) Naturally I did not suspect that Mr. Terman was the editor of that section of the army report which I quoted against him. What reason or right had I to assume he had forgotten what he himself had edited?

(2) Mr. Terman will find no insistence in my reply that the "fourteen-year-old" statement about the army rests on the Stanford Binet scale. I do insist, first, that the statement conflicts with the Stanford Binet scale, and that, therefore, "either the army tests are unreliable or the Stanford norm is incorrect"; secondly, I insist that since mental age is based on an average, no standard adult mental age can be established independently of the average of the army as a whole which will measure the mental age of the army as a whole.

This point is clearly made on page 785 of the Army report. I cite this passage to Mr. Terman, once again not daring to suspect that he edited it.

(3) What Mr. Terman has to say about the effect of the time limit is very gratifying. It is in complete agreement with what I argued, and it is an absolute repudiation of the generalizations of Mr. Lothrop Stoddard which I was criticizing.

Note that Mr. Terman admits that "the one purpose of the classification was to get the *relative* position." This is equivalent to an admission that the army tests did not measure the intelligence of the American Nation. It is an admission that the talk about the percentage of "A" men is a misunderstanding of the army results. For the relative order of abilities is no measure of the strength of those abilities.

Therefore I can agree with Mr. Terman (and said this in my articles) that it was irrelevant what time limits were used if the one purpose was to fix the order of abilities. But the time limit is highly relevant, because it decisively changes the result, if the classifications as "A" men, etc., are to be taken as a measurement of American intelligence. I am happy to say that Mr. Terman now admits this.

WALTER LIPPMANN

New York City

Testing for Competence Rather Than for "Intelligence"

David C. McClelland

THE TESTING MOVEMENT IN THE UNITED STATES HAS BEEN A SUCCESS, if one judges success by the usual American criteria of size, influence, and profitability. Intelligence and aptitude tests are used nearly everywhere by schools, colleges, and employers. It is a sign of backwardness not to have test scores in the school records of children. The Educational Testing Service alone employs about 2,000 people, annually administers Scholastic Aptitude Tests to thousands of aspirants to college, and makes enough money to support a large basic research operation. Its tests have tremendous power over the lives of young people by stamping some of them "qualified" and others "less qualified" for college work. Until recent "exceptions" were made (over the protest of some), the tests have served as a very efficient device for screening out black, Spanish-speaking, and other minority applicants to colleges. Admissions officers have protested that they take other qualities besides test achievements into account in granting admission, but careful studies by Wing and Wallach (1971) and others have shown that this is true only to a very limited degree.

Why should intelligence or aptitude tests have all this power? What justifies the use of such tests in selecting applicants for college entrance or jobs? On what assumptions is the success of the movement based? They deserve careful examination before we go on rather blindly promoting the use of tests as instruments of power over the lives of many Americans.

The key issue is obviously the *validity* of so-called intelligence tests. Their use could not be justified unless they were valid, and it is my conviction that the evidence for their validity is by no means so overwhelming as most of us, rather unthinkingly, had come to think it was. In point of fact, most of us just believed the results that the testers gave us, without subjecting them to the kind of fierce skepticism that greets, for example, the latest attempt to show that ESP exists. My objectives are to review skeptically the main lines of evidence for the validity of intelligence and aptitude tests and to draw some inferences from this review as to new lines that testing might take in the future.

Let us grant at the outset that brain-damaged or retarded people do less well on intelligence tests than other people. Wechsler (1958) initially used this criterion to validate his instrument, although it has an obvious weakness: brain-damaged people do less well on almost *any* test so that it is hard to argue that something unique called "lack of intelligence" is responsible for the deficiency in test scores. The multimethod, multitrait criterion has not been applied here.

Tests Predict Grades in School

The games people are required to play on aptitude tests are similar to the games teachers require in the classroom. In fact, many of Binet's original tests were taken from exercises that teachers used in French schools. So it is scarcely surprising that aptitude test scores are correlated highly with grades in school. The whole Scholastic Aptitude Testing movement rests its case largely on this single undeniable fact. Defenders of intelligence testing, like McNemar (1964), often seem to be suggesting that this is the only kind of validity necessary. McNemar remarked that "the manual of the Differential Aptitude Test of the Psychological Corporation contains a staggering total of 4,096, yes I counted 'em, validity coefficients." What more could you ask for, ladies and gentlemen? It was not until I looked at the manual myself (McNemar certainly did not enlighten me) that I confirmed my suspicion that almost every one of those "validity" coefficients involved predicting grades in courses—in other words, performing on similar types of tests.

So what about grades? How valid are they as predictors? Researchers have in fact had great difficulty demonstrating that grades in school are related to any other behaviors of importance—other than doing well on aptitude tests. Yet the general public—including many psychologists and most college officials—simply has been unable to believe or accept this fact. It seems so self-evident to educators that those who do well in their classes must go on to do better in life that they systematically have disregarded evidence to the contrary that has been accumulating for some time. In the early 1950s, a committee of the Social Science Research Council of which I was chairman looked into the matter and concluded that while grade level attained seemed related to future measures of success in life, performance within grade was related only slightly. In other words, being a high school or college graduate gave one a credential that opened up certain higher level jobs, but the poorer students in high school or college did as well in life as the top students. As a college teacher, I found this hard to believe until I made a simple check. I took the top eight students in a class in the late 1940s at Wesleyan University where I was teaching—all straight A students—and contrasted what they were doing in the early 1960s with what eight really poor students were doing—all of whom were getting barely passing averages in college (C – or below). To my great surprise, I could not distinguish the two lists of men fifteen to eighteen years later. There were lawyers, doctors, research scientists, and college and high school teachers in both groups. The only difference I noted was that those with better grades got into better law or medical schools, but even with this supposed advantage they did not have notably more successful careers as compared with the poorer students who had had to be satisfied with "second-rate" law and medical schools at the outset. Doubtless the C – students could not get into even second-rate law and medical schools under the stricter admissions testing standards of today. Is that an advantage for society?

Such outcomes have been documented carefully by many researchers (cf. Hoyt, 1965) both in Britain (Hudson, 1960) and in the United States. Berg (1970), in a book suggestively titled *Education and Jobs: The Great Training Robbery,* has summarized studies showing that neither amount of education nor grades in school are related to vocational success as a factory worker, bank teller, or air traffic controller. Even for highly intellectual jobs like scientific researcher, Taylor et al. (1963) have shown that superior on-the-job performance is related in no way to better grades in college. The average college grade for the top third in research success was 2.73 (about B –), and for the bottom third, 2.69 (also

B —). Such facts have been known for some time. They make it abundantly clear that the testing movement is in grave danger of perpetuating a mythological meritocracy in which none of the measures of merit bears significant demonstrable validity with respect to any measures outside of the charmed circle. Psychologists used to say as a kind of an "in" joke that intelligence is what the intelligence tests measure. That seems to be uncomfortably near the whole truth and nothing but the truth. But what's funny about it, when the public took us more seriously than we did ourselves and used the tests to screen people out of opportunities for education and high-status jobs? And why call excellence at these test games intelligence?

Even further, why keep the best education for those who are already doing well at the games? This in effect is what the colleges are doing when they select from their applicants those with the highest Scholastic Aptitude Test scores. Isn't this like saying that we will coach especially those who already can play tennis well? One would think that the purpose of education is precisely to improve the performance of those who are not doing very well. So when psychologists predict on the basis of the Scholastic Aptitude Test who is most likely to do well in college, they are suggesting implicitly that these are the "best bets" to admit. But in another sense, if the colleges were interested in proving that they could educate people, high-scoring students might be poor bets because they would be less likely to show improvement in performance. To be sure, the teachers want students who will do well in their courses, but should society allow the teachers to determine who deserves to be educated, particularly when the performance of interest to teachers bears so little relation to any other type of life performance?

Do Intelligence Tests Tap Abilities That Are Responsible for Job Success?

Most psychologists think so; certainly the general public thinks so (Cronbach, 1970, p. 300), but the evidence is a whole lot less satisfactory than one would think it ought to be to justify such confidence.

Thorndike and Hagen (1959), for instance, obtained 12,000 correlations between aptitude test scores and various measures of later occupational success on over 10,000 respondents and concluded that the number of significant correlations did not exceed what would be expected by chance. In other words, the tests were invalid. Yet psychologists go on using them, trusting that the poor validities must be due to restriction in range due to the fact that occupations do not admit individuals with lower scores. But even here it is not clear whether the characteristics required for entry are, in fact, essential to success in the field. One might suppose that finger dexterity is essential to being a dentist, and require a minimum test score for entry. Yet, it was found by Thorndike and Hagen (1959) to be related negatively to income as a dentist! Holland and Richards (1965) and Elton and Shevel (1969) have shown that no consistent relationships exist between scholastic aptitude scores in college students and their *actual accomplishments* in social leadership, the arts, science, music, writing, and speech and drama.

Yet what are we to make of Ghiselli's (1966, p. 121) conclusions, based on a review of fifty years of research, that general intelligence tests correlate .42 with trainability and .23 with proficiency across all types of jobs? Each of these correlations is based on over 10,000 cases. It is small wonder that psychologists believe intelligence tests are valid predictors of job success. Unfortunately, it is impossible to evaluate Ghiselli's conclusion, as he does not cite his sources and he does not state exactly how job proficiency was measured for each of his correlations. We can draw some conclusions from his results, however, and we can make a good guess that job proficiency often was measured by supervisors' ratings or by such indirect indicators of supervisors' opinions as turnover, promotion, salary increases, and the like.

What is interesting to observe is that intelligence test correlations with proficiency in higher-status jobs are regularly higher than with proficiency in lower-status jobs (Ghiselli, 1966, pp. 34, 78). Consider the fact that intelligence test scores correlate −.08 with proficiency as a canvasser or solicitor and .45 with proficiency as a stock and bond salesman. This should be a strong clue as to what intelligence tests are getting at, but most observers have overlooked it or simply assumed that it takes more general ability to be a stock and bond salesman than a canvasser. But these two jobs differ also in social status, in the language, accent, clothing, manner, and connections by education and family necessary for success in the job. The basic problem with many job proficiency measures for validating ability tests is that they depend heavily on the *cre-*

dentials the man brings to the job—the habits, values, accent, interests, etc.—that mean he is acceptable to management and to clients. Since we also know that social-class background is related to getting higher-ability test scores (Nuttall and Fozard, 1970), as well as to having the right personal credentials for success, *the correlation between intelligence test scores and job success often may be an artifact,* the product of their joint association with class status. Employers may have a right to select bond salesmen who have gone to the right schools because they do better, but psychologists do not have a right to argue that it is their *intelligence* that makes them more proficient in their jobs.

We know that correlation does not equal causation, but we keep forgetting it. Far too many psychologists still report average-ability test scores for high- and low-prestige occupations, inferring incorrectly that this evidence shows it takes more of this type of brains to perform a high-level than a low-level job. For instance, Jensen (1972, p. 9) wrote recently:

> Can the I.Q. tell us anything of practical importance? Is it related to our commonsense notions about mental ability as we ordinarily think of it in connection with educational and occupational performance? Yes, indeed, and there is no doubt about it. . . . The I.Q. obtained after 9 or 10 years of age also predicts final adult occupational status to almost as high a degree as it predicts scholastic performance. . . . The *average* I.Q. of persons within a particular occupation is closely related to that occupation's standing in terms of average income and the amount of prestige accorded to it by the general public.

He certainly leaves the impression that it is "mental ability as we ordinarily think of it" that is responsible for this association between average IQ scores and job prestige. But the association can be interpreted as meaning, just as reasonably, that it takes more *pull,* more opportunity, to get the vocabulary and other habits required by those in power from incumbents of high-status positions. Careful studies that try to separate the *credential* factor from the *ability* factor in job success have been very few in number.

Ghiselli (1966) simply did not deal with the problem of what the criteria of job proficiency may mean for validating the tests. For example, he reported a correlation of .27 between intelligence test scores and proficiency as a policeman or a detective (p. 83), with no attention given to the very important issues involved in how a policeman's performance is to be evaluated. Will supervisors' ratings do? If so, it discriminates against black policemen (Baehr et al., 1968) because white supervisors regard them as inferior. And what about the public? Shouldn't their opinion as to how they are served by the police be part of the criterion? The most recent careful review (Kent and Eisenberg, 1972) of the evidence relating ability test scores to police performance concluded that there is no stable, significant relationship. Here is concrete evidence that one must view with considerable skepticism the assumed relation of intelligence test scores to success on the job.

One other illustration may serve to warn the unwary about accepting uncritically simple statements about the role of ability, as measured by intelligence tests, in life outcomes. It is stated widely that intelligence promotes general adjustment and results in lower neuroticism. For example, Anderson (1960) reported a significant correlation between intelligence test scores obtained from boys in 1950, age 14–17, and follow-up ratings of general adjustment made five years later. Can we assume that intelligence promotes better adjustment to life as has been often claimed? It sounds reasonable until we reflect that the "intelligence" test is a test of ability to do well in school (to take academic type tests), that many of Anderson's sample were still in school or getting started on careers, and that those who are not doing well in school or getting a good first job because of it are likely to be considered poorly adjusted by themselves and others. Here the test has become part of the criterion and has introduced the correlation artificially. In case this sounds like special reasoning, consider the fact, not commented on particularly by Anderson, that the same correlation between "intelligence" test scores and adjustment in girls was an insignificant .06. Are we to conclude that intelligence does *not* promote adjustment in girls? It would seem more reasonable to argue that the particular ability tested, here associated with scholastic success, is more important to success (and hence adjustment) for boys than for girls. But this is a far cry from the careless inference that intelligence tests tap a general ability to adapt successfully to life's problems because high-IQ children (read "men") have better mental health (Jensen, 1972).

To make the point even more vividly, suppose you are a ghetto resident in the Roxbury section of Boston. To qualify for being a policeman you have to take a three-hour-long general intelligence test in which you must know the meaning of words like "quell," "pyromaniac," and "lexicon." If you do not know enough of those words or cannot play analogy games with them, you do not qualify and must be satisfied with some such job as being a janitor for which an intelligence test is not required yet by the Massachusetts Civil Service Commission. You, not unreasonably, feel angry, upset, and unsuccessful. Because you do not know those words, you are considered to have low intelligence, and since you consequently have to take a low-status job and are unhappy, you contribute to the celebrated correlations of low intelligence with low occupational status and poor adjustment. Psychologists should be ashamed of themselves for promoting a view of general intelligence that has encouraged such a testing program, particularly when there is no solid evidence that significantly relates performance on this type of intelligence test with performance as a policeman.

The Role of Power in Controlling Life-outcome Criteria

Psychologists have been, until recently, incredibly naïve about the role of powerful interests in controlling the criteria against which psychologists have validated their tests. Terman felt that his studies had proved conclusively that "giftedness," as he measured it with psychological tests, was a key factor in life success. By and large, psychologists have agreed with him. Kohlberg et al. (1970), for instance, in a recent summary statement concluded that Terman and Oden's (1947) study "indicated the gifted were more successful occupationally, maritally, and socially than the average group, and were lower in 'morally deviant' forms of psychopathology (e.g., alcoholism, homosexuality)." Jensen (1972, p. 9) agreed:

> One of the most convincing demonstrations that I.Q. is related to "real life" indicators of ability was provided in a classic study by Terman and his associates at Stanford University. . . . Terman found that for the most part these high-I.Q. children in later adulthood markedly excelled the general population on every indicator of achievement that was examined: a higher level of education completed; more scholastic honors and awards; higher occupational status; higher income; production of more articles, books, patents and other signs of creativity; more entries in *Who's Who*; a lower mortality rate; better physical and mental health; and a lower divorce rate. . . . Findings such as these establish beyond a doubt that I.Q. tests measure characteristics that are obviously of considerable

> importance in our present technological society. To say that the kind of ability measured by intelligence tests is irrelevant or unimportant would be tantamount to repudiating civilization as we know it.

I do not want to repudiate civilization as we know it, or even to dismiss intelligence tests as irrelevant or unimportant, but I do want to state, as emphatically as possible, that Terman's studies do *not* demonstrate unequivocally that it is the kind of ability measured by the intelligence tests that is responsible for (i.e., causes) the greater success of the high-IQ children. Terman's studies *may* show only that the rich and powerful have more opportunities, and therefore do better in life. And if that is even possibly true, it is socially irresponsible to state that psychologists have established "beyond a doubt" that the kind of ability measured by intelligence tests is essential for high-level performance in our society. For, by current methodological standards, Terman's studies (and others like them) were naïve. No attempt was made to equate for *opportunity* to be successful occupationally and socially. His gifted people clearly came from superior socioeconomic backgrounds to those he compared them with (at one point all men in California, including day laborers). He had no unequivocal evidence that it was "giftedness" (as reflected in his test scores) that was responsible for the superior performance of his group. It would be *as* legitimate (though also not proven) to conclude that sons of the rich, powerful, and educated were apt to be more successful occupationally, maritally, and socially because they had more material advantages. To make the point in another way, consider the data in table 1,

Table 1. *Numbers of Students in Various IQ and SES Categories (Sixth Grade) and Percentage Subsequently Going to College*

IQ	*Socioeconomic status*			
	High	*% to college*	*Low*	*% to college*
High	51	71	57	23
Low	33	18	96	5

Note. $\chi^2 = 11.99$, $p < .01$, estimated tetrachoric $r = .35$, SES × IQ. (Table adapted from Havighurst et al., 1962. Copyright by Wiley, 1962.)

which are fairly representative of findings in this area. They were obtained by Havighurst et al. (1962) from a typical town in Middle America. One observes the usual strong relationship between social class and IQ and between IQ and college-going—which leads on to occupational success. The traditional interpretation of such findings is that more stupid children come from the lower classes because their parents are also stupid which explains why they are lower class. A higher proportion of children with high IQ go to college because they are more intelligent and more suited to college study. This is as it should be because IQ predicts academic success. The fact that more intelligent people going to college come more often from the upper class follows naturally because the upper classes contain more intelligent people. So the traditional argument has gone for years. It seemed all very simple and obvious to Terman and his followers.

However, a closer look at table 1 suggests another interpretation that is equally plausible, though not more required by the data than the one just given. Compare the percentages going to college in the "deviant" boxes—high socioeconomic status and low IQ versus high IQ and low socioeconomic status. It appears to be no more likely for the bright children (high IQ) from the lower classes to go to college (despite their high aptitude for it) than for the "stupid" children from the upper classes. Why is this? An obvious possibility is that the bright but poor children do not have the money to go to college, or they do not want to go, preferring to work or do other things. In the current lingo, they are "disadvantaged" in the sense that they have not had access to the other factors (values, aspirations, money) that promote college-going in upper-class children. But now we have an alternative explanation of college-going—namely, socioeconomic status which seems to be as good a predictor of this type of success as ability. How can we claim that ability as measured by these tests is the critical factor in college-going? Very few children, even with good test-taking ability, go to college if they are from poor families. One could argue that they are victims of oppression: they do not have the opportunity or the values that permit or encourage going to college. Isn't it likely that the same oppressive forces may have prevented even more of them from learning to play school games well at all?

Belonging to the power elite (high socioeconomic status) not only helps a young man go to college and get jobs through contacts his family has, it also gives him easy access as a child to the credentials that permit him to get into certain occupations. Nowadays, those credentials include the words and word-game skills used in Scholastic Aptitude Tests. In the Middle Ages, they required knowledge of Latin for the learned professions of law, medicine, and theology. Only those young men who could read and write Latin could get into those occupations, and if tests had been given in Latin, I am sure they would have shown that professionals scored higher in Latin than men in general, that sons who grew up in families where Latin was used would have an advantage in those tests compared to those in poor families where Latin was unknown, and that these men were more likely to get into the professions. But would we conclude we were dealing with a general ability factor? Many a ghetto resident must or should feel that he is in a similar position with regard to the kind of English he must learn in order to do well on tests, in school, and in occupations today in America. I was recently in Jamaica where all around me poor people were speaking an English that was almost incomprehensible to me. If I insisted, they would speak patiently in a way that I could understand, but I felt like a slow-witted child. I have wondered how well I would do in Jamaican society if this kind of English were standard among the rich and powerful (which, by the way, it is not), and, therefore, required by them for admission into their better schools and occupations (as determined by a test administered perhaps by the Jamaican Testing Service). I would feel oppressed, not less intelligent, as the test would doubtless decide I was because I was so slow of comprehension and so ignorant of ordinary vocabulary.

When Cronbach (1970) concluded that such a test "is giving realistic information on the presence of a handicap," he is, of course, correct. But psychologists should recognize that it is those in power in a society who often decide what is a handicap. We should be a lot more cautious about accepting as ultimate criteria of ability the standards imposed by whatever group happens to be in power.

Does this mean that intelligence tests are invalid? As so often when you examine a question carefully in psychology, the answer depends on what you mean. Valid for what? Certainly they are valid for predicting who will get ahead in a number of prestige jobs where credentials are important. So is white skin: it too is a valid predictor of job success in prestige jobs. But no one would argue that white skin per se is an ability factor. Lots of the celebrated correlations between so-called intelligence test scores and success can lay no greater claim to representing an ability factor.

Valid for predicting success in school? Certainly, because school success depends on taking similar types of tests. Yet, neither the tests nor school grades seem to have much power to predict real competence in many life outcomes, aside from the advantages that credentials convey on the individuals concerned.

Are there *no* studies which show that general intelligence test scores predict competence with all of these other factors controlled? I can only assert that I have had a very hard time finding a good carefully controlled study of the problem because testers simply have not worked very hard on it: they have believed so much that they were measuring true competence that they have not bothered to try to prove that they were. Studies do exist, of course, which show significant positive correlations between special test scores and job-related skills. For example, perceptual speed scores are related to clerical proficiency. So are tests of vocabulary, immediate memory, substitution, and arithmetic. Motor ability test scores are related to proficiency as a vehicle operator (Ghiselli, 1966). And so on. Here we are on the safe and uncontroversial ground of using tests as criterion samples. But this is a far cry from inferring that there is a general ability factor that enables a person to be more competent in anything he tries. The evidence for this general ability factor turns out to be contaminated heavily by the power of those at the top of the social hierarchy to insist that the skills they have are the ones that indicate superior adaptive capacity.

Where Do We Go from Here?

Criticisms of the testing movement are not new. The Social Science Research Council Committee on Early Identification of Talent made some of these same points nearly fifteen years ago (McClelland et al., 1958). But the beliefs on which the movement is based are held so firmly that such theoretical or empirical objections have had little impact up to now. The testing movement continues to grow and extend into every corner of our society. It is unlikely that it can be simply stopped, although minority groups may have the political power to stop it. For the tests are clearly discriminatory against those who have not been exposed to the culture, entrance to which is guarded by the tests. What hopefully can happen is that testers will recognize what is going on and attempt to redirect their energies in a sounder direction. The report of the special committee on testing to the College Entrance Examination Board (1970) is an important sign that changes in thinking are occurring—if only they can be implemented at a practical level. The report's gist is that a wider array of talents should be assessed for college entrance and reported as a profile to the colleges. This is a step in the right direction if everyone keeps firmly in mind that the criteria for establishing the "validity" of these new measures really ought to be not *grades in school,* but "grades in life" in the broadest theoretical and practical sense.

But now I am on the spot. Having criticized what the testing movement has been doing, I feel some obligation to suggest alternatives. How would I do things differently or better? I do not mind making suggestions, but I am well aware that some of them are as open to criticism on other grounds as the procedures I have been criticizing. So I must offer them in a spirit of considerable humility, as approaches that at least some people might be interested in pursuing who are discouraged with what we have been doing. My goal is to brainstorm a bit on how things might be different, not to present hard evidence that my proposals are better than what has been done to date. How would one test for competence, if I may use that word as a symbol for an alternative approach to traditional intelligence testing?

1. *The best testing is criterion sampling.* The point is so obvious that it would scarcely be worth mentioning, if it had not been obscured so often by psychologists like McNemar (1964) and Jensen (1972) who tout a general intelligence factor. If you want to know how well a person can drive a car (the criterion), sample his ability to do so by giving him a driver's test. Do not give him a paper-and-pencil test for following directions, a general intelligence test, and so on. As noted above, there is ample evidence that tests which sample job skills will predict proficiency on the job.

Academic skill tests are successful precisely because they involve criterion sampling for the most part. As already pointed out, the Scholastic Aptitude Test taps skills that the teacher is looking for and will give high grades for. No one could object if it had been recognized widely that this was *all* that was going on when aptitude tests were used to predict who would do well in school. Trouble started only when people assumed that these skills had some more general validity, as implied in the use of words like intelligence.

Yet, even a little criterion analysis would show that there are almost no occupations or life situations that require a person to do word analogies, choose the most correct of four alternative meanings of a word, and the like.

Criterion sampling means that testers have got to get out of their offices where they play endless word and paper-and-pencil games and into the field where they actually analyze performance into its components. If you want to test who will be a good policeman, go find out what a policeman does. Follow him around, make a list of his activities, and sample from that list in screening applicants. Some of the job sampling will have to be based on theory as well as practice. If policemen generally discriminate against blacks, that is clearly not part of the criterion, because the law says that they must not. So include a test which shows the applicant does not discriminate. Also sample the vocabulary he must use to communicate with the people he serves since his is a position of interpersonal influence—and not the vocabulary that men who have never been on a police beat think it is proper to know. And do not rely on supervisors' judgments of who are the better policemen because that is not, strictly speaking, job analysis but analysis of what people think involves better performance. Baehr et al. (1968), for instance, found that black policemen in Chicago who were rated high by their superiors scored high on the Deference scale of the Edwards Personal Preference Test. No such relationship appeared for white policemen. In other words, if you wanted to be considered a good cop in Chicago and you were black, you had to at least talk as if you were deferent to the white power system. Any psychologist who used this finding to pick black policemen would be guilty of improper job analysis, to put it as mildly as possible.

Criterion sampling, in short, involves both theory and practice. It requires real sophistication. Early testers knew how to do it better than later testers because they had not become so caught up in the ingrown world of "intelligence" tests that simply were validated against each other. Testers of the future must relearn how to do criterion sampling. If someone wants to know who will make a good teacher, they will have to get videotapes of classrooms, as Kounin (1970) did, and find out how the behaviors of good and poor teachers differ. To pick future businessmen, research scientists, political leaders, prospects for a happy marriage, they will have to make careful behavioral analyses of these outcomes and then find ways of sampling the adaptive behavior in advance. The task will not be easy. It will require new psychological skills not ordinarily in the repertoire of the traditional tester. What is called for is nothing less than a revision of the role itself—moving it away from word games and statistics toward behavioral analysis.

2. *Tests should be designed to reflect changes in what the individual has learned.* It is difficult, if not impossible, to find a human characteristic that cannot be modified by training or experience, whether it be an eye blink or copying Kohs' block designs. To the traditional intelligence tester, this fact has been something of a nuisance because he has been searching for some unmodifiable, unfakeable index of innate mental capacity. He has reacted by trying to keep secret the way his tests are scored so that people will not learn how to do them better, and by selecting tests, scores on which are stable from one administration to the next. Stability is supposed to mean that the score reflects an innate aptitude that is unmodified by experience, but it could also mean that the test is simply insensitive to important changes in what the person knows or can do. That is, the skill involved may be so specialized, so unrelated to general experience, that even though the person has learned a lot, he performs the same in this specialized area. For example, being able to play a word game like analogies is apparently little affected by a higher education, which is not so surprising since few teachers ask their students to do analogies. Therefore, being able to do analogies is often considered a sign of some innate ability factor. Rather, it might be called an achievement so specialized that increases in general wisdom do not transfer to it and cause changes in it. And why should we be interested in such specialized skills? As we have seen, they predictably do not seem to correlate with any life-outcome criteria except those that involve similar tests or that require the credentials that a high score on the test signifies.

It seems wiser to abandon the search for pure ability factors and to select tests instead that are valid in the sense that scores on them change as the person grows in experience, wisdom, and ability to perform effectively on various tasks that life presents to him. Thus, the second principle of the new approach to testing becomes a corollary of the first. If one begins by using as tests samples of life-outcome behaviors, then one way of determining whether those tests are valid is to observe that the person's ability to perform them increases as his competence in the life-outcome behavior increases. For example, if excellence in a policeman is defined partly in terms of being evenhanded toward all minority groups, then a test of fair-mindedness (or lack of ethnocentrism) might be used to select policemen and also should reflect growth in fair-mindedness as a police recruit develops on the job. One of the hidden prejudices of psychology, borrowed from the notion of fixed inherited aptitudes, is that any trait, like racial prejudice, is unmodifiable by training. Once a bigot, always a bigot. There is no solid evidence that this trait or any other human trait cannot be changed. So it is worth insisting that a new test should be designed especially to reflect growth in the characteristic it assesses.

3. *How to improve on the characteristic tested should be made public and explicit.* Such a principle contrasts sharply with present practice in which psychologists have tried hard—backed up by the APA Ethics Committee—to keep answers to many of their tests a secret lest people practice and learn how to do better on them or fake high scores. Faking a high score is impossible if you are performing the criterion behavior, as in tests for reading, spelling, or driving a car. Faking becomes possible the more indirect the connection is between the test behavior and the criterion behavior. For example, in checking out hundreds of items for predicting flight training success, it may turn out that something like playing the piano as a boy has diagnostic validity. But no one knows exactly why: perhaps it has something to do with mechanical ability, perhaps with a social class variable, or with conscientiousness in practicing. The old-fashioned tester could not care less what the reason was as long as the item worked. But he had to be very careful about security because men who wanted to become pilots easily could report they had played the piano if they knew such an answer would help them be selected. If playing the piano actually helped people become better pilots—which no psychologist bothered to check out in World War II—then it might make some sense to make this known and encourage applicants to learn to play. That would be very like the criterion-sampling approach to testing proposed here, in which the person tested is told how to improve on the characteristic for which he will be tested.

Or to take another example, doing analogies is a task that predicts grades in school fairly well. Again no one knows quite why because schoolwork ordinarily does not involve doing analogies. So psychologists have had to be security conscious for fear that if students got hold of the analogies test answers, they might practice and become good at analogies and "fake" high aptitude. What is meant by faking here is that doing well on analogies is *not* part of the criterion behavior (getting good grades), or else it could hardly be considered faking. Rather, the test must have some indirect connection with good grades, so that doing well on it through practice destroys its predictive power: hence the high score is a "fake." The person can do analogies but that does not mean any longer that he will get better grades. Put this way, the whole procedure seems like a strange charade that testers have engaged in because they did not know what was going on, behaviorally speaking, and refused to take the trouble to find out as long as the items "worked." How much simpler it is, both theoretically and pragmatically, to make explicit to the learner what the criterion behavior is that will be tested. Then psychologist, teacher, and student can collaborate openly in trying to improve the student's score on the performance test. Certain school achievement tests, of course, follow this model. In the Iowa Test of Basic Skills, for instance, both pupil and teacher know how the pupil will be tested on spelling, reading, or arithmetic, how he should prepare for the test, how the tests will be scored, etc. What is proposed here is that *all* tests should follow this model. To do otherwise is to engage in power games with applicants over the secrecy of answers and to pretend knowledge of what lies behind correlations, which does not in fact exist.

4. *Tests should assess competencies involved in clusters of life outcomes.* If we abandon general intelligence or aptitude tests, as proposed, and move toward criterion sampling based on job analysis, there is the danger that the tests will become extremely specific to the criterion involved. For example, Project ABLE (Gagné, 1965) has identified over 50 separate skills that can be assessed for the exit level of millman apprentice (job family: woodworker and related occupations). They include skills like "measures angles," "sharpens tools and planes," and "identifies sizes and types of fasteners using gauges and charts." This approach has all of the characteristics of the new look in testing so far proposed: the tests are criterion samples; improvement in skill shows up in the tests; how to pass them is public knowledge; and both teacher and pupil can collaborate to improve test performance. However, what one ends up with is hundreds, even thousands, of specific tests for dozens of different occupations. For some purposes, it may be desirable to assess competencies that are more generally useful in clusters of life

outcomes, including not only occupational outcomes but social ones as well, such as leadership, interpersonal skills, etc. Project ABLE has been excellent at identifying the manual skills involved in being a service station attendant, but so far, it has been unable to get a simple index of whether or not the attendant is pleasant to the customers.

Some of these competencies may be rather traditional cognitive ones involving reading, writing, and calculating skills. Others should involve what traditionally have been called personality variables, although they might better be considered competencies. Let me give some illustrations.

(*a*) *Communication skills.* Many jobs and most interpersonal situations require a person to be able to communicate accurately by word, look, or gesture just what he intends or what he wants done. Writing is one simple way to test these skills. Can the person put together words in a way that makes immediate good sense to the reader? Word-game skills do not always predict this ability, as is often assumed. I will never forget an instance of a black student applicant for graduate school at Harvard who scored in something like the fifth percentile in the Miller Analogies Test, but who obviously could write and think clearly and effectively as shown by the stories he had written as a reporter in the college paper. I could not convince my colleagues to admit him despite the fact that he had shown the criterion behavior the Analogies Test is supposed to predict. Yet if he were admitted as a psychologist, he would be writing papers in the future, not doing analogies for his colleagues. It is amazing to me how often my colleagues say things like: "I don't care how well he can write. Just look at those test scores." Testers may shudder at this, and write public disclaimers, but what practically have they done to stop the spread of this blind faith in test scores?

In Ethiopia in 1968 we were faced with the problem of trying to find out how much English had been learned by high school students who had been taught by American Peace Corps volunteers. The usual way of doing this there, as elsewhere, is to give the student a "fill in the blanks," multiple-choice objective test to see whether the student knows the meaning of words, understands correct grammatical forms, etc. We felt that this left out the most important part of the criterion behavior: the ability to use English to communicate. So we asked students to write brief stories which we then coded objectively, not for grammatical or spelling correctness, but for complexity of thought which the student was able to express correctly in the time allotted. This gave a measure of English fluency that predictably did correlate with occupational success among Ethiopian adults and also with school success, although curiously enough it was significantly negatively related to a word-game skill (English antonyms) that more nearly approximates the usual test of English competence (Bergthold, 1969).

Important communication skills are nonverbal. When the proverbial Indian said, "White man speak with forked tongue," he doubtless meant, among other things, that what the white man was saying in words did not jibe with what he was doing or expressing nonverbally. The abilities to know what is going on in a social setting and to set the correct emotional tone for it are crucial life-outcome criteria. Newmeyer (1970), for instance, has found a way to measure success at enacting certain emotions so that others receive them correctly and to measure success at receiving the correct emotions over various enactors. He found that black boys at a certain age were consistently better than white boys at this particular kind of communication skill, which is a far more crucial type of criterion behavior than most paper-and-pencil tests sample.

(*b*) *Patience,* or response delay as phychologists would call it, is a human characteristic that seems essential for many life outcomes. For instance, it is desirable for many service occupations where clients' needs and demands can be irritating. It would seem particularly desirable in a policeman who has the power and authority to do great damage to people who irritate him. Kagan, Pearson, and Welch (1966) have shown that it is an easily measured human characteristic that is relatively stable over time and can be taught directly.

(*c*) *Moderate goal setting* is important in achievement-related games, as I have explained fully elsewhere (McClelland, 1961). In most life situations, it is distinctly preferable to setting goals either too high or too low, which leads more often to failure. Many performance situations have been devised which measure the tendency to set moderate, achievable goals and help the person learn how to set more realistic goals in the future (Alschuler et al., 1970; McClelland and Winter, 1969).

(*d*) *Ego development.* Many scholars (*see* Erikson, 1950; Loevinger, 1970; White, 1959) have reasoned that there is a general kind of competence which develops with age and to a higher level in some people than in others. Costa (1971a) recently has developed a Thematic Apperception Test code for ego development which appears to have many of the aspects sought in the new measurement direction proposed here. The thought characteristics sampled represent criterion behavior in the sense that at stage 1, for example, the person is thinking at a passive conformist level, whereas at stage 4, he represents people in his stories as taking initiative on behalf of others (a more developed competency). The score on this measure predicts very well which junior or high school students will be perceived by their teachers as more competent (even when correlations with intelligence and grade performance are removed), and furthermore, a special kind of education in junior high school moves students up the ego development scale significantly. That is, training designed to develop a sense of initiative produced results that were reflected sensitively in this score. Pupils and teachers can collaborate in increasing this kind of thinking which ought to prepare students for competent action in many spheres of life.

5. *Tests should involve operant as well as respondent behavior.* One of the greatest weaknesses of nearly all existing tests is that they structure the situation in advance and demand a response of a certain kind from the test taker. They are aimed at assessing the capacity of a person to make a certain kind of response or choice. But life outside of tests seldom presents the individual with such clearly defined alternatives as "Which dog is most likely to bite?" or "Complete the following number series: 1 3 6 10 15 —," or "Check the word which is most similar in meaning to lexicon." If we refer to these latter behaviors as respondents in the sense that the stimulus situation clearly is designed to evoke a particular kind of response, then life is much more apt to be characterized by operant responses in the sense that the individual spontaneously makes a response in the absence of a very clearly defined stimulus. This fact probably explains why most existing tests do not predict life-outcome behaviors. Respondents generally do not predict operants. To use a crude example, a psychologist might assess individual differences in the *capacity* to drink beer, but if he used this measure to predict actual beer consumption over time, the chances are that the relationship would be very low. How much beer a person can drink is not related closely to how much he does drink.

Testers generally have used respondent behaviors to save time in scoring answers and to get higher test-retest reliability. That is, the person is more likely to give the same response in a highly structured situation than in an unstructured one that allows him to emit any behavior. Yet, slavishly pursuing these goals has led to important lacks in validity of the tests because life simply is not that structured, and often does not permit one to choose between defined-in-advance responses. The *n* Achievement measure, which is an operant in the sense that the subject emits responses (tells stories) under only very vague instructions, has predicted over a twelve-to-fourteen-year period in three different samples those who will drift into entrepreneurial business occupations (McClelland, 1965). Here an operant is predicting an operant—the tendency to think spontaneously about doing better all the time predicts a series of spontaneous acts over time which leads the individual into an entrepreneurial occupation. But predicting from operants to respondents or vice versa does not work, at least for men (McClelland, 1966). The *n* Achievement score is not related to grades or academic test scores (respondent measures), nor do grades relate to entering entrepreneurial occupations (*see* McClelland, 1961).

Even within fairly structured test situations, it is possible to allow for more operant behavior than has been the usual practice. Not long ago, we tried to find an existing performance test on which a person with high *n* Achievement ought to do well because such a test might be a useful substitute for the Thematic Apperception Test storytelling measure in certain situations. Theoretically, such a test should permit operant behavior in which the individual generates a lot of alternatives for solving a problem in search of the most efficient solution. But to our surprise, we could find no such test. Tests of divergent thinking existed that counted the number of operants (e.g., original uses for a paper clip) an individual could come up with, but they did not require the person to find the best alternative. Most other tests simply required the person to find the one correct answer the test maker had built into the item. What was needed were test items to which there were many correct answers, among which one was better than others in terms of some criteria of efficiency that the person would have to apply. This task seemed more lifelike to us and certainly more like the type of behavior characteristic of people with high *n* Achievement. So we invented an Airlines Scheduling Test (Bergthold, 1969) in which the person is faced with a number of problems of getting a passenger from City A to City B by such and such a time at minimum expenditure in time, energy, money, and discomfort. From schedules provided, several alternative routes and connections can be generated (if the test taker is energetic enough to think them up) that will solve the problem, but one is clearly the most efficient. The test has promise in that it correlates with the *n* Achievement score at a low level. But the main point is that it requires more lifelike operant behavior in generating alternative solutions and therefore it should have more predictive power to a variety of situations in which what the person is expected to do is not so highly structured as in standard respondent tests.

6. *Tests should sample operant thought patterns to get maximum generalizability to various action outcomes.* As noted already, the movement toward defining behavioral objectives in occupational testing can lead to great specificity and huge inventories of small skills that have little general predictive power. One way to get around this problem is to focus on defining thought codes because, almost by definition, they have a wider range of applicability to a variety of action possibilities. That is, they represent a higher order of behavioral abstraction than any given act itself which has not the capacity to stand for other acts the way a word does. And in empirical fact, this is the way it has worked out. The *n* Achievement score—an operant thought measure—has many action correlates from goal setting and occupational styles to color and time-span preferences (McClelland, 1961) which individually have little power as "actones" to predict each other. A more recent example is provided by an operant thought measure of power motivation which has very low positive correlations with four action characteristics: drinking, gambling, accumulating prestige supplies, and confessing to having many aggressive impulses that are not acted on (McClelland et al., 1972). These action characteristics are completely unrelated to each other so that they would be unlikely to come out on the same dimension in a factor analysis. But what is particularly interesting is that they appear to be alternative outlets for the power drive because the power motivation score correlates much higher with the maximum expression of *any one* of these alternatives than it does with any one alone or with the sum of standard scores on all of them. The thought characteristic—here the desire to "have impact," to make a big splash—is the higher order abstraction that gives the test predictive power for alternative ways of making a big splash in action—by gambling, drinking, and so on. The tester of the future is likely to get farther in finding generaliz-

able competencies of characteristics across life outcomes if he starts by focusing on thought patterns rather than by trying to infer what thoughts must lie behind the clusters of action that come out in various factors in the traditional trait analysis.

However, I have been arguing for this approach for over twenty years, and as far as I can see, the testing movement has been affected little by my eloquence. Why? There are lots of reasons: People keep insisting that the *n* Achievement score is invalid because it will not predict grades in school—which is ironic since it was designed precisely to predict life outcomes and not grades in school. Or they argue it does not predict all types of achievement (Klinger, 1966)—when, of course, it is not supposed to, on theoretical grounds. But the practical problem (outside the tedium of content coding) is the unreliability of operant thought measures. Many of them are unreliable, though not all. Costa's (1971b) ego-development score has a test-retest stability coefficient over a year of .66, N = 223. Unreliability is a fatal defect if the goal of testing is to *select* people, let us say, with high *n* Achievement. For rejected applicants could argue that they had been excluded improperly or that they might have high scores the next time they took the test, and the psychologist would have no good defense. One could just imagine beleaguered psychologists trying to defend themselves against irate parents whose children had not gotten into a preferred college because their *n* Achievement scores were too low.

But the emphasis in the new testing movement should be as much on evaluating educational progress as it is on identifying fixed characteristics for selection purposes. The operant thought measures are certainly reliable enough for the former objective. The educator can use them to assess whether a certain class or an innovative approach to teaching has tended, on the average, to promote ego development in thought as assessed by Costa's measure. The educator does not care *which particular child* is high in the measure since he does not plan to use the measure to select the child for special treatment. So its unreliability does not matter. He, as an administrator, can use the test information to decide whether the goals of the school are being forwarded by one educational approach or another. In a sense, the very unreliability of the thought measures may be a virtue if they encourage educators to stop thinking only about selection and start thinking more about evaluating educational progress.

Does this mean that test reliability is always unimportant? Not at all. Sometimes it will be important to diagnose deficiencies reliably that are to be made up. On other occasions tests will have to be used to pick out those most likely to be able to do a particular job well. So something will have to be done about reliability. Thus, a man with a high *n* Achievement score is a better bet for a sales job than a man with a low *n* Achievement score, but the measure of *n* Achievement from content coding of thought samples is not very defensible for selection purposes because it is unreliable. In this instance, the thought code can be used as the criterion against which more reliable performance measures can be validated. For example, the Airlines Scheduling Test score is reliable, and if it turns out to be related consistently to the *n* Achievement score based on thought sampling, it can be used as a substitute for the latter in selection. In fact, the thought codes can be considered devices for finding the clusters of action patterns that can be measured more reliably to get indexes of various competency domains central to various life outcomes. For example, if it turns out that an elevated socialized power (*s* Power) score (McClelland et al., 1972) characterizes successful policemen more than unsuccessful ones—as would be expected—then the action correlates of socialized power, such as capacity to lead or be influential in social groups, can be used to select potentially good policemen. The *s* Power score itself could not be so used because it is unreliable and "fakeable" if you learn the scoring system, but it is essential as a validating criterion for more reliable measures because its wide network of empirical and theoretical relationships helps find the action characteristics that will be useful for selection purposes.

While the six principles just enumerated for the new testing movement may affect occupational testing, the fact remains that testing has had its greatest impact in the schools and currently is doing the worst damage in that area by falsely leading people to believe that doing well in school means that people are more competent and, therefore, more likely to do well in life because of some real ability factor. Concretely, what would an organization like the Educational Testing Service do differently if it were to take these six principles into account? As a start, it might have to drop the term intelligence from its vocabulary and speak of scholastic achievement tests that are more or less content specific. The noncontent-specific achievements (formerly called "aptitudes") do predict test-taking and symbol manipulation competencies, and these competencies are central to certain life-outcome criteria—like making up tests for others to pass or being proficient as a clerk (Ghiselli, 1966). But it is a serious practical and theoretical error to label them general intelligence on the basis of evidence now available.

Once the innate intelligence philosophy is discarded, it becomes apparent that the role of such a testing service is to report to schools a profile of scholastic and nonscholastic achievements in a number of different areas. Then, in the case of selection, it is for the college to decide whether it has the educational programs that will promote growth in given areas of low performance. If performance is already high, say, in mathematics, then the college probably can produce little improvement in that area and should ask itself in what other areas it can educate such a student, as shown by his lower levels of accomplishment at the outset. The profile particularly should include measures of such general characteristics as ego development or moral development (Kohlberg and Turiel, 1971) based on thought samples, because these general competencies ought to be improved by higher educational systems anyway.

The profile of achievements should be reported not only at entrance but at various points throughout the schooling to give teachers, administrators, and students feedback on whether growth in desired characteristics actually is occurring. Test results then become a device for helping students and teachers redesign the teaching-learning process to obtain mutually agreed-on objectives. Only then will educational testing turn from the sentencing procedure it now is into the genuine service it purports to be.

References

ALSCHULER, A. S., TABOR, D., and MCINTYRE, J. 1970. *Teaching Achievement Motivation.* Middletown, Conn.: Educational Ventures.

ANDERSON, J. E. 1960. "The Prediction of Adjustment over Time." In *Personality Development in Children,* ed. I. Iscoe and H. Stevenson. Austin: University of Texas Press.

BAEHR, M. E., FURCON, J. E., and FROEMEL, E. C. 1968. *Psychological Assessment of Patrolman Gratifications in Relation to Field Performance.* Washington, D.C.: Government Printing Office.

BERG, I. E. 1970. *Education and Jobs: The Great Training Robbery.* New York: Praeger Publishers.

BERGTHOLD, G. D. 1969. "The Impact of Peace Corps Teachers on Students in Ethiopia." Unpublished doctoral dissertation, Harvard University.

COLLEGE ENTRANCE EXAMINATION BOARD. 1970. *Report, Special Committee on Testing.* Princeton: Educational Testing Service.

COSTA, P. 1971a. "Introduction to the Costa Ego Development Manual." Department of Social Relations, Harvard University. Mimeographed.

———. 1971b. "Working Papers on Ego Development Validation Research." Harvard University. Mimeographed.

CRONBACH, L. J. 1970. *Essentials of Psychological Testing.* 3rd ed. New York: Harper & Row.

ELTON, C. F., and SHEVEL, L. R. 1969. *Who Is Talented? An Analysis of Achievement.* Research Report no. 31. Iowa City: American College Testing Program.

ERIKSON, E. H. 1950. *Childhood and Society.* New York: W. W. Norton & Co.

GAGNÉ, R. M. 1965. *The Conditions of Learning.* New York: Holt, Rinehart & Winston.

GHISELLI, E. E. 1966. *The Validity of Occupational Aptitude Tests.* New York: John Wiley & Sons.

HAVIGHURST, R. J., BOWMAN, P. H., LIDDLE, G. P., MATTHEWS, C. V., and PIERCE, J. V. 1962. *Growing Up in River City.* New York: John Wiley & Sons.

HOLLAND, J. L., and RICHARDS, J. M., JR. 1965. *Academic and Nonacademic Accomplishment: Correlated or Uncorrelated?* Research Report no. 2. Iowa City: American College Testing Program.

HOYT, D. P. 1965. *The Relationship between College Grades and Adult Achievement, a Review of the Literature.* Research Report no. 7. Iowa City: American College Testing Program.

HUDSON, L. 1960. "Degree Class and Attainment in Scientific Research." *British Journal of Psychology* 51: 67–73.

JENSEN, A. R. 1972. "The Heritability of Intelligence." *Saturday Evening Post* 244, no. 2:9, 12, 149.

KAGAN, J., PEARSON, L., and WELCH, L. 1966. "The Modifiability of an Impulsive Tempo." *Journal of Educational Psychology* 57: 359–365.

KENT, D. A., and EISENBERG, T. 1972. "The Selection and Promotion of Police Officers." *Police Chief,* February 20–29.

KLINGER, E. 1966. "Fantasy Need Achievement As a Motivational Construct." *Psychological Bulletin* 66: 291–308.

KOHLBERG, L., LACROSSE, J., and RICKS, D. 1970. "The Predictability of Adult Mental Health from Childhood Behavior." In *Handbook of Child Psychopathology,* ed. B. Wolman. New York: McGraw-Hill Book Co.

KOHLBERG, L., and TURIEL, E. 1971. *Moralization Research, the Cognitive Development Approach.* New York: Holt, Rinehart & Winston.

KOUNIN, J. S. 1970. *Discipline and Group Management in Classrooms.* New York: Holt, Rinehart & Winston.

LOEVINGER, J., and WESSLER, R. 1970. *Measuring Ego Development.* San Francisco: Jossey-Bass, 2 vols.

McClelland, D. C. 1961. *The Achieving Society.* New York: Van Nostrand Reinhold Co.

———. 1965. "*N* Achievement and Entrepreneurship: A Longitudinal Study." *Journal of Personality and Social Psychology* 1: 398–392.

———. 1966. "Longitudinal Trends in the Relation of Thought to Action." *Journal of Consulting Psychology* 30: 479–483.

———. 1972. "Education for Competence." In *Proceedings of the 1971 FOLEB Conference,* ed. H. Heckhausen and W. Edelstein. Berlin, Germany: Institut für Bildungsforschung in der Max-Planck-Gesellschaft.

McClelland, D. C., Baldwin, A. L., Bronfenbrenner, U., and Strodtbeck, F. L. 1958. *Talent and Society.* Princeton: D. Van Nostrand Co.

McClelland, D. C., Davis, W. N., Kalin, R., and Wanner, H. E. 1972. *The Drinking Man.* New York: Free Press.

McClelland, D. C., and Winter, D. G. 1969. *Motivating Economic Achievement.* New York: Free Press.

McNemar, Q. 1964. "Lost: Our Intelligence? Why?" *American Psychologist* 19: 871–882.

Newmeyer, J. A. 1970. "Creativity and Non-verbal Communication in Pre-Adolescent White and Black Children." Unpublished doctoral dissertation, Harvard University.

Nuttall, R. L., and Fozard, T. L. 1970. "Age, Socioeconomic Status and Human Abilities." *Aging and Human Development* 1: 161–169.

Taylor, C., Smith, W. R., and Ghiselin, B. 1963. "The Creative and other Contributions of One Sample of Research Scientists." In *Scientific Creativity: Its Recognition and Development,* ed. C. W. Taylor and F. Barron. New York: John Wiley & Sons.

Terman, L. M., and Oden, M. H. 1947. *The Gifted Child Grows Up.* Stanford: Stanford University Press.

Thorndike, R. L., and Hagen, E. 1959. *10,000 Careers.* New York: John Wiley & Sons.

Wechsler, D. 1958. *The Measurement and Appraisal of Adult Intelligence.* 4th ed. Baltimore: Williams & Wilkins.

White, R. W. 1959. "Motivation Reconsidered: The Concept of Competence." *Psychological Review* 66: 297–333.

Wing, C. W. Jr., and Wallach, M. A. 1971. *College Admissions and the Psychology of Talent.* New York: Holt, Rinehart & Winston.

PART II

Genetic Component of IQ Differences

Introduction

THIS SECTION DEALS WITH VARIOUS ASPECTS OF THE FOLLOWING QUESTION: Is there a genetic contribution to IQ differences? The main issues dealt with are:

- What sort of evidence can we have for heritability of a characteristic?
- What is the nature of the actual evidence on which heritability estimates are based?
- Is heritability of any interest?
- What is the relationship between heritability and other concepts of inheritance?
- What is the connection between heritability and changeability?
- What is the connection between heritability within a racial group and genotypic differences between races?
- Can we have evidence of a genotypic component of race differences?

Of the eleven pieces included, the first seven are concerned largely with genetic differences between races, while the last four are mainly about heritability within a given population.

Richard C. Lewontin's "Race and Intelligence" contains the first (and the clearest) criticism of Arthur R. Jensen's use of the claimed high heritability of IQ within the white population to support the claim that blacks have a lower average genotypic IQ. Lewontin also clarifies the rather loose relationship of the heritability of a characteristic to its changeability, and he comments on a wide variety of issues such as the arbitrariness of arguments for "general intelligence" and the desirability of organizing society "meritocratically."

In the course of restating those of his claims that he believes to be most important, Jensen says he did not state that one could *prove* that blacks have lower genotypic IQ based on a high heritability among whites and the lower IQ of blacks. He says, rather, that the latter two facts make it *probable* that blacks are lower in genotypic IQ. Lewontin replies that available data suggest that IQ has not been very highly selected for in whites; nonetheless, genetic drift may have produced genotypic IQ differences between races; current information, however, does not make it probable that races differ in genotypic IQ, nor do current data suggest the direction of the difference if there is one. Lewontin's article and his reply to Jensen contain a number of new (1974) footnotes.

Also included in this section is Sandra Scarr-Salapatek's article, which is a review of Jensen's original article, Herrnstein's article "IQ," and H. J. Eysenck's *The IQ Argument: Race, Intelligence and Education.*[1] She condemns Eysenck's book as "an uncritical popularization of Jensen's ideas," though she also commends it for pointing out that characteristics can often be changed markedly even if they have high heritability. Scarr-Salapatek agrees with much of what Herrnstein says, while rejecting his much-discussed conclusion that we are moving toward a society in which there will be no substantial intergenerational mobility between classes. She goes on to argue against Jensen that no conclusions about genotypic racial differences can now be drawn, though indirect evidence of such differences is available in principle. She also points out that even if IQ has high heritability, redistribution of existing types of environments can be expected to change IQ substantially, while the use of new types of environmental intervention could change IQ distributions radically.

J. M. Thoday's "Limitations to Genetic Comparison of Populations" argues that in present circumstances, no one can know whether there is a genotypic component to racial IQ differences, though nonetheless, there is a method for ascertaining the existence (but not the direction) of genotypic racial differences. Thoday discusses the kind of data that are usually used to settle such questions in nonhuman genetics and explains why such techniques cannot be applied to human racial differences. He then goes on to point out a number of confusions often involved in arguments for racial genotypic differences; he makes a number of distinctions that are important to understanding the issues involved in reasoning about genotypic racial differences.

Thoday's "Educability and Group Differences" is a review of the book by Arthur Jensen with the same title. Jensen's book is an attempt to show that 50 to 75 percent of the IQ differences between blacks and whites are due to racial genetic differences. While agreeing with Jensen that the heritability of IQ is high within the white population, Thoday argues that there are a number of errors of reasoning and fact in Jensen's argument for racial differences in genotypic IQ. Thoday states, for example, that Jensen's argument based on blacks' regression toward a lower mean than whites is fallacious, and that Jensen misrepresents data about Australian aborigines who have varying degrees of caucasian ancestry.

Parts of Jerry Hirsch's article, "Behavior-Genetic Analysis and Its Biosocial Consequences," are devoted to placing behavior genetics in the context of its racist origins. He describes the "typological way of thought" exemplified by Francis Galton's development of a composite photograph of "the Jewish type" and Karl Pearson's (the eminent English biostatistician) claim of Jewish inferiority. Hirsch also analyzes the concepts of heritability, norm of reaction, and genotype-environment interaction, and criticizes Jensen's assumption that the higher the heritability of a characteristic, the less its potential for environmental control.

Lewontin's "The Analysis of Variance and the Analysis of Causes" claims that analysis of variance of human characteristics into genetic and environmental components is of no interest. He argues that analysis of variance can (in most circumstances) provide only a very local picture of the relative contributions of genotype and environment, and that this picture is likely to be a poor guide to the relation between genotype and environment when the structure of the environment changes. Thus, analysis of variance is likely to yield misleading expectations about the effect of environmental changes.

David Layzer's "Science or Superstition? A Physical Scientist Looks at the IQ Controversy" (which has been extensively revised for this volume) argues that IQ measurements do not satisfy the formal requirements of the theory of polygenic inheritance, and thus that estimates of the heritability of IQ are meaningless. He also argues that there is no reason to believe IQ is an index of capacity to acquire higher cognitive skills, and that no foreseeable experimental methods can test hypotheses of genotypic IQ differences among ethnic groups.

"Heredity, Intelligence, Politics, and Psychology: I" is part of a talk Leon J. Kamin gave at many universities in 1972–1973. It is a critical study of some of the data and reasoning that comprise the bases for the claim that IQ has a heritability of 75 percent in Western white populations. Kamin says, for example, that Sir Cyril Burt's studies of identical twins reared apart report correlation coefficients that are the same to three decimal places even though the sample of twins grew from twenty-one pairs to fifty-three pairs of twins. He argues that this strange consistency as well as many equally strange inconsistencies throw a great deal of doubt on Burt's data. Jensen, who relied heavily on the Burt studies in arguing for high heritability, has since conceded, in effect, that many of Kamin's criticisms of Burt's studies are substantially correct.[2]

Arthur S. Goldberger's "Mysteries of the Meritocracy" is a dissection of Barbara Burks's classic heritability study and of R. J. Herrnstein's use of the study. Goldberger concludes that Herrnstein substantially misrepresents the study, and that, in addition, Burks's conclusion that heritability of IQ is high is not supported by her data.

Notes

1 H. J. Eysenck, *The IQ Argument: Race, Intelligence and Education* (New York: Library Press, 1971).

2 A. R. Jensen, "Kinship Correlations Reported by Sir Cyril Burt," *Behavior Genetics* 4, no. 1 (March 1974) : 1–28.

Race and Intelligence

Richard C. Lewontin

In the spring of 1653, Pope Innocent X condemned a pernicious heresy which espoused the doctrines of "total depravity, irresistible grace, lack of free will, predestination and limited atonement." That heresy was Jansenism and its author was Cornelius Jansen, Bishop of Ypres.

In the winter of 1968, the same doctrine appeared in the *Harvard Educational Review*. That doctrine is now called "Jensenism" by the *New York Times* magazine and its author is Arthur R. Jensen, professor of educational psychology at the University of California at Berkeley. It is a doctrine as erroneous in the twentieth century as it was in the seventeenth. I shall try to play the Innocent.

Jensen's article, "How Much Can We Boost IQ and Scholastic Achievement?" created such a furor that the *Review* reprinted it along with critiques by psychologists, theorists of education, and a population geneticist under the title "Environment, Heredity and Intelligence." The article first came to my attention when, at no little expense, it was sent to every member of the National Academy of Sciences by the eminent white Anglo-Saxon inventor, William Shockley, as part of his continuing campaign to have the Academy study the effects of interracial mating. It is little wonder that the *New York Times* found the matter newsworthy, and that Professor Jensen has surely become the most discussed and least read essayist since Karl Marx. I shall try, in this article, to display Professor Jensen's argument, to show how the structure of his argument is designed to make his point, and to reveal what appear to be deeply embedded assumptions derived from a particular world view, leading him to erroneous conclusions. I shall say little or nothing about the critiques of Jensen's article, which would require even more space to criticize than the original article itself.

The Position

Jensen's argument consists essentially of an elaboration on two incontrovertible facts, a causative explanation and a programmatic conclusion. The two facts are that black people perform, on the average, more poorly than whites on standard IQ tests, and that special programs of compensatory education so far tried have not had much success in removing this difference. His causative explanation for these facts is that IQ is highly heritable, with most of the variation among individuals arising from genetic rather than environmental sources. His programmatic conclusion is that there is no use in trying to remove the difference in IQ by education since it arises chiefly from genetic causes and the best thing that can be done for black children is to capitalize on those skills for which they are biologically adapted. Such a conclusion is so clearly at variance with the present egalitarian consensus and so clearly smacks of a racist elitism, whatever its merit or motivation, that a very careful analysis of the argument is in order.

The article begins with the pronouncement: "Compensatory education has been tried and it apparently has failed." A documentation of that failure and a definition of compensatory education are left to the end of the article for good logical and pedagogical reasons. Having caught our attention by whacking us over the head with a two-by-four, like that famous trainer of mules, Jensen then asks:

> What has gone wrong? In other fields, when bridges do not stand, when aircraft do not fly, when machines do not work, when treatments do not cure, despite all the conscientious efforts on the part of many persons to make them do so, one begins to question the basic assumptions, principles, theories, and hypotheses that guide one's efforts. Is it time to follow suit in education?

Who can help but answer that last rhetorical question with a resounding "Yes"? What thoughtful and intelligent person can avoid being struck by the intellectual and empirical bankruptcy of educational psychology as it is practiced in our mass educational systems? The innocent reader will immediately fall into close sympathy with Professor Jensen, who, it seems, is about to dissect educational psychology and show it up as a prescientific jumble without theoretic coherence or prescriptive competence. But the innocent reader will be wrong. For the rest of Jensen's article puts the blame for the failure of his science not on the scientists but on the children. According to him, it is not that his science and its practitioners have failed utterly to understand human motivation, behavior, and development, but simply that the damn kids are ineducable.

The unconscious irony of his metaphor of bridges, airplanes, and machines has apparently been lost on him. The fact is that, in the twentieth century, bridges do stand, machines do work, and airplans do fly because they are built on clearly understood mechanical and hydrodynamic principles which even moderately careful and intelligent engineers can put into practice. In the seventeenth century that was not the case, and the general opinion was that men would never succeed in their attempts to fly because flying was impossible. Jensen proposes that we take the same view of education and that, in the terms of his metaphor, fallen bridges be taken as evidence of the unbridgeability of rivers. The alternative explanation, that educational psychology is still in the seventeenth century, is apparently not part of his philosophy.

This view of technological failure as arising from ontological rather than epistemological sources is a common form of apology at many levels of practice. Anyone who has dealt with plumbers will appreciate how many things "can't be fixed" or "weren't meant to be used like that." Physicists tell me that their failure to formulate an elegant general theory of fundamental particles is a result of there not being any underlying regularity to be discerned. How often men, in their overweening pride, blame nature for their own failures. This professionalist bias, that if a problem were soluble it would have been solved, lies at the basis of Jensen's thesis which can only be appreciated when seen in this light.

Having begun with the assumption that IQ cannot be equalized, Jensen now goes on to why not. He begins his investigation with a discussion of the "nature of intelligence," by which he means the way in which intelligence is defined by testing and the correlation of intelligence test scores with scholastic and occupational performance. A very strong point is made that IQ testing was developed in a Western industrialized society specifically as a prognostication of success in that society by the generally accepted criteria. He makes a special point of noting that psychologists' notions of status and success have a high correlation with those of the society at large, so that it is entirely reasonable that tests created by psychologists will correlate highly with conventional measures of success. One might think that this argument, that IQ testing is "culture bound," would militate against Jensen's general thesis of the biological and specifically genetical basis of IQ differences. Indeed, it is an argument often used against IQ testing for so-called "deprived" children, since it is supposed that they have developed in a subculture that does not prepare them for such tests. What role does this "environmentalist" argument play in Jensen's thesis? Is it simply evidence of his total fairness and objectivity? No. Jensen has seen, more clearly than most, that the argument of the specific cultural origins of IQ testing, and especially the high correlation of these tests with occupational status, cuts both ways. For if the poorer performance of blacks on IQ tests has largely genetic rather than environmental causes, then it follows that blacks are also genetically handicapped for other high status components of Western culture. That is, what Jensen is arguing is that differences between cultures are, in large part, genetically determined and that IQ testing is simply one manifestation of those differences.

In this light, we can also understand his argument concerning the existence of "general intelligence" as measured by IQ tests. Jensen is at some pains to convince his readers that there is a single factor, *g*, which, in factor analysis of various intelligence tests, accounts for a large fraction of the variance of scores. The existence of such a factor, while not critical to the argument, obviously simplifies it, for then IQ tests would really be testing for "something" rather than just being correlated with scholastic and occupational performance. While Jensen denies that intelligence should be reified, he comes perilously close to doing so in his discussion of *g*.

Without going into factor analysis at any length, I will point out only that factor analysis does not give a unique result for any given set of data. Rather, it gives an infinity of possible results among which the investigator chooses according to his tastes and preconceptions of the models he is fitting. One strategy in factor analysis is to pack as much weight as possible into one factor, while another is to distribute the weights over as many factors as possible as equally as possible. Whether one chooses one of these or some other depends upon one's model, the numerical analysis only providing the weights appropriate for each model. Thus, the impression left by Jensen that factor analysis somehow naturally or ineluctably isolates one factor with high weight is wrong.

"True Merit"?

In the welter of psychological metaphysics involving concepts of "crystallized" as against "fluid" intelligence, "generalized" intelligence, "intelligence" as opposed to "mental ability," there is some danger of losing sight of Jensen's main point: IQ tests are culture bound and there is good reason that they should be, because they are predictors of culture-bound activities and values. What is further implied, of course, is that those who do not perform well on these tests are less well suited for high status and must paint barns rather than pictures. We read that "We have to face it: the assortment of persons into occupational roles simply is not 'fair' in any absolute sense. The best we can hope for is that true merit, given equality of opportunity, act as a basis for the natural assorting process." What a world view is there revealed. The most rewarding places in society shall go to those with "true merit" and that is the best we can hope for. Of course, Professor Jensen is safe since, despite the abject failure of educational psychology to solve the problems it has set itself, that failure does not arise from lack of true merit on the part of psychologists but from the natural intransigence of their human subjects.

Having established that there are differences among men in the degree to which they are adapted to higher status and high satisfaction roles in Western society, and having stated that education has not succeeded in removing these differences, Jensen now moves on to their cause. He raises the question of "fixed" intelligence and quite rightly dismisses it as misleading. He introduces us here to what he regards as the two real issues. "The first issue concerns the genetic basis of individual differences in intelligence; the second concerns the stability or constancy of the IQ through the individual's lifetime." Jensen devotes some three-quarters of his essay to an attempt to demonstrate that IQ is developmentally rather stable, being, to all intents and purposes, fixed after the age of eight, and that most of the variation in IQ among individuals in the population has a genetic rather than environmental basis. Before looking in detail at some of these arguments, we must again ask where he is headed. While Jensen argues strongly that IQ is culture bound, he wishes to argue that it is not environmentally determined. This is a vital distinction. IQ is culture bound in the sense that it is related to performance in a Western industrial society. But the determination of the ability to perform culturally defined tasks might itself be entirely genetic. For example, a person suffering from a genetically caused deaf-mutism is handicapped to different extents in cultures requiring different degrees of verbal performance, yet his disorder did not have an environmental origin.

Jensen first dispenses with the question of developmental stability of IQ. Citing Benjamin Bloom's survey of the literature, he concludes that the correlation between test scores of an individual at different ages is close to unity after the age of eight. The inference to be drawn from this fact is, I suppose, that it is not worth trying to change IQ by training after that age. But such an inference cannot be made. All that can be said is that, given the usual progression of educational experience to which most children are exposed, there is sufficient consistency not to cause any remarkable changes in IQ. That is, a child whose educational experience (in the broad sense) may have ruined his capacity to perform by the age of eight is not likely to experience an environment in his later years that will do much to alter those capacities. Indeed, given the present state of educational theory and practice, there is likely to be a considerable reinforcement of early performance. To say that children do not change their IQ is not the same as saying they cannot. Moreover, Jensen is curiously silent on the lower correlation and apparent plasticity of IQ at younger ages, which is, after all, the chief point of Bloom's work.

The Genetic Argument

The heart of Jensen's paper is contained in his long discussion of the distribution and inheritance of intelligence. Clearly, he feels that here his main point is to be established. The failure of compensatory education, the developmental stability of IQ, the obvious difference between the performance of blacks and whites can be best understood, he believes, when the full impact of the findings of genetics is felt. In his view, insufficient attention has been given by social scientists to the findings of geneticists, and I must agree with him. Although there are exceptions, there has been a strong professional bias toward the assumption that human behavior is infinitely plastic, a bias natural enough in men whose professional commitment is to changing behavior. It is as a reaction to this tradition, and as a natural outcome of his confrontation with the failure of educational psychology, that Jensen's own opposite bias flows, as I have already claimed.

The first step in his genetical argument is the demonstration that IQ scores are normally distributed or nearly so. I am unable to find in his paper any explicit statement of why he regards this point as so important. From repeated references to Sir Francis Galton, filial regression, mutant genes, a few major genes for exceptional ability and assortative mating, it gradually emerges that an underlying normality of the distribution appears to Jensen as an important consequence of genetic control of IQ. He asks: "is intelligence itself—not just our measurements of it—really normally distributed?" Apparently, he believes that if intelligence, quite aside from measurement, were really normally distributed, this would demonstrate its biological and genetical status. Aside from a serious epistemological error involved in the question, the basis for his concern is itself erroneous. There is nothing in genetic theory that requires, or even suggests, that a phenotypic character should be normally distributed, even when it is completely determined genetically. Depending upon the degree of dominance of genes, interaction between them, frequencies of alternative alleles at the various gene loci in the population, and allometric growth relations between various parts of the organism transforming primary gene effects, a character may have almost any unimodal distribution and under some circumstances even a multimodal one.

After establishing the near-normality of the curve of IQ scores, Jensen goes directly to a discussion of the genetics of continuously varying characters. He begins by quoting, with approbation, E. L. Thorndike's maxim: "In the actual race of life, which is not to get ahead, but to get ahead of somebody, the chief determining factor is heredity." This quotation, along with many others used by Jensen, shows a style of argument that is not congenial to natural scientists; however, it may be a part of other disciplines. There is a great deal of appeal to authority and the acceptance of the empirically unsubstantiated opinions of eminent authorities as a kind of relevant evidence. We hear of "three eminent geneticists," or "the most distinguished exponent [of genetical methods], Sir Cyril Burt." The irrelevance of this kind of argument is illustrated precisely by the appeal to E. L. Thorndike, who, despite his eminence in the history of psychology, made the statement quoted by Jensen in 1905, when nothing was known about genetics outside of attempts to confirm Mendel's paper. Whatever the eventual truth of his statement turns out to be, Thorndike made it out of his utter ignorance of the genetics of human behavior, and it can only be ascribed to the sheer prejudice of a Methodist Yankee.

Heritability

To understand the main genetical argument of Jensen, we must dwell, as he does, on the concept of heritability. We cannot speak of a trait being molded by heredity, as opposed to environment. Every character of an organism is the result of a unique interaction between the inherited genetic information and the sequence of environments through which the organism has passed during its development. For some traits, the variations in environment have little effect, so that once the genotype is known, the eventual form of the organism is pretty well specified. For other traits, specification of the genetic make-up may be a very poor predictor of the eventual phenotype because even the smallest environmental effects may affect the trait greatly. But for all traits, there is a many-many relationship between gene and character and between environment and character. Only by a specification of both the genotype and the environmental sequence can the character be predicted. Nevertheless, traits do vary in the degree of their genetic determination; this degree can be expressed, among other ways, by their heritabilities.

The distribution of character values, say, IQ scores, in a population arises from a mixture of a large number of genotypes. Each genotype in the population does not have a unique phenotype corresponding to it because the different individuals of that genotype have undergone somewhat different environmental sequences in their development. Thus, each genotype has a distribution of IQ scores associated with it. Some genotypes are more common in the population so their distributions contribute heavily to determining the overall distribution, while others are rare and make little contribution. The total variation in the population, as measured by the variance, results from the variation between the mean IQ scores of the different genotypes and the variation around each genotypic mean. The heritability of a measurement is defined as the ratio of the variance due to the differences between the genotypes to the total variance in the population. If this heritability were 1.0, it would mean that all the variation in the population resulted from differences between genotypes but that there was no environmentally caused variation around each genotype mean. On the other hand, a heritability of 0.0 would mean that there was no genetic variation because all individuals were effectively identical in their genes, and that all the variation in the population arose from environmental differences in the development of the different individuals.

Defined in this way, heritability is not a concept that can be applied to a trait in general, but only to a trait in a particular population, in a particular set of environments. Thus, different populations may have more or less genetic variation for the same character. Moreover, a character may be relatively insensitive to environment in a particular environmental range, but be extremely sensitive outside this range. Many such characters are known, and it is the commonest kind of relation between character and environment. Finally, some genotypes are more sensitive to environmental fluctuation than others so that two populations with the same genetic variance but different genotypes, and living in the same environments, may still have different heritabilities for a trait.

The estimation of heritability of a trait in a population depends on measuring individuals of known degrees of relationship to each other and comparing the observed correlation in the trait between relatives with the theoretical correlation from genetic theory. There are two difficulties that arise in such a procedure. First, the exact theoretical correlation between relatives, except for identical twins, cannot be specified unless there is detailed knowledge of the mode of inheritance of the character. A first-order approximation is possible, however, based upon some simplifying assumptions, and it is unusual for this approximation to be badly off.

A much more serious difficulty arises because relatives are correlated, not only in their heredities, but also in their environments. Two siblings are much more alike in the sequence of environments in which they developed than are two cousins or two unrelated persons. As a result, there will be an overestimate of the heritability of a character, arising from the added correlation between relatives from environmental similarities. There is no easy way to get around this bias in general, so that great weight must be put on peculiar situations in which the ordinary environmental correlations are disturbed. That is why so much emphasis is placed, in human genetics, on the handful of cases of identical twins raised apart from birth, and the much more numerous cases of totally unrelated children raised in the same family. Neither of these cases is completely reliable, however, since twins separated from birth are nevertheless likely to be raised in families belonging to the same socioeconomic, racial, religious, and ethnic categories, while unrelated children raised in the same family may easily be treated rather more differently than biological siblings. Despite these difficulties, the weight of evidence from a variety of correlations between relatives puts the heritability of IQ in various human populations between .6 and .8.* For reasons of his argument, Jensen prefers the higher value but it is not worth quibbling over. Volumes could be written on the evaluation of heritability estimates for IQ, and one can find a number of faults with Jensen's treatment of the published data. However, it is irrelevant to questions of race and intelligence and to questions of the failure of compensatory education, whether the heritability of IQ is .4 or .8, so I shall accept Jensen's rather high estimate without serious argument.

The description I have given of heritability, its application to a specific population in a specific set of environments, and the difficulties in its accurate estimation are all discussed by Jensen. While the emphasis he gives to various points differs from mine, and his

* This last sentence, taken out of context, has sometimes been misunderstood to mean that I, in fact, accept the estimate of heritability of 60–80%. Nothing could be further from the truth. First, there is no such thing as *the* heritability of IQ, since heritability of a trait is different in different populations at different times. Second, the data on which the estimate of 80% for Caucasian populations is based, are themselves of very doubtful status. Jensen leans heavily upon the twin studies of Burt, to whom he appeals over and over, and whom he characterizes as "the most distinguished exponent [of genetical methods]." But it now appears that Burt's publications contain a large number of highly suspicious discrepancies and coincidences, and that Burt manipulated the data in very unobjective ways. These discrepancies were first pointed out by Professor Leon Kamin in an address to the American Psychological Association (reprinted in this book) and now more thoroughly documented in his book, *The Science and Politics of IQ* (Potomac, Md.: Erlbaum Associates, 1974). Kamin's argument was so devastating that Jensen himself has now discounted the entire corpus of Burt's published work on heritability of IQ ("Kinship Correlations Reported by Sir Cyril Burt," *Behavior Genetics* 4, no. 1 [March 1974]: 1–28). Kamin has also shown that other studies of twins suffer from serious biases, and the general conclusion of his book is that there is currently insufficient evidence to assign *any* nonzero heritability to IQ in any population.

estimate of heritability is on the high side, he appears to have said in one way or another just about everything that a judicious man can say. The very judiciousness of his argument has been disarming to geneticists especially, and they have failed to note the extraordinary conclusions that are drawn from these reasonable premises. Indeed, the logical and empirical hiatus between the conclusions and the premises is especially striking and thought-provoking in view of Jensen's apparent understanding of the technical issues.

The first conclusion concerns the cause of the difference between the IQ distributions of blacks and whites. On the average, over a number of studies, blacks have a distribution of IQ scores whose mean is about 15 points—about 1 standard deviation—below whites. Taking into account the lower variance of scores among blacks than among whites, this difference means that about 11 percent of blacks have IQ scores above the mean white score (as compared with 50 percent of whites) while 18 percent of whites score below the mean black score (again, as compared to 50 percent of blacks). If, according to Jensen, "gross socioeconomic factors" are equalized between the tested groups, the difference in means is reduced to 11 points. It is hard to know what to say about overlap between the groups after this correction, since the standard deviations of such equalized populations will be lower. From these and related observations, and the estimate of .8 for the heritability of IQ (in white populations, no reliable estimate existing for blacks), Jensen concludes that

> all we are left with are various lines of evidence, no one of which is definitive alone, but which, viewed altogether, make it a not unreasonable hypothesis that genetic factors are strongly implicated in the average Negro-white intelligence difference. The preponderance of evidence is, in my opinion, less consistent with a strictly environmental hypothesis than with a genetic hypothesis, which, of course, does not exclude the influence of environment on its interaction with genetic factors.

Anyone not familiar with the standard litany of academic disclaimers ("not unreasonable hypothesis," "does not exclude," "in my opinion") will, taking this statement at face value, find nothing to disagree with since it says nothing. To contrast a "strictly environmental hypothesis" with "a genetic hypothesis which . . . does not exclude the influence of the environment" is to be guilty of the utmost triviality. If that is the only conclusion he means to come to,

Jensen has just wasted a great deal of space in the *Harvard Educational Review*. But of course, like all cant, the special language of the social scientist needs to be translated into common English. What Jensen is saying is: "It is pretty clear, although not absolutely proved, that most of the difference in IQ between blacks and whites is genetical." This, at least, is not a trivial conclusion. Indeed, it may even be true. However, the evidence offered by Jensen is irrelevant.

Is It Likely?

How can that be? We have admitted the high heritability of IQ and the reality of the difference between the black and the white distributions. Moreover, we have seen that adjustment for gross socio-economic level still leaves a large difference. Is it not then likely that the difference is genetic? No. It is neither likely nor unlikely. There is no evidence. The fundamental error of Jensen's argument is to confuse heritability of a character within a population with heritability of the difference between two populations. Indeed, between two populations, the concept of heritability of their difference is meaningless. This is because a variance based upon two measurements has only one degree of freedom and so cannot be partitioned into genetic and environmental components. The genetic basis of the difference between two populations bears no logical or empirical relation to the heritability within populations and cannot be inferred from it, as I will show in a simple but realistic example. In addition, the notion that eliminating what appear a priori to be major environmental variables will serve to eliminate a large part of the environmentally caused difference between the populations is biologically naïve. In the context of IQ testing, it assumes that educational psychologists know what the major sources of environmental difference between black and white performance are. Thus, Jensen compares blacks with American Indians whom he regards as far more environmentally disadvantaged. But a priori judgments of the importance of different aspects of the environment are valueless, as every ecologist and plant physiologist knows. My example will speak to that point as well.

Let us take two completely inbred lines of corn. Because they are completely inbred by self-fertilization, there is no genetic variation in either line, but the two lines will be genetically different from each other. Let us now plant seeds of these two inbred lines in flowerpots with ordinary potting soil, one seed of each line to a pot. After they have germinated and grown for a few weeks we will measure the height of each plant. We will discover variation in height from plant to plant. Because each line is completely inbred, the variation in height within lines must be entirely environmental, a result of variation in potting conditions from pot to pot. Then the heritability of plant height in both lines is 0.0. But there will be an average difference in plant height between lines that arises entirely from the fact that the two lines are genetically different. Thus the difference between lines is entirely genetical even though the heritability of height is 0!

Now let us do the opposite experiment. We will take two handsful from a sack containing seed of an open-pollinated variety of corn. Such a variety has lots of genetic variation in it. Instead of using potting soil, however, we will grow the seed in vermiculite watered with a carefully made up nutrient, Knop's solution, used by plant physiologists for controlled growth experiments. One batch of seed will be grown on complete Knop's solution, but the other will have the concentration of nitrates cut in half and, in addition, we will leave out the minute trace of zinc salt that is part of the necessary trace elements (30 parts per billion). After several weeks we will measure the plants. Now we will find variation within seed lots which is entirely genetical since no environmental variation within lots was allowed. Thus heritability will be 1.0. However, there will be a radical difference between seed lots which is ascribable entirely to the difference in nutrient levels. Thus, we have a case where heritability within populations is complete, yet the difference between populations is entirely environmental!

But let us carry our experiment to the end. Suppose we do not know about the difference in the nutrient solutions because it was really the carelessness of our assistant that was involved. We call in a friend who is a very careful chemist and ask him to look into the matter for us. He analyzes the nutrient solutions and discovers the obvious—only half as much nitrates in the case of the stunted plants. So we add the missing nitrates and do the experiment again. This time, our second batch of plants will grow a little larger but not much, and we will conclude that the difference between the lots is genetic since equalizing the large difference in nitrate level had so little effect. But, of course, we would be wrong, for it is the missing trace of zinc that is the real culprit. Finally, it should be pointed out that it took many years before the importance of minute trace elements in plant physiology was worked out because ordinary laboratory glassware will leach out enough of many trace elements to let plants grow normally. Should educational psychologists study plant physiology?

Having disposed, I hope, of Jensen's conclusion that the high heritability of IQ and the lack of effect of correction for gross socioeconomic class are presumptive evidence for the genetic basis of the difference between blacks and whites, I will turn to his second erroneous conclusion. The article under discussion began with the observation, which he documents, that compensatory education for the disadvantaged (blacks, chiefly) has failed. The explanation offered for the failure is that IQ has a high heritability and that therefore the difference between the races is also mostly genetical. Given that the racial difference is genetical, then environmental change and educational effort cannot make much difference and cannot close the gap very much between blacks and whites. I have already argued that there is no evidence one way or the other about the genetics of interracial IQ differences. To understand Jensen's second error, however, we will suppose that the difference is indeed genetical. Let it be entirely genetical. Does this mean that compensatory education, having failed, must fail? The supposition that it must arises from a misapprehension about the fixity of genetically determined traits. It was thought at one time that genetic disorders, because they were genetic, were incurable. Yet we now know that inborn errors of metabolism are indeed curable if their biochemistry is sufficiently well understood and if deficient metabolic products can be supplied exogenously. Yet in the normal range of environments, these inborn errors manifest themselves irrespective of the usual environmental variables. That is, even though no environment in the normal range has an effect on the character, there may be special environments, created in response to our knowledge of the underlying biology of a character, which are effective in altering it.

But we do not need recourse to abnormalities of development to see this point. Jensen says that "there is no reason to believe that the IQs of deprived children, given an environment of abundance, would rise to a higher level than the already privileged children's IQs." It is empirically wrong to argue that, if the richest environment experience we can conceive does not raise IQ substantially, that we have exhausted the environmental possibilities. In the seventeenth century, the infant mortality rates were many times their present level at all socioeconomic levels. Using what was then the normal range of environments, the infant mortality rate of the highest socioeconomic class would have been regarded as the limit below which one could not reasonably expect to reduce the death rate. But changes in sanitation, public health, and disease control—changes which are commonplace to us now but would have seemed incredible to a man of the seventeenth century—have reduced the infant mortality rates of disadvantaged urban Americans well below those of even the richest members of seventeenth-century society. The argument that compensatory education is hopeless is equivalent to saying that changing the form of the seventeenth-century gutter would not have a pronounced effect on public sanitation. What compensatory education will be able to accomplish when the study of human behavior finally emerges from its prescientific era is anyone's guess. It will be most extraordinary if it stands as the sole exception to the rule that technological progress exceeds by manyfold what even the most optimistic might have imagined.

The real issue in compensatory education does not lie in the heritability of IQ or in the possible limits of educational technology. On the reasonable assumption that ways of significantly altering mental capacities can be developed if it is important enough to do so, the real issue is what the goals of our society will be. Do we want to foster a society in which the "race of life" is "to get ahead of somebody" and in which true merit, be it genetically or environmentally determined, will be the criterion of men's earthly reward? Or do we want a society in which every man can aspire to the fullest measure of psychic and material fulfillment that social activity can produce? Professor Jensen has made it fairly clear to me what sort of society he wants.

I oppose him.

Race and the Genetics of Intelligence

A Reply to Lewontin

Arthur R. Jensen

PROFESSOR LEWONTIN (*Bulletin of the Atomic Scientists,* MARCH 1970, pp. 1–8) has likened my article "How Much Can We Boost IQ and Scholastic Achievement?" (*Harvard Educational Review,* Winter 1969) to the "pernicious heresy . . . of total depravity, irresistable grace, lack of free will, predestination and limited atonement" attributed to Bishop Jansen in the seventeenth century. Lewontin goes on to claim that the same doctrine is now called "Jensenism" (a term coined by the *Wall Street Journal*), and that Jensenism is "as erroneous in the twentieth century as it was in the seventeenth." Lewontin proposes to play the role of Pope Innocent X (who denounced Bishop Jansen in 1653) by holding up and condemning his own version, incomplete and distorted, of Jensenism.

Thus Lewontin sets the stage for the *ad hominem* flavor of the rest of his paper. His role may resemble that of Pope Innocent's in trying to put down what he perceives as a heresy, but readers of Lewontin's piece may be reminded of a closer ecclesiastical parallel in Bishop Wilberforce, who, in debating evolution with T. H. Huxley, resorted to commenting that Darwin's physiognomy bore a simian resemblance; and he begged to know of Huxley, "was it through his grandfather or grandmother that he claimed his descent from a monkey?" Thus we see Lewontin, albeit in a milder vein, referring to Edward L. Thorndike (probably America's greatest psychologist and a pioneer in twin studies of the heritability of intelligence) as a "Methodist Yankee" and to William Shockley (a Nobel Laureate in physics, author of some three hundred scientific articles, and winner of numerous scientific awards and distinctions) as "the eminent white Anglo-Saxon inventor." (True, Shockley has eighty-five patented inventions, including the junction transistor.) If Lewontin is trying to be uncomplimentary, it is interesting to see the labels he picks for this.

In connection with Lewontin's reference to Shockley, an error of fact calls for correction. Shockley has not urged the Academy to study "the effects of interracial mating." This is a distortion of Shockley's aim, which is to see the Academy openly encourage scientific inquiry into the genetics of human abilities and proclivities, including their racial aspects. Lewontin's approach makes it appear to me that he views the problems of criticizing my article as that of making a case for the "good guys" versus the "bad guys," and he wants there to be no doubt in the reader's mind that he is very much one of the good guys. Thus he finally makes it perfectly clear in the last few sentences of his article that he opposes me mainly for ideological reasons and not on scientific or technical grounds.

A Persistent Question

Lewontin's statement that "Jensen has made it fairly clear to me what sort of society he wants" is not based on knowledge that Lewontin has of my social or political philosophy. It is a subjective surmise reflecting Lewontin's antipathy for anyone who would raise the question of genetic racial intelligence differences in an obviously nonpolitical, scholarly context. The question of whether the observed racial differences in mental abilities and scholastic performance involve genetic as well as environmental factors is indeed tabooed. Nevertheless, it is a persistent question. My belief is that scientists in the appropriate disciplines must face the question and not repeatedly sweep it back under the rug. In the long run, the safest and sanest thing we can urge is intensive, no-holds-barred inquiry in the best tradition of science.

Before proceeding with comments on specific technical points in Lewontin's paper, it would be well to put them in proper perspective by giving a capsule summary of what my article was about.

Survey Findings

First, I reviewed some of the evidence and the conclusions of a nationwide survey and evaluation of the large, federally funded compensatory education programs made by the United States Commission on Civil Rights, which concluded that these special programs had produced no significant improvement in the measured intelligence or scholastic performance of the disadvantaged children whose educational achievements these programs were specifically intended to improve. The massive evidence presented by the Civil Rights Commission suggests to me that merely applying more of the same approach to compensatory education on a still larger scale is not at all likely to lead to the desired results, namely, increasing the benefits of public education to the disadvantaged. The well-documented fruitlessness of these well-intentioned compensatory programs indicates the importance of now questioning the assumptions, theories and practices on which they were based.

I agree with Lewontin that these assumptions, theories and practices—espoused over the past decade by the majority of educators, social and behavioral scientists—are bankrupt. I do not blame the children who fail to benefit from these programs, as Lewontin would have his readers think. A large part of the failure, I believe, has resulted from the failure and reluctance of the vast majority of the educational establishment, aided and abetted by social scientists, to take seriously the problems of individual differences in developmental rates, patterns of ability, and learning styles. The prevailing philosophy has been that all children are basically very much alike—they are all "average children"—in their mental development and capabilities, and that the only causes of the vast differences that show up as they go through school are due to cultural factors and home influences that mold the child even before he enters kindergarten. By providing the culturally disadvantaged with some of the cultural amenities enjoyed by middle-class children for a period of a year or two before they enter school, we are told, the large differences in scholastic aptitude would be minimized, and the schools could go on thereafter treating all children very much alike and expect nearly all to perform as average children for their grade in school.

It hasn't worked. And educators are now beginning to say, "Let's really look at individual differences and try to find a variety of instructional methods and differentiated programs that will accommodate these differences." Whatever their causes may be, it now seems certain that they are not so superficial as to be erased by a few months of "cultural enrichment," "verbal stimulation," and the like. I have pointed out that some small-scale experimental intervention programs, which gear specific instructional methods to developmental differences, have shown more promise of beneficial results than the large-scale programs based on a philosophy of general cultural enrichment and a multiplication of the resources in already existing programs for the average child.

The Opportunities

One of the chief obstacles to providing differentiated educational programs for children with different patterns of abilities, aside from the lack of any detailed technical knowledge as to how to go about this most effectively, is the fact that children in different visibly identifiable subpopulations probably will be disproportionately represented in different instructional programs. This highly probable consequence of taking individual differences really seriously is misconstrued by some critics as inequality of opportunity. But actually, one child's opportunity can be another's defeat. To me, equality of opportunity does not mean uniform treatment of all children, but equality of opportunity for a diversity of educational experiences and services. If we fail to take account either of innate or acquired differences in abilities and traits, the ideal of equality of educational opportunity can be interpreted so literally as to be actually harmful, just as it would be harmful for a physician to give all his patients the same medicine.

I know personally of many instances in which children with educational problems were denied the school's special facilities for dealing with such problems (small classes, specialist teachers, tutorial help, diagnostic services), not because the children did not need this special attention or because the services were not available to the school, but simply because the children were black and no one wanted to single them out as being different or in need of special attention. So instead, white middle-class children with similar educational problems get nearly all the attention and special treatment, and most of them benefit from it. No one objects, because this is not viewed by anyone as "discrimination." But some school districts have been dragged into court for trying to provide similar facilities for minority children with educational problems. In these actions, the well-intentioned plaintiffs undoubtedly viewed themselves as the good guys. Many children, I fear, by being forced into the educational mold of the "average child" from grade 1 on, are soon "turned off" on school learning and have to pay the consequences in frustration and defeat both in school and in the world of work for which their schooling has not prepared them.

I do not advocate abandoning efforts to improve the education of the disadvantaged. I urge increased emphasis on these efforts, in the spirit of experimentation, expanding the diversity of approaches and improving the rigor of evaluation in order to boost our chances of discovering the methods that will work best.

Learning and IQ

My article also dealt with my theory of two broad categories of mental abilities, which I call intelligence (or abstract reasoning ability) and associative learning ability. These types of ability appear to be distributed differently in various social classes and racial groups. While large racial and social class differences are found for intelligence, there are practically negligible differences among these groups in associative learning abilities, such as memory span and serial and paired-associate rote learning.

Research should be directed at delineating still other types of abilities and at discovering how the particular strengths of each individual's pattern of abilities can be most effectively brought to bear on school learning and on the attainment of occupational skills. By pursuing this path, I believe we can discover the means by which the reality of individual differences need not mean educational rewards for some children and utter frustration and defeat for others.

Intelligence

I pointed out that IQ tests evolved to predict scholastic performance in largely European and North American middle-class populations around the turn of the century. They evolved to measure those abilities most relevant to the curriculum and type of instruction, which, in turn, were shaped by the pattern of abilities of the children the schools were then intended to serve.

IQ or abstract reasoning ability is thus a selection of just one portion of the total spectrum of human mental abilities. This aspect of mental abilities measured by IQ tests is important to our society, but is obviously not the only set of educationally or occupationally relevant abilities. Other mental abilities have not yet been adequately measured; their distributions in various segments of the population have not been adequately determined; and their educational relevance has not been fully explored.

I believe a much broader assessment of the spectrum of abilities and potentials, and the investigation of their utilization for educational achievement will be an essential aspect of improving the education of children regarded as disadvantaged.

Inheritance

Much of my paper was a review of the methods and evidence that led me to the conclusion that individual differences in intelligence —that is, IQ—are predominantly attributable to genetic differences, with environmental factors contributing a minor portion of the variance among individuals. The heritability of the IQ—that is, the percentage of individual differences variance attributable to genetic factors—comes out to about 80 percent, the average value obtained from all relevant studies now reported.

These estimates of heritability are based on tests administered to European and North American populations and cannot properly be generalized to other populations. I believe we need similar heritability studies in minority populations if we are to increase our understanding of what our tests measure in these populations and how these abilities can be most effectively used in the educational process.

Class Differences

Although the full range of IQ and other abilities is found among children in every socioeconomic stratum in our population, it is well established that IQ differs, on the average, among children from different social class backgrounds. The evidence, some of which I referred to in my article, indicates to me that some of this IQ difference is attributable to environmental differences and some of it is attributable to genetic differences among social classes—largely as a result of differential selection of the parent generations for different patterns of ability.

I have not yet met or read a modern geneticist who disputes this interpretation of the evidence. In the view of geneticist C. O. Carter: "Sociologists who doubt this show more ingenuity than judgment." At least three sociologists who are students of this problem —Pitirim Sorokin, Bruce Eckland, and Otis Dudley Duncan—all agree that selective factors in social mobility and assortative mating have resulted in a genetic component in social class intelligence differences. As Eckland points out, this conclusion holds within socially defined racial groups but cannot properly be generalized among racial groups, since barriers to upward mobility have undoubtedly been quite different for various racial groups.

Race Differences

I have always advocated dealing with persons as individuals, each in terms of his own merits and characteristics, and I am opposed to according treatment to persons solely on the basis of their race, color, national origin, or social class background. But I am also opposed to ignoring or refusing to investigate the causes of the well-established differences among racial groups in the distribution of educationally relevant traits, particularly IQ.

I believe that the causes of observed differences in IQ and scholastic performance among different ethnic groups is, scientifically, still an open question, an important question and a researchable one. I believe that official statements such as: "It is a demonstrable fact that the talent pool in any one ethnic group is substantially the same as in any other ethnic group" (United States Office of Education, 1966), and "Intelligence potential is distributed among Negro infants in the same proportion and pattern as among Icelanders or Chinese, or any other group" (United States Department of Labor, 1965) are without scientific merit. They lack any factual basis and must be regarded only as hypotheses.

The fact that different racial groups in this country have widely separated geographic origins and have had quite different histories which have subjected them to different selective social and economic pressures make it highly likely that their gene pools differ for some genetically conditioned behavioral characteristics, including intelligence or abstract reasoning ability. Nearly every anatomical, physiological, and biochemical system investigated shows racial differences. Why should the brain be any exception? The reasonableness of the hypothesis that there are racial differences in genetically conditioned behavioral characteristics, including mental abilities, is not confined to the poorly informed, but has been expressed in writings and public statements by geneticists such as K. Mather, C. D. Darlington, R. A. Fisher, and Francis Crick, to name a few.

In my article, I indicated several lines of evidence which support my assertion that a genetic hypothesis is not unwarranted. The fact that we still have only inconclusive results with respect to this hypothesis does not mean that the opposite of the hypothesis is true. Yet some social scientists speak as if this were the case and have even publicly censured me for suggesting an alternative to purely environmental hypothesis of intelligence differences. Scientific investigation proceeds most effectively by means of what Platt has called "strong inference," pitting alternative hypotheses that lead to different predictions against one another and then putting the predictions to an empirical test.

Most environmentalist theories are so inadequate that they often fail to explain even the facts they were devised to account for. In this area, psychologists, sociologists, and anthropologists have not followed the usual methods of scientific investigation, which consist, in part, in testing rival hypotheses in such a way that empirical evidence can disconfirm either one or the other, or both. There has been only one acceptable hypothesis—the environmentalists'—and research has consisted largely of endless enumeration of subtler and subtler environmental differences among subpopulations and of showing their psychological, educational, and sociological correlates, without even asking if genetic factors are in any way implicated at any point in the correlational network. Social scientists, for the most part, simply decree, on purely ideological grounds, that all races are identical in the genetic factors that condition various behavioral traits, including intelligence. Most environmental hypotheses proposed to account for intelligence differences among racial groups, therefore, have not had to stand up to scientific tests of the kind that other sciences have depended upon for the advancement of knowledge. Until genetic, as well as environmental, hypotheses are seriously considered in our search for causes, it is virtually certain that we will never achieve a scientifically acceptable answer to the question of racial differences in intellectual performance.

Dysgenic Trends

Lewontin does not comment on my article's pointing to a problem which is socially more important than the question of racial differences per se, namely, the high probability of dysgenic trends in our urban slums. At least 16 percent of black children (as compared with less than 2 percent of white children) in our nation's schools are mentally retarded by the criterion of IQs under 70 and scholastic performance commensurate with this level of ability. The figure is much higher in "inner city" schools, and these children come from the largest families. How much of this retardation is attributable to genetic factors and how much to environmental influences, we do not know. It is my position that we should try to find out. What hope is there for improving this condition, and for ameliorating the frustration and suffering obviously implied by these facts if we do not discover the causes? Some of the causes are undoubtedly environmental, nutritional, pre- and perinatal, and cultural, and my article includes sections on all these factors. But I also suggest that genetic hypotheses (which, of course, do not exclude the effects of environment) be considered in our efforts to understand these conditions.

Census data show markedly higher birth rates among the poorest segments of the Negro population than among successful, middle-class Negroes. This social class differential in birth rate appears to be much greater in the Negro than in the white population. That is, the educationally and occupationally least able among Negroes have a higher reproductive rate than their white counterparts, and the most able segment of the Negro population has a lower reproductive rate than its white counterpart.

If social class intelligence differences within the Negro population have a genetic component, as in the white population, the condition I have described could create and widen the genetic intelligence differences between Negroes and whites. The social and educational implications of this trend, if it exists and persists, are enormous. The problem obviously deserves thorough investigation by social scientists and geneticists and should not be ignored or superficially dismissed as a result of well-meaning wishful thinking. The possible consequences of our failure seriously to study these questions may well be viewed by future generations as our society's greatest injustice to Negro Americans.

Specific Comments

I agree with Lewontin that much of educational psychology and educational practices are still in the seventeenth century, especially as regards recognition of individual and group differences. Just as the seventeenth-century alchemists tried to transmute base metals into gold, the twentieth-century alchemists in our schools would like to make all children conform to their concept of the average child, so that all can be taught the same things in the same way at the same pace.

Lewontin seems to believe that anything is possible, given sufficient technological implementation. But reality does not bow to technology. Technology depends upon a correct assessment of reality. With all our technological progress in the physical sciences since the seventeenth century, we have not yet produced the philosopher's stone that can change base metals into gold. Though this was the most highly sought goal of the forerunners of modern chemistry, it was abandoned as soon as scientists discovered the actual nature of matter. Scientific inquiry took the place of wishful thinking. So tremendous technological capabilities were never brought to bear on this prescientific goal of discovering the philosopher's stone. Yet men have found other ways to create wealth, ways compatible with reality.

Lewontin points out that "to say that children do not change their IQ is not the same as saying they cannot." I have never said anything to the contrary, but I would point out that no one knows how to change IQs appreciably, and in those few children in whom true large shifts in IQ are found, either there is no explanation or the explanation involves changes in physiological and biochemical factors. Except in the case of children reared in almost total social isolation, there is no known psychological or educational treatment that systematically will boost IQs more than the few points' gain that comes from direct practice in taking the tests. In writing about the high heritability of intelligence, I have stated: "This is not to say, however, that as yet undiscovered biological, chemical, or psychological forms of intervention in the genetic or developmental processes could not diminish the relative importance of heredity as a determinant of intellectual differences."

Although Lewontin dislikes E. L. Thorndike's statement ("In the actual race of life, which is not to get ahead but to get ahead of somebody, the chief determining factor is heredity"), it should be noted that the statement is found in an empirical paper by Thorndike based on twin correlations. The statement thus was not made out of "utter ignorance," and, in fact, it still emphasizes a most important point about heritability—that the genes do not fix an absolute level of performance but determine differences among individuals given equal opportunity.

Lewontin states that "one can find a number of faults with Jensen's treatment of the published data" pertaining to the heritability of IQ. I assume they are not very important faults, if existent at all, or Lewontin surely would have enumerated them. (A number of highly qualified geneticists have reviewed my treatment of quantitative genetics in the article and have found no faults with it.) I point out that heritability estimates for IQ range between about 0.6 and 0.9. Lewontin thinks I prefer the "higher" estimate of 0.8. I don't prefer it; I simply find that 0.81 turns out to be the average heritability value based on all the data which has been reported in the literature, and I have made a most thorough survey. Surely no one at all familiar with the relevant literature could reasonably argue that the evidence leads to conclusions significantly at variance with those in my article: that heredity is about twice as important as environment in accounting for IQ differences in the populations on which the heritability of IQ has been investigated.

The main thrust of Lewontin's argument, as he sees it, actually attacks only a straw man set up by himself: the notion that heritability of a trait within a population does not prove that genetic factors are involved in the mean difference between two different populations on the same trait. I agree. But nowhere in my *Harvard Educational Review* discussion of race differences do I propose this line of reasoning, nor have I done so in any other writings. I do, however, discuss many other lines of evidence which I believe are more consistent with a hypothesis that genetic factors are involved in the average Negro-white IQ differences than with purely environmental theories.

But let us further consider Lewontin's statement that heritability (i.e., proportion of variance attributable to genetic factors) within populations is irrelevant to the question of genetic differences between populations. Theoretically, this is true: It is possible to have genetic differences within populations and no genetic differences between populations which differ phenotypically; conversely, it is possible to have zero heritability within populations and complete genetic determination of the mean difference between populations. Therefore, heritability coefficients obtained within populations, no matter how high, cannot prove the existence of a genetic difference between populations. All this follows strictly from the quantitative logic of estimating heritability, and Lewontin has given some good concrete examples of this logic in the case of plant physiology. But it is necessary to distinguish between the possible and the probable, and between proof in the sense of mathematical tautology and the probabilistic statements that result from hypothesis testing in empirical science. The real question is not whether a heritability estimate, by its mathematical logic, can prove the existence of a genetic difference between two groups, but whether there is any probabilistic connection between the magnitude of the heritability and the magnitude of group differences. Given two populations (A and B) whose means on a particular characteristic differ by x amount, and given the heritability (h_A^2 and h_B^2) of the characteristic in each of the two populations, the probability that the two populations differ from one another genotypically as well as phenotypically is some monotonically increasing function of the magnitudes of h_A^2 and h_B^2. Such probabilistic statements are commonplace in all branches of science. It seems that only when we approach the question of genetic race differences do some scientists talk as though only one of two probability values is possible, either 0 or 1. The possibility for scientific advancement in any field would be in a sorry state if this restriction were a universal rule. Would Lewontin maintain, for example, that there would be no difference in the probability that two groups differ genetically where h^2 for the trait in question is 0.9 in each group as against the case where h^2 is 0.1? Pygmies average under five feet in height; the Watusis average over six feet. The fact that the heritability of physical stature is close to 0.9 does not prove that all the difference is not caused by environmental factors, but it is more probable that genetic factors may be involved in the difference than would be the probability in the case of a group difference in the amount of scarification (body markings) which very likely has a heritability close to zero. Since pygmies and Watusis live in very different environments, why should we not bet on the proposition that their difference in mean height is attributable entirely to environment? In short, the high heritability of height suggests a reasonable hypothesis. We would then look for other lines of evidence to test the hypothesis—for example, comparing the heights of pygmy orphans from birth in the Watusis tribe and vice versa; of pygmies and Watusis living in highly similar environments and eating the same foods; of the offspring of pygmy and Watusis matings; and so on. We can proceed similarly in studying group differences in behavioral characteristics. Within-group heritability estimates thus can give us probabilistic clues as to which characteristics are most likely to show genetic differences between groups when investigated through all other available lines of evidence. If a genetic hypothesis of Negro-white differences in intelligence is not plausible to Lewontin, he does not tell us why, nor does he offer a more plausible hypothesis. Lewontin merely shows his bias when he repeatedly says I am "wrong" and "in error," instead of saying why he disagrees with the tenability of the hypothesis I have proposed to account for the data.

Negro and Indian

The comparison I drew between Negro and American Indian children in IQ and scholastic performance was perfectly valid. It shows that despite greater environmental disadvantage, as assessed by twelve different indices, the Indian children, on the average, exceeded the Negro in IQ and achievement. But I did not pick the environmental indices. The sociologists picked them. They are those environmental factors most often cited by social scientists as the cause of the Negroes' poor performance on IQ tests and in school work. Does not the fact that another group which rates even lower than the Negro on these environmental indices (Indians are as far below Negroes as Negroes are below whites), yet displays better intellectual performance, bring into question the major importance attributed to these environmental factors by sociologists? Or should we grant immunity from empirical tests to sociological theories when they are devised to explain racial differences?

There is an understandable reluctance to come to grips scientifically with the problem of race differences in intelligence—to come to grips with it, that is to say, in the same way that scientists would approach the investigation of any other phenomenon. This reluctance is manifested in a variety of "symptoms" found in most writings and discussions of the psychology of race differences, particularly differences in mental ability. These include a tendency to remain on the remotest fringes of the subject; to sidestep central questions; to blur the issues; and tolerate a degree of vagueness in definitions, concepts, and inferences that would be unseemly in any other realm of scientific discourse. The writings express an unwarranted degree of skepticism about reasonably well-established quantitative methods and measurements. They deny or belittle already generally accepted facts—accepted, that is, when brought to bear on inferences outside the realm of race differences—and demand practically impossible criteria of certainty before even seriously proposing or investigating genetic hypotheses, as contrasted with extremely uncritical attitudes toward purely environmental hypotheses. There is a failure to distinguish clearly between scientifically answerable aspects of the question and the moral, political, and social policy issues; a tendency to beat dead horses and to set up straw men on what is represented as the genetic side of the argument. We see appeals to the notion that the topic is either really too unimportant to be worthy of scientific curiosity or too complex, or too difficult, or that it is forever impossible for any kind of research to be feasible, or that answers to key questions are fundamentally "unknowable" in any scientifically acceptable sense. Finally, there is complete denial of intelligence and race as realities, or as quantifiable attributes, or as variables capable of being related to one another and there follows, ostrichlike, dismissal of the subject altogether.

These tendencies will be increasingly overcome the more widely and openly the subject is discussed among scientists and scholars. As some of the taboos against the public discussion of the topic fall away, the issues will become clarified on a rational basis. We will come to know better just what we do and do not yet know about the subject, and we will be in a better position to deal with it objectively and constructively. I believe my article has made a substantial contribution toward this goal. It has provoked serious thought and discussion among leaders in genetics, psychology, sociology, and education concerned with these important fundamental issues and their implications for public education. I expect that my work will stimulate further relevant research as well as efforts to apply the knowledge gained thereby to educationally and socially beneficial purposes.

In my view, society will benefit most if scientists and educators treat these problems in the spirit of scientific inquiry rather than as a battlefield upon which one or another preordained ideology may seemingly triumph.

Further Remarks on Race and the Genetics of Intelligence

Richard C. Lewontin

PROFESSOR JENSEN HAS, UNDERSTANDABLY, RESPONDED AT SOME length to my analysis of his article "How Much Can We Boost IQ and Scholastic Achievement?" In large part, his response only reinforces many of the points I made about his original article, but he does raise some new and very interesting issues. I shall try to deal with his reply as briefly as I can.

Jensen's overall objection is that my article makes liberal use of ad hominem argumentation in an attempt to establish myself as a good guy attacking a bad guy. Thus, Professor Jensen establishes himself as a dispassionate scientist who, having written an objective empirical scientific paper, is attacked on nonscientific, ideological grounds. But Jensen is wrong in two respects. There is no ad hominem argument in my article. I confess to one episode of self-caricature when I compared my role to that of a Pope denouncing a heresy, and to a rather vulgar attempt to have some fun with Dr. Shockley by describing him in reverse racist terms, but in no case is my argument about Dr. Jensen's paper made on any grounds but its merit and logic. Indeed, my remarks about E. L. Thorndike are the best demonstration of that. In what Jensen informs us is "an empirical paper based on twin correlations," Thorndike makes the remarkable statement that "In the actual race of life, which is not to get ahead, but to get ahead of somebody, the chief determining factor is heredity." That maxim is a conjunction of a socioeconomic prejudice about the nature of human relations and a scientific statement with completely inadequate theoretical and experimental basis. The paper in question appeared in 1905, thirteen years before Fisher's paper establishing the statistical theory on which heritabilities are estimated, ten years before Fisher worked out the sampling distribution of the correlation coefficient, and five years before Morgan's chromosome theory of inheritance. In an attempt to explain how "America's greatest psychologist" could have made such an obviously unscientific statement, I postulated that it was a prejudice that might be expected from the son of a New England Methodist clergyman. I did not attempt by that hypothesis to discredit Thorndike's statement. It discredits itself.

But more important, Jensen's article is not an objective empirical scientific paper which stands or falls on the correctness of his calculation of heritability. It is, rather, a closely reasoned ideological document springing, as I have shown, from deep-seated professionalist bias and permeated, like Thorndike's work, with an elitist and competitive world view. While Jensen's original article gave many instances of this world view, of which I quoted a few, his reply to my analysis provides yet fresh evidence. Thus, we read of the attempt to equalize children's school performance as being analogous to the attempt to "transmute base metals into gold." Jensen speaks of "particular strengths in each individual's pattern of abilities" as if he regarded those differences in a valuefree way, objectively. Yet a little later, he discusses "dysgenic trends" among blacks. How revealing is rhetoric.

Weight of Authority

As in his original article, Jensen in his reply relies heavily on the weight of authority as relevant evidence. We hear of a "Nobel Laureate in physics," or "three sociologists who are students of this problem" and who "all agree," "geneticists such as K. Mather, C. D. Darlington, R. A. Fisher, and Francis Crick, to name a few," and finally, "a number of highly qualified geneticists" who have reviewed his "treatment of quantitative genetics and have found no fault with it." Well, I am a very highly qualified geneticist whose field is the study of genetic variation in natural populations, and I found a few faults. For example, the estimate of heritability by ratios of correlation differences are upwardly biased by environmental correlations which may be considerable. (One of those other "highly qualified geneticists" also points this out in his comments on Jensen's article in the *Harvard Educational Review*, so Jensen's phalanx of authorities is not quite unbroken.) Moreover, heritabilities, being ratios, should not be averaged in the usual way. No standard error is given for Jensen's estimate of heritability. No examination of the sensitivity of heritability estimates to different genetic models is given. This is important if there is a lot of dominance variance. And so on. But if authority is evidence, what do we do when authorities disagree? We might take a vote, but I do not think Jensen would favor that technique any more than I would, especially in view of the fact that the membership at last year's meeting of the National Academy voted almost unanimously not to consider the question of race and the inheritance of IQ. (They have since reversed themselves.)

Jensen has parried my major scientific thrust at his thesis by saying that I have demanded an unrealistic level of proof. I share with Jensen an impatience with that smart aleck who is always telling us our evidence is only circumstantial and we have not really proved our point. But that is not my case at all. I think there is an honest misunderstanding here, not simply a polemical question. If two populations have very low heritability for a trait but differ from each other on the average, there are three possibilities. Each population may have been highly inbred, in which case the genetic component of the differences between them may be very high. Each population may have been subject to a different force of natural selection, again causing them each to be nearly homozygous, so that again the difference between populations might be chiefly genetic. Finally, both populations might be highly variable genetically, in which case the populations almost certainly owe their observed difference almost entirely to environment.* One cannot assign a priori probabilities (or likelihoods, better) to these

* An underlying assumption of Jensen's argument is that the observed *direction* of difference between black and white populations reveals an underlying average genetic difference *in the same direction*. But this is an unjustified assumption. Even if there were a nontrivial genetic difference between blacks and whites for genes influencing IQ, it does not follow that elimination of environmental differences would result in simply narrowing the gap in performance. It could as easily result in producing a gap in the *opposite direction*, with blacks outperforming whites on the average. Which group would have the better average performance in a uniform environment depends upon the particular form of relationship between environment and phenotype characteristic of the genes in question (there is absolutely no evidence of any kind on this issue) and on the particular set of environments chose for the populations.

three situations. In any common-sense meaning of the word, they have equal likelihoods, since all three circumstances occur quite frequently in the history of species. What about the reverse situation, the one applicable to our problem? If two populations have high heritabilities for a character and there is an average difference between them, is that difference mostly genetical? One possibility is that the populations differ genetically because of a previous history of differential selection of a type that causes genetic variation to be stabilized. Another possibility is that the populations may differ genetically because of historical accidents of genetic sampling (genetic drift) without differential selection. A third possibility is that the populations are genetically much alike but live in environments that differ from each other in some critical limiting factor. All of these occur in nature, and again no a priori likelihoods can be fairly assigned to them.

For the race problem, however, we can say something because of other information. The first possibility is quite unlikely because the result of selection would be the elimination of additive genetic variance, leaving only dominance and interaction variance. But Burt's data, quoted by Jensen, show that 48 percent of the variance in IQ is additive genetic variance. This is a high figure for a quantitative trait in general, and absurdly high for any trait that has long been under natural selection. It appears that IQ has been selectively neutral, at least over much of our species history.* The second and third possibilities are more or less equally likely explanations of the situation in man, and I would not care to bet the educational future of any children on one or the other.

* The most fundamental and general proposition in the population genetics of quantitative characters is Fisher's "Fundamental Theorem of Natural Selection," which states that changes in a character under selection are equal to the *additive* genetic variance in the character, and which has a corollary that during the process of selection, additive genetic variance will disappear (because the frequencies of genotypes change during selection), leaving behind only nonadditive variance. This loss of additive variance is generally a process that occurs more slowly than the change in mean itself, so that small changes in mean will not necessarily exhaust the additive variance. It is sometimes erroneously claimed that the presence of nonadditive variance is an evidence that selection *has* gone on. (*See* J. Jinks and D. Fulker, "Comparison of the Biometrical, Genetical, MAVA, and Classical Approaches to the Analysis of Human Behavior," *Psychological Bulletin* 73 [1970]: 311–349.) This is based on the hypothesis that in the very early stages of evolution of a species, all the variance from a character is additive, and that nonadditive variance evolves by the modi-

Future IQs

Jensen remarks that I have said nothing about dysgenic trends among blacks which he regards as "socially more important than the question of racial differences per se." Apparently, Jensen believes that lower IQ blacks are out-breeding higher IQ blacks so that the average difference between blacks and whites will become even greater than it is. The evidence for this is indirect and is of the form: lower socioeconomic classes have more children than higher ones; lower socioeconomic classes have lower IQ scores; IQ score is highly heritable; therefore, IQ will decrease. Such eminent geneticists as C. D. Darlington and R. A. Fisher used to make this argument about social classes among whites too, but they were proved wrong by a then unknown human geneticist who teaches at Grand Valley State College. Carl Bajema (whose work is quoted by Jensen) showed that the old story that lower IQ classes out-breed higher IQ classes was the erroneous result of an egregious statistical blunder: They forgot to count women who had no children! In fact, women with low IQs have much bigger families when they have a family, but many fewer of them have families. The result is that the reproductive rate of the highest IQ classes is actually the highest. This information does not exist for blacks and all the information quoted by Jensen about blacks is of the pre-Bajema biased variety.

I would like to end my contribution to this controversy by returning to my original point. Jensen has spent a great deal of energy on the question of whether there is a genetic difference between blacks and whites in IQ. He believes this to be an important social question and not simply a matter of vulgar curiosity. But suppose the difference between the black and white IQ distributions were completely genetic: What program for social action flows from that fact? Should all black children be given a different education from all white children, even the 11 percent who are better than the average white child? Should all black men be unskilled laborers and all black women clean other women's houses?

---

fication of gene action during evolution. This is only a speculation, and even if the speculation were true, the rate of evolution of nonadditive variation by the modification of gene action has been shown by Fisher and Wright to be necessarily much slower than the rate of exhaustion of additive variance during the course of selection.

Jensen says he believes in the primacy of the individual; yet, he is deeply concerned with the genetic causation of group differences. Why? Because, he says:

> Since much of the current thinking behind civil rights, fair employment, and equality of educational opportunity appeals to the fact that there is a disproportionate representation of different racial groups in various levels of the educational, occupational, and socioeconomic hierarchy, we are forced to examine all possible reasons for this inequality. . . .

Not True

Nonsense. Does Jensen really believe that all the fuss about civil rights has occurred because someone noticed that blacks were underrepresented in college classes? It is simply not true that "we are forced to examine all possible reasons for this inequality." What we are morally obliged to do is to eliminate blackness per se as a cause of unequal treatment and for that program we have no need of genetics.

But that program cannot be accomplished unless we challenge a yet deeper flaw in Jensen's scheme. Putting questions of race quite aside, we must expose the fallacy that, because human behavior is chiefly genetically determined at present, it must always be so and ought always to be so. Children are different. They are different at birth and different when they reach school. But what Jensen continues to misunderstand is that whether those differences are genetical, maternal, obstetrical, or Oedipal, the decision about what role each child is to play eventually in society and what rewards he will receive, is social. At present, our society is truly one in which "the race is to get ahead of somebody" and nothing suits that dog-eat-dog philosophy better than the notion that winners, like heroes, are born, not made. But that is a social attitude, not an ineluctable biological result. In answer to Professor Jensen's rhetorical question "How Much Can We Boost IQ and Scholastic Achievement?" I say "As much or as little as our social values may eventually demand."

Unknowns in the IQ Equation

Sandra Scarr-Salapatek

IQ SCORES HAVE BEEN REPEATEDLY ESTIMATED TO HAVE A LARGE heritable component in United States and Northern European white populations.[1] Individual differences in IQ, many authors have concluded, arise far more from genetic than from environmental differences among people in these populations, at the present time, and under present environmental conditions. It has also been known for many years that white lower-class and black groups have lower IQs, on the average, than white middle-class groups. Most behavioral scientists comfortably "explained" these group differences by appealing to obvious environmental differences between the groups in standards of living, educational opportunities, and the like. But recently, an explosive controversy has developed over the heritability of between-group differences in IQ, the question at issue being: If individual differences within the white population as a whole can be attributed largely to heredity, is it not plausible that the average differences between social class groups and between racial groups also reflect significant genetic differences? Can the former data be used to explain the latter?

To propose genetically based racial and social class differences is anathema to most behavioral scientists, who fear any scientific confirmation of the pernicious racial and ethnic prejudices that abound in our society. But now that the issue has been openly raised and has been projected into the public context of social and educational policies, a hard scientific look must be taken at what is known and at what inferences can be drawn from that knowledge.

The public controversy began when A. R. Jensen, in a long paper in the *Harvard Educational Review*, persuasively juxtaposed data on the heritability of IQ and the observed differences between groups. Jensen suggested that current large-scale educational attempts to raise the IQs of lower-class children, white and black, were failing because of the high heritability of IQ. In a series of papers and rebuttals to criticism, in the same journal and elsewhere,[2] Jensen put forth the hypothesis that social class and racial differences in mean IQ were due largely to differences in the gene distributions of these populations. At least, he said, the genetic-differences hypothesis was no less likely, and probably more likely, than a simple environmental hypothesis to explain the mean difference of 15 IQ points between blacks and whites[3] and the even larger average IQ differences between professionals and manual laborers within the white population.

Jensen's articles have been directed primarily at an academic audience. An article by Richard Herrnstein in the *Atlantic*[4] and a book (first published in England) by H. J. Eysenck[5] have brought the argument to the attention of the wider lay audience. Both Herrnstein and Eysenck agree with Jensen's genetic-differences hypothesis as it pertains to individual differences and to social class groups, but Eysenck centers his attention on the genetic explanation of racial group differences, which Herrnstein only touches on. Needless to say, many other scientists will take issue with them.

Eysenck's Racial Thesis

Eysenck has written a popular account of the race, social class, and IQ controversy in a generally inflammatory book. The provocative title and the disturbing cover picture of a forlorn black boy are clearly designed to tempt the lay reader into a pseudobattle between Truth and Ignorance. In this case, Truth is genetic-environmental interactionism[6] and Ignorance is naïve environmentalism. For the careful reader, the battle fades out inconclusively as Eysenck admits that scientific evidence to date does not permit a clear choice of the genetic-differences interpretation of black inferiority on intelligence tests. A quick reading of the book, however, is sure to leave the reader believing that scientific evidence today strongly supports the conclusion that United States blacks are genetically inferior to whites in IQ.

The basic theses of the book are as follows:

1. IQ is a highly heritable characteristic in the United States white population and probably equally heritable in the U.S. black population.
2. On the average, blacks score considerably lower than whites on IQ tests.
3. United States blacks are probably a nonrandom, lower-IQ sample of native African populations.
4. The average IQ difference between blacks and whites probably represents important genetic differences between the races.
5. Drastic environmental changes will have to be made to improve the poor phenotypes that United States blacks now achieve.

The evidence and nonevidence that Eysenck cites to support his genetic hypothesis of racial differences make a curious assortment. Audrey Shuey's review[7] of hundreds of studies showing mean phenotypic differences between black and white IQs leads Eysenck to conclude:

> All the evidence to date suggests the strong and indeed overwhelming importance of genetic factors in producing the great variety of intellectual differences which we observe in our culture, and much of the difference observed between certain racial groups. This evidence cannot be argued away by niggling and very minor criticisms of details which do not really throw doubts on the major points made in this book [p. 126].

To "explain" the genetic origins of these mean IQ differences he offers these suppositions:

> White slavers wanted dull beasts of burden, ready to work themselves to death in the plantations, and under those conditions intelligence would have been counterselective. Thus there is every reason to expect that the particular sub-sample of the Negro race which is constituted of American Negroes is not an unselected sample of Negroes, but has been selected throughout history according to criteria which would put the highly intelligent at a disadvantage. The inevitable outcome of such selection would of course be a gene pool lacking some of the genes making for higher intelligence [p. 42].

Other ethnic minorities in the U.S. are also, in his view, genetically inferior, again because of the selective migration of lower IQ genotypes:

> It is known [*sic*] that many other groups came to the U.S.A. due to pressures which made them very poor samples of the original populations. Italians, Spaniards, and Portuguese, as well as Greeks, are examples where the less able, less intelligent were forced through circumstances to emigrate, and where their American progeny showed significantly lower IQ's than would have been shown by a random sample of the original population [p. 43].

Although Eysenck is careful to say that these are not established facts (because no IQ tests were given to the immigrants or non-immigrants in question?), the tone of his writing leaves no doubt about his judgment. There is something in this book to insult almost everyone except WASPs and Jews.

Despite his conviction that United States blacks are genetically inferior in IQ to whites, Eysenck is optimistic about the potential effects of radical environmental changes on the present array of Negro IQ phenotypes. He points to the very large IQ gains produced by intensive one-to-one tutoring of black urban children with low-IQ mothers, contrasting large environmental changes and large IQ gains in intensive programs of this sort with insignificant environmental improvements and small IQ changes obtained by Headstart and related programs. He correctly observes that, whatever the heritability of IQ (or, it should be added, of any characteristic), large phenotypic changes may be produced by creating appropriate, radically different environments never before encountered by those genotypes. On this basis, Eysenck calls for further research to determine the requisites of such environments.

Since Eysenck comes to this relatively benign position regarding potential improvement in IQs, why, one may ask, is he at such pains to "prove" the genetic inferiority of blacks? Surprisingly, he expects that new environments, such as that provided by intensive educational tutoring, will not affect the black-white IQ differential, because black children and white will probably profit equally from such treatment. Since many middle-class white children already have learning environments similar to that provided by tutors for the urban black children, we must suppose that Eysenck expects great IQ gains from relatively small changes in white, middle-class environments.

This book is an uncritical popularization of Jensen's ideas without the nuances and qualifiers that make much of Jensen's writing credible or at least responsible. Both authors rely on Shuey's review, but Eysenck's way of doing it is to devote some twenty-five pages to quotes and paraphrases of her chapter summaries. For readers to whom the original Jensen article is accessible, Eysenck's book is a poor substitute; although he defends Jensen and Shuey, he does neither a service.

It is a maddeningly inconsistent book filled with contradictory caution and incaution; with hypotheses stated both as hypotheses and as conclusions; with both accurate and inaccurate statements on matters of fact. For example, Eysenck thinks evoked potentials offer a better measure of "innate" intelligence than IQ tests. But on what basis? Recently, F. B. Davis[8] has failed to find any relationship whatsoever between evoked potentials and either IQ scores or scholastic achievement, to which intelligence is supposed to be related. Another example is Eysenck's curious use of data to support a peculiar line of reasoning about the evolutionary inferiority of blacks: First, he reports that African and United States Negro babies have been shown to have precocious sensorimotor development by white norms (the difference, by several accounts, appears only in gross motor skills and even there is slight). Second, he notes that by three years of age, United States white exceed United States black children in mean IQ scores. Finally, he cites a (very slight) negative correlation, found in an early study, between sensorimotor intelligence in the first year of life and later IQ. From exaggerated statements of these various data, he concludes:

> These findings are important because of a very general view in biology according to which the more prolonged the infancy the greater in general are the cognitive or intellectual abilities of the species. This law appears to work even within a given species [p. 79].

Eysenck would apparently have us believe that Africans and their relatives in the United States are less highly evolved than Caucasians, whose longer infancy is related to later higher intelligence. I am aware of no evidence whatsoever to support a within-species relationship between longer infancy and higher adult capacities.

The book is carelessly put together, with no index; few references, and those not keyed to the text; and long, inadequately cited quotes that carry over several pages without clear beginnings and ends. Furthermore, considering the gravity of Eysenck's theses, the book has an occasional jocularity of tone that is offensive. A careful book on the genetic hypothesis, written for a lay audience, would have merited publication. This one, however, has been publicly disowned as irresponsible by the entire editorial staff of its London publisher, New Society. But never mind, the American publisher has used that and other condemnations to balance the accolades and make its advertisement[9] of the book more titillating.

Herrnstein's Social Thesis

Thanks to Jensen's provocative article, many academic psychologists who thought IQ tests belonged in the closet with the Rorschach inkblots have now explored the psychometric literature and found it to be a trove of scientific treasure. One of these is Richard Herrnstein, who from a Skinnerian background has become an admirer of intelligence tests—a considerable leap from shaping the behavior of pigeons and rats. In contrast to Eysenck's book, Herrnstein's popular account in the *Atlantic* of IQ testing and its values is generally responsible, if overly enthusiastic in parts.

Herrnstein unabashedly espouses IQ testing as "psychology's most telling accomplishment to date," despite the current controversy over the fairness of testing poor and minority-group children with IQ items devised by middle-class whites. His historical review of IQ test development, including tests of general intelligence and multiple abilities, is interesting and accurate. His account of the validity and usefulness of the tests centers on the fairly accurate prediction that can be made from IQ scores to academic and occupational achievement and income level. He clarifies the pattern of relationship between IQ and these criterion variables: High IQ is a necessary but not sufficient condition for high achievement, while low IQ virtually assures failure at high academic and occupational levels. About the usefulness of the tests, he concludes:

> An IQ test can be given in an hour or two to a child, and from this infinitesimally small sample of his output, deeply important predictions follow—about schoolwork, occupation, income, satisfaction with life, and even life expectancy. The predictions are not perfect, for other factors always enter in, but no other single factor matters as much in as many spheres of life [p. 53].

One must assume that Herrnstein's enthusiasm for intelligence tests rests on population statistics, not on predictions for a particular child, because many children studied longitudinally have been shown to change IQ scores by 20 points or more from childhood to adulthood. It is likely that extremes of giftedness and retardation can be sorted out relatively early by IQ tests, but what about the 95 percent of the population in between? Their IQ scores may vary from dull to bright normal for many years. Important variations in IQ can occur up to late adolescence.[10] On a population basis, Herrnstein is correct; the best early predictors of later achievement are ability measures taken from age five on. Predictions are based on correlations, however, which are not sensitive to absolute changes in value, only to rank orders. This is an important point to be discussed later.

After reviewing the evidence for average IQ differences by social class and race, Herrnstein poses the nature-nurture problem of "which is primary" in determining phenotypic differences in IQ. For racial groups, he explains, the origins of mean IQ differences are indeterminate at the present time because we have no information from heritability studies in the black population or from other, unspecified, lines of research which could favor primarily genetic or primarily environmental hypotheses. He is thoroughly convinced, however, that individual differences and social-class differences in IQ are highly heritable at the present time, and are destined, by environmental improvements, to become even more so:

> If we make the relevant environment much more uniform [by making it as good as we can for everyone], then an even larger proportion of the variation in IQ will be attributable to the genes. The average person would be smarter, but intelligence would run in families even more obviously and with less regression toward the mean than we see today [p. 58].

For Herrnstein, society is, and will be even more strongly, a meritocracy based largely on inherited differences in IQ. He presents a "syllogism" (p. 58) to make his message clear:

1. If differences in mental abilities are inherited, and,
2. If success requires those abilities, and,
3. If earnings and prestige depend on success,
4. Then social standing (which reflects earnings and prestige) will be based to some extent on inherited differences among people.

Five "corollaries" for the future predict that the heritability of IQ will rise; that social mobility will become more strongly related to inherited IQ differences; that most bright people will be gathered in the top of the social structure, with the IQ dregs at the bottom; that many at the bottom will not have the intelligence needed for new jobs; and that the meritocracy will be built not just on inherited intelligence but on all inherited traits affecting success, which will presumably become correlated characters. Thus from the successful realization of our most precious, egalitarian, political, and social goals, there will arise a much more rigidly stratified society, a "virtual caste system" based on inborn ability.

To ameliorate this effect, society may have to move toward the socialist dictum, "From each according to his abilities, to each according to his needs," but Herrnstein sees complete equality of earnings and prestige as impossible because high-grade intelligence is scarce and must be recruited into those critical jobs that require it by the promise of high earnings and high prestige. Although garbage collecting is critical to the health of the society, almost anyone can do it; to waste high-IQ persons on such jobs is to misallocate scarce resources at society's peril.

Herrnstein points to an ironic contrast between the effects of caste and class systems. Castes, which established artificial hereditary limits on social mobility, guarantee the inequality of opportunity that preserves IQ heterogeneity at all levels of the system. Many bright people are arbitrarily kept down and many unintelligent people are artificially maintained at the top. When arbitrary bounds on mobility are removed, as in our class system, most of the bright rise to the top and most of the dull fall to the bottom of the social system, and IQ differences between top and bottom become increasingly hereditary. The greater the environmental equality, the greater the hereditary differences between levels in the social structure. The thesis of egalitarianism surely leads to its antithesis in a way that Karl Marx never anticipated.

Herrnstein proposes that our best strategy, in the face of increasing biological stratification, is publicly to recognize genetic human differences but to reallocate wealth to a considerable extent. The IQ have-nots need not be poor. Herrnstein does not delve into the psychological consequences of being publicly marked as genetically inferior.

Does the evidence support Herrnstein's view of hereditary social classes, now or in some future Utopia? Given his assumptions about the high heritability of IQ, the importance of IQ to social mobility, and the increasing environmental equality of rearing and opportunity, hereditary social classes are to some extent inevitable. But one can question the limits of genetic homogeneity in social class groups and the evidence for his syllogism at present.

Is IQ as highly heritable throughout the social structure as Herrnstein assumes? Probably not. In a recent study of IQ heritability in various racial and social class groups,[11] I found much lower proportions of genetic variance that would account for aptitude differences among lower-class than among middle-class children, in both black and white groups. Social disadvantage in prenatal and postnatal development can substantially lower phenotypic IQ and reduce the genotype-phenotype correlation. Thus, average phenotypic IQ differences between the social classes may be considerably larger than the genotypic differences.

Are social classes largely based on hereditary IQ differences now? Probably not as much as Herrnstein believes. Since opportunities for social mobility act at the phenotypic level, there still may be considerable genetic diversity for IQ at the bottom of the social structure. In earlier days, arbitrary social barriers maintained genetic variability throughout the social structure. At present, individuals with high phenotypic IQs are often upwardly mobile; but inherited wealth acts to maintain genetic diversity at the top, and nongenetic biological and social barriers to phenotypic development act to maintain a considerable genetic diversity of intelligence in the lower classes.

As P. E. Vernon has pointed out,[12] we are inclined to forget that the majority of gifted children in recent generations have come from working-class, not middle-class, families. A larger percentage of middle-class children are gifted, but the working and lower classes produce gifted children in larger numbers. How many more disadvantaged children would have been bright if they had had middle-class gestation and rearing conditions?

I am inclined to think that intergenerational class mobility will always be with us, for three reasons. First, since normal IQ is a polygenic characteristic, various recombinations of parental genotypes will always produce more variable genotypes in the offspring than in the parents of all social class groups, especially the extremes. Even if both parents, instead of primarily the male, achieved social class status based on their IQs, recombinations of their genes would always produce a range of offspring, who would be upwardly or downwardly mobile relative to their families of origin.

Second, since, as Herrnstein acknowledges, factors other than IQ—motivational, personality, and undetermined—also contribute to success or the lack of it, high IQs will always be found among lower-class adults, in combination with schizophrenia, alcoholism, drug addiction, psychopathy, and other limiting factors. When recombined in offspring, high IQ can readily segregate with facilitating motivational and personality characteristics, thereby leading to upward mobility for many offspring. Similarly, middle-class parents will always produce some offspring with debilitating personal characteristics which lead to downward mobility.

Third, for all children to develop phenotypes that represent their best genotypic outcome (in current environments) would require enormous changes in the present social system. To improve and equalize all rearing environments would involve such massive intervention as to make Herrnstein's view of the future more problematic than he seems to believe.

Race as Caste

Races are castes between which there is very little mobility. Unlike the social class system, where mobility based on IQ is sanctioned, the racial caste system, like the hereditary aristocracy of medieval Europe and the caste system of India, preserves within each group its full range of genetic diversity of intelligence. The Indian caste system was, according to Dobzhansky,[13] a colossal genetic failure—or success, according to egalitarian values. After the abolition of castes at independence, Brahmins and untouchables were found to be equally educable despite—or because of—their many generations of segregated reproduction.

While we may tentatively conclude that there are some genetic IQ differences between social class groups, we can make only wild speculations about racial groups. Average phenotypic IQ differences between races are not evidence for genetic differences (any more than they are evidence for environmental differences). Even if the heritabilities of IQ are extremely high in all races, there is still no warrant for equating within-group and between-group heritabilities.[14] There are examples in agricultural experiments of within-group differences that are highly heritable but between-group differences that are entirely environmental. Draw two random samples of seeds from the same genetically heterogeneous population. Plant one sample in uniformly good conditions, the other in uniformly poor conditions. The average height difference between the populations of plants will be entirely environmental, although the individual differences in height within each sample will be entirely genetic. With known genotypes for seeds and known environments, genetic and environmental variances between groups can be studied. But racial groups are not random samples from the same population, nor are members reared in uniform conditions within each race. Racial groups are of unknown genetic equivalence for polygenic characteristics like IQ, and the differences in environments within and between the races may have as yet unquantified effects.

There is little to be gained from approaching the nature-nurture problem of race differences in IQ directly.[15] Direct comparisons of estimated within-group heritabilities and the calculation of between-group heritabilities require assumptions that few investigators are willing to make, such as that all environmental differences are quantifiable, that differences in the environments of blacks and whites can be assumed to affect IQ in the same way in the two groups, and that differences in environments between groups can be "statistically controlled." A direct assault on race differences in IQ is vulnerable to many criticisms.

Indirect approaches may be less vulnerable. These include predictions of parent-child regression effects and admixture studies. Regression effects can be predicted to differ for blacks and whites if the two races indeed have genetically different population means. If the population mean for blacks is 15 IQ points lower than that of whites, then the offspring of high-IQ black parents should show greater regression (toward a lower population mean) than the offspring of whites of equally high IQ. Similarly, the offspring of low-IQ black parents should show less regression than those of white parents of equally low IQ. This hypothesis assumes that assortative mating for IQ is equal in the two races, which could be empirically determined but has not been studied as yet. Interpretable results from a parent-child regression study would also depend upon careful attention to intergenerational environmental changes, which could be greater in one race than the other.

Studies based on correlations between degree of white admixture and IQ scores *within* the black group would avoid many of the pitfalls of between-group comparisons. If serological genotypes can be used to identify persons with more and less white admixture, and if estimates of admixture based on blood groups are relatively independent of visible characteristics like skin color, then any positive correlation between degree of admixture and IQ would suggest genetic racial differences in IQ. Since blood groups have not been used directly as the basis of racial discrimination, positive findings would be relatively immune from environmentalist criticisms. The trick is to estimate individual admixture reliably. Several loci which have fairly different distributions of alleles in contemporary African and white populations have been proposed.[16] No one has yet attempted a study of this sort.

h^2 *and Phenotype*

Suppose that the heritabilities of IQ differences within all racial and social class groups were .80, as Jensen estimates, and suppose that the children in all groups were reared under an equal range of conditions. Now, suppose that racial and social class differences in mean IQ still remained. We would probably infer some degree of genetic difference between the groups. So what? The question now turns from a strictly scientific one to one of science and social policy.

As Eysenck, Jensen, and others have noted, eugenic and euthenic strategies are both possible interventions to reduce the number of low-IQ individuals in all populations. Eugenic policies could be advanced to encourage or require reproductive abstinence by people who fall below a certain level of intelligence. The Reeds[17] have determined that one-fifth of the mental retardation among whites of the next generation could be prevented if no mentally retarded persons of this generation reproduced. There is no question that a eugenic program applied at the phenotypic level of parents' IQ would substantially reduce the number of low-IQ children in the future white population. I am aware of no studies in the black population to support a similar program, but some proportion of future retardation could surely be eliminated. It would be extremely important, however, to sort out genetic and environmental sources of low IQ both in racial and in social class groups before advancing a eugenic program. The request or demand that some persons refrain from any reproduction should be a last resort, based on sure knowledge that their retardation is caused primarily by genetic factors and is not easily remedied by environmental intervention. Studies of the IQ levels of adopted children with mentally retarded natural parents would be most instructive, since some of the retardation observed among children of retarded parents may stem from the rearing environments provided by the parents.

In a pioneering study of adopted children and their adoptive and natural parents, Skodak[18] reported greater *correlations* of children's IQs with their natural than with their adoptive parents' IQs. This statement has been often misunderstood to mean that the children's *levels* of intelligence more closely resembled their natural parents', which is completely false. Although the rank order of the children's IQs resembled that of their mothers' IQs, the children's IQs were higher, being distributed, like those of the adoptive parents, around a mean above 100, whereas their natural mothers' IQs averaged only 85. The children, in fact, averaged 21 IQ points higher than their natural mothers. If the (unstudied) natural fathers' IQs averaged around the population mean of 100, the mean of the children's would be expected to be 94, or 12 points lower than the mean obtained. The unexpected boost in IQ was presumably due to the better social environments provided by the adoptive families. Does this mean that phenotypic IQ can be substantially changed?

Even under existing conditions of child rearing, phenotypes of children reared by low-IQ parents could be markedly changed by giving them the same rearing environment as the top IQ group provide for their children. According to DeFries,[19] if children whose parents average 20 IQ points below the population mean were reared in environments such as usually are provided only by parents in the top .01 percent of the population, these same children would average 5 points *above* the population mean instead of 15 points below, as they do when reared by their own families.

Euthenic policies depend upon the demonstration that different rearing conditions can change phenotypic IQ sufficiently to enable most people in a social class or racial group to function in future society. I think there is great promise in this line of research and practice, although its efficacy will depend ultimately on the cost and feasibility of implementing radical intervention programs. Regardless of the present heritability of IQ in any population, phenotypes can be changed by the introduction of new and different environments. (One merit of Eysenck's book is the attention he gives to this point.) Furthermore, it is impossible to predict phenotypic outcomes under very different conditions. For example, in the Milwaukee Project,[20] in which the subjects are ghetto children whose mothers' IQs are less than 70, intervention began soon after the children were born. Over a four-year period Heber has intensively tutored the children for several hours every day and has produced an enormous IQ difference between the experimental group (mean IQ 127) and a control group (mean IQ 90). If the tutored children continue to advance in environments which are radically different from their homes with retarded mothers, we shall have some measure of the present phenotypic range of reaction[21] of children whose average IQs might have been in the 80 to 90 range. These data support Crow's comment on h^2 in his contribution to the *Harvard Educational Review* discussion (p. 158):

> It does not directly tell us how much improvement in IQ to expect from a given change in the environment. In particular, it offers no guidance as to the consequences of a new kind of environmental influence. For example, conventional heritability measures for height show a value of nearly 1. Yet, because of unidentified environmental influences, the mean height in the United States and in Japan has risen by a spectacular amount. Another kind of illustration is provided by the discovery of a cure for a hereditary disease. In such cases, any information on prior heritability may become irrelevant. Furthermore, heritability predictions are less dependable at the tails of the distribution.

To illustrate the phenotypic changes that can be produced by radically different environments for children with clear genetic anomalies, Rynders[22] has provided daily intensive tutoring for Down's syndrome infants. At the age of two, these children have average IQs of 85 while control-group children, who are enrolled in a variety of other programs, average 68. Untreated children have even lower average IQ scores.

The efficacy of intervention programs for children whose expected IQs are too low to permit full participation in society depends on their long-term effects on intelligence. Early childhood programs may be necessary but insufficient to produce functioning adults. There are critical research questions yet to be answered about euthenic programs, including what kinds, how much, how long, how soon, and toward what goals?

Does h^2 *Matter?*

There is growing disillusionment with the concept of heritability, as it is understood and misunderstood. Some who understand it very well would like to eliminate h^2 from human studies for at least two reasons. First, the usefulness of h^2 estimates in animal and plant genetics pertains to decisions about the efficacy of selective breeding to produce more desirable phenotypes. Selective breeding does not apply to the human case, at least so far. Second, if important phenotypic changes can be produced by radically different environments, then, it is asked, who cares about the heritability of IQ? Morton[23] has expressed these sentiments well:

> Considerable popular interest attaches to such questions as "is one class or ethnic group innately superior to another on a particular test?" The reasons are entirely emotional, since such a difference, if established, would serve as no better guide to provision of educational or other facilities than an unpretentious assessment of phenotypic differences.

I disagree. The simple assessment of phenotypic performance does not suggest any particular intervention strategy. Heritability estimates can have merit as indicators of the effects to be expected from various types of intervention programs. If, for example, IQ tests, which predict well to achievements in the larger society, show low heritabilities in a population, then it is probable that simply providing better environments which now exist will improve average performance in that population. If h^2 is high but environments sampled in that population are largely unfavorable, then (again) simple environmental improvement will probably change the mean phenotypic level. If h^2 is high and the environments sampled are largely favorable, then novel environmental manipulations are probably required to change phenotypes, and eugenic programs may be advocated.

The most common misunderstanding of the concept "heritability" relates to the myth of fixed intelligence: If h^2 is high, this reasoning goes, then intelligence is genetically fixed and unchangeable at the phenotypic level. This misconception ignores the fact that h^2 is a population statistic, bound to a given set of environmental conditions at a given point in time. Neither intelligence nor h^2 estimates are fixed.

It is absurd to deny that the frequencies of genes for behavior may vary between populations. For individual differences within populations, and for social-class differences, a genetic hypothesis is almost a necessity to explain some of the variance in IQ, especially among adults in contemporary white populations living in average or better environments. But what Jensen, Shuey, and Eysenck (and others) propose is that genetic racial differences are necessary to account for the current phenotypic differences in mean IQ between populations. That may be so, but it would be extremely difficult, given current methodological limitations, to gather evidence that would dislodge an environmental hypothesis to account for the same data. And to assert, despite the absence of evidence, and in the present social climate, that a particular race is genetically disfavored in intelligence is to scream "FIRE! . . . I think" in a crowded theater. Given that so little is known, further scientific study seems far more justifiable than public speculations.

Notes

I thank Philip Salapatek, Richard Weinberg, I. I. Gottesman, and Leonard I. Heston for their critical reading of this paper. They are not in any way responsible for its content, however.

1. For a review of studies, *see* L. Erlenmeyer-Kimling and L. F. Jarvik, "Genetics and Intelligence," *Science* 142 (1963):1477–1479. Heritability is the ratio of genetic variance to total phenotypic variance. For human studies, heritability is used in its broad sense of total genetic variance–total phenotypic variance.
2. The *Harvard Educational Review* compilation, *Environment, Heredity, and Intelligence*, Reprint Series no. 2 (Cambridge, Mass., 1969), includes Jensen's paper "How Much Can We Boost IQ and Scholastic Achievement?" comments on it by J. S. Kagan, J. McV. Hunt, J. F. Crow, C. Bereiter, D. Elkind, L. J. Cronbach, and W. F. Brazziel, and a rejoinder by Jensen. *See also* A. R. Jensen, "Can We and Should We Study Race Differences?" in *Compensatory Education: A National Debate*, ed. J. Hellmuth, vol. 3, *The Disadvantaged Child* (New York: Brunner-Mazel, 1970).
3. P. L. Nichols, Ph.D. thesis, University of Minnesota, 1970. Nichols reports that in two large samples of black and white children, seven-year WISC IQ scores showed the same means and distributions for the two racial groups, once social-class variables were equated. These results are unlike those of several other studies, which found that matching socioeconomic status did not create equal means in the two racial groups (A. Shuey, *The Testing of Negro Intelligence* [New York: Social Science Press, 1966]; A. B. Wilson, *Racial Isolation in the Public Schools*, vol. 2 [Washington, D.C.: Government Printing Office, 1967]). In Nichols' samples, prenatal and postnatal medical care was equally available to blacks and whites, which may have contributed to the relatively high IQ scores of the blacks in these samples.
4. R. Herrnstein ,"IQ," *Atlantic Monthly* September 1971, pp. 44–64.
5. H. J. Eysenck, *The IQ Argument: Race, Intelligence and Education* (New York: Library Press, 1971).
6. By interaction, Eysenck means simply $P = G + E$, or "heredity and environ-

ment acting together to produce the observed phenotype" (p. 111). He does not mean what most geneticists and behavior geneticists mean by interaction; that is, the *differential* phenotypic effects produced by various combinations of genotypes and environments, as in the interaction term of analysis-of-variance statistics. Few thinking people are not interactionists in Eysenck's sense of the term, because that is the only way to get the organism and the environment into the same equation to account for variance in any phenotypic trait. How much of the phenotypic variance is accounted for by each of the terms in the equation is the real issue.

7. A. Shuey, *The Testing of Negro Intelligence* (New York: Social Science Press, 1966), pp. 499–519.
8. F. B. Davis, *The Measurement of Mental Capacity Through Evoked-Potential Recordings* (Greenwich, Conn.: Educational Records Bureau, 1971). "As it turned out, no evidence was found that the latency periods obtained . . . displayed serviceable utility for predicting school performance or level of mental ability among pupils in preschool through grade 8" (p. v).
9. *New York Times*, October 8, 1971, p. 41.
10. J. Kagan and H. A. Moss, *Birth to Maturity* (New York: John Wiley & Sons, 1962).
11. S. Scarr-Salapatek, "Race, Social Class, and IQ," *Science* 174 (1971):1285–1295.
12. P. E. Vernon, *Intelligence and Cultural Environment* (London: Methuen & Co., 1969).
13. T. Dobzhansky, *Mankind Evolving* (New Haven, Conn.: Yale University Press, 1962), pp. 234–238.
14. J. Thoday, "Limitations to Genetic Comparisons of Populations," *Journal of Biosocial Science*, suppl. 1 (1969), pp. 3–14. (Reprinted in this volume.)
15. L. L. Cavalli-Sforza and W. F. Bodmer, *The Genetics of Human Populations* (San Francisco: W. H. Freeman & Co., 1971), pp. 753–804. They propose that the study of racial differences is useless and not scientifically supportable at the present time.
16. T. E. Reed, "Caucasian Genes in American Negroes," *Science* 165 (1969): 762; *American Journal of Human Genetics* 21 (1969):1; C. MacLean and P. L. Workman, paper presented at a meeting of the American Society of Human Genetics, Indianapolis, 1970. (Studies of the sort suggested show no evidence of genotypic IQ differences between blacks and whites. See J. Loehlin, S. Vandenberg, and R. Osborne, "Blood Group Genes and Negro-White Ability Differences." *Behavior Genetics* 3 (1973):263–270. Eds.)
17. E. W. Reed and S. C. Reed, *Mental Retardation: A Family Study* (Philadelphia: W. B. Saunders Co., 1965); *Social Biology*, suppl. 18 (1971), p. 42.
18. M. Skodak and H. M. Skeels, "A Final Follow-up Study of One Hundred Adopted Children," *Journal of Genetic Psychology* 75 (1949):85–125.
19. J. C. DeFries, paper for the COBRE Research Workshop on Genetic En-

dowment and Environment in the Determination of Behavior, Rye, N.Y., October 3–8, 1971.

20. R. Heber, *Rehabilitation of Families at Risk for Mental Retardation*, Regional Rehabilitation Center, University of Wisconsin (Madison, 1969). S. P. Strickland, "Can Slum Children Learn?" *American Education* 7 (1971):3.

21. I. I. Gottesman, "Biogenetics of Race and Class," in *Social Class, Race, and Psychological Development*, ed. M. Deutsch, I. Katz, and A. R. Jensen (New York: Holt, Rinehart & Winston, 1968), pp. 11–51.

22. J. Rynders, personal communication, November 1971.

23. N. E. Morton, paper for the COBRE Research Workshop on Genetic Endowment and Environment in the Determination of Behavior, Rye, N.Y., October 3–8, 1971.

Limitations to Genetic Comparison of Populations

J. M. Thoday

THE DISCUSSION OF BIOLOGICAL AND SOCIAL ASPECTS OF RACE IS, OF course, a tender topic, in dealing with which I feel strongly that we must all take especial care to express our ignorance when we are ignorant. Especially is this so in the most sensitive area of the topic, the question whether or not differences between racial groups in so-called mental abilities have an innate genetic component or not. I hope to convince you that nobody knows the answer to this question, and that in present circumstances it is impossible to know or even to foresee with certainty that it will ever be possible to know. Hence, whoever speaks as if he knows that such a racial difference is genetic, or whoever speaks as if he knows that such a racial difference is not genetic, is showing a bias that cannot be justified by the facts. Both they who say the differences are genetic and they who say they are environmental or cultural are equally prejudiced.

Of course, races differ genetically in some respects. This is what race means, though zoologists use the word rather little, preferring to use the more precise term "subspecies" when appropriate, and the less precise term "local population" when appropriate. Either of these can mean what I take it race means to anthropologists: a race is a population of a species originally living in a different region and recognizably different from other populations of that species, and it is implicit that some of the differences will be genetic. Further, the more clearly the differences are genetic, the more they are used as diagnostic criteria of race.

That races differ genetically, then, is not the controversial question. Controversy arises when people believe and act as if, because races differ genetically, they may assume that all racial differences are necessarily genetic. The most controversial area, of course, involves observed differences in performance at psychological "ability" tests.

What we must concern ourselves with, therefore, is the criteria by which it may be possible to test whether, for example, a particular observed IQ difference between two human populations belonging to two ethnic groups is genetic or part genetic or purely environmental or cultural in origin.

Now the criteria needed to assess the genetic differences between two populations differ according to the particular variables we are interested in. First and most critical is the question whether the character we are dealing with varies discontinuously or continuously within each population. IQ varies continuously. Second is the question whether heritability is 100 percent or less in each population. The heritability of IQ (that is, the proportion of IQ variance that arises from genetic variety) is less than 100 percent. Thirdly, when heritability is less than 100 percent, that is to say when part of the variation within the population is not genetic but environmental in causation, we have the vital question, Is the environmental variation in whole or part culturally caused, and if part is culturally caused, is the culture socially inherited? Part of IQ variation is culturally caused, and we need have little doubt that part of the cultural variation is socially inherited.

Discontinuous Variables

There is little or no difficulty in assessing the degree to which a race difference in a straightforward* discontinuous variable has genetic causes. For such differences we are in a position, if the alternative kinds of individual are frequent enough, to say exactly how the relevant biological inheritance works and to say that this individual has these genes and that individual those genes. We may then de-

* I am leaving out of consideration threshold phenomena, important though they are, where an underlying continuity can give rise to an observed discontinuity as, for example, with many kinds of disease resistance where the discontinuity is imposed at the environmental level. Such characters involve the same problems as continuous variables.

termine the gene and genotype frequencies of different populations and describe the genetic differences distinguishing them quite precisely. This has been widely done with blood groups and various other biological differences, and different populations and different races have been shown to differ in gene frequencies.

Such characters, then, present us with little difficulty and enable us to make unequivocal statements about differences between races. These statements, however, should be understood properly. They are not usually of the kind that this race has such and such a gene and that race some allelic (alternative) gene. They are of the kind, the frequency of this gene in this population is higher than in that population. They make, therefore, no statement which we can apply to individuals: we cannot say without specific investigation that this white man differs in blood group from that American Indian. We should bear this in mind when thinking about less easily handled characters, especially when they involve people's emotive attitudes. We do not deal in real life with racial types. The concept is a dangerous one in this field; it involves abstractions that have no reality. Populations of a species do not differ absolutely, but in the relative frequency of differing genotypes.

Continuous Variables

While discontinuous variables such as blood groups present us with little difficulty, continuous variables such as IQ are a different matter, for it is not possible with these to identify specific genotypes and it is therefore not possible to determine gene frequencies. Furthermore, there are always environmental as well as genetic causes of variation. We may measure the relative importance of environmental and genetic causes of variation or heritability within a population; and if the heritabilities are very high, that is, the variation is almost entirely a consequence of genetic variety, we may know more than if they are low. But even if they are high, as with fingerprint ridge counts, we are already in difficulties with population comparisons, for *there is no warrant for equating within-group heritabilities and between-group heritabilities.*

Since this is the core of my argument, I propose to consider it in a little detail, to indicate how the experimental biologist solves the problem with nonhuman, that is, experimentally tractable material, tell what kinds of results he may get, and show why the techniques used cannot be used for races of man.

Transplant Experiments

The sort of problem we are confronted with is a commonplace in plant ecology. The populations of the same species of a plant that are to be found in different environments are often different. For example, Coxfoot grass plants growing in shady habitats often have broader leaves than those growing in open habitats. There is a well-recognized shade form. The question the ecogeneticist has to ask is whether the shade form is a consequence of development in the shady environment or whether it is a result of the evolution of a different genotype specially adapted to the shady environment. Now the only way he can answer this question is to transplant seeds, plants, or cuttings from both habitats to some common environment, say, in his experimental garden, and grow them together so as to compare them in a common environment.

Turesson did this with the shade and open-habitat forms of Coxfoot. The results depended upon the geographical origin of the shade form used. Shade forms from some areas preserved their shade form in the experimental plots: these shade forms were genetically different from the open-habitat form. Shade forms from other habitats lost their shade form in the experimental plots so that there was no evidence that they differed genetically from the open-habitat populations. We shall, however, see later that this lack of evidence that there is a genetic difference cannot be taken as evidence that there is not a genetic difference.

Now this experiment has lessons for us. The same apparent difference between shade form and open-habitat form can sometimes have purely environmental causes and sometimes purely genetic causes, and only transplant experiments will tell us which is the cause in any particular case.

Another plant experiment is equally revealing. Populations of goldenrod to be found in different places differ from one another. The plants in southern Sweden in shaded habitats are taller than those to be found in exposed habitats in northern Norway. These have been transplanted to experimental environment chambers and grown in high- or low-light intensities with the results given in table 1.

Here we see that the different populations from the different habitats differ genetically, but that the environment has profound effects also, the different genotypes reacting in opposite ways to the environmental variable. We also see a trap we might readily fall

Table 1. Solidago virgaurea—*height of plant*

Natural habitat of plants	*Environment of test (light intensity)*	
	Low	*High*
Exposed, northern Norway	Dwarf	Medium
Shaded, southern Sweden	Tall	Medium

into. Grown in high-light intensity the two genotypes look alike so that we get no evidence that they differ genetically. As I pointed out, this is not evidence that they are genetically alike. In fact, their reaction to low-light intensity shows them to be different. If they had only been grown in the high-light intensity, we might have reached the erroneous conclusion that they were genetically alike

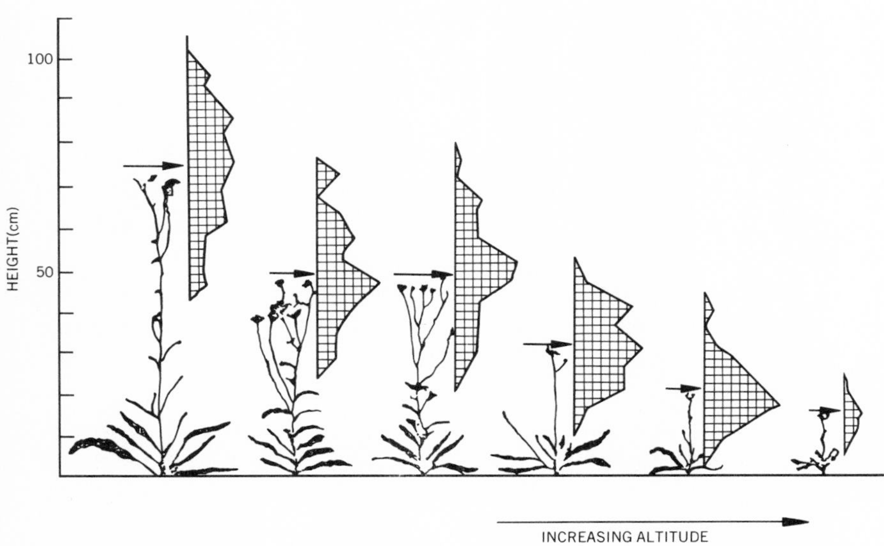

Figure 1. *The figure shows an average plant and the range of the distribution of heights in samples of plants taken from each of several populations from different altitudes and raised together in a common environment. The distinct populations are clearly genetically different yet the range in each is such that there is considerable overlap. Transplanting nonrandom samples could clearly have either exaggerated or negated the real genetic difference between the populations. (After Clausen* et al., *1940.)*

and that their differences in the wild were purely environmental. The lessons of this result are, I think, obvious.

My final lesson from plants is illustrated in figure 1, which shows some of the results of the classic work of Clausen, Keck, and Hiesey (1940) in California. This concerns not only the average results for a population but its variability in transplant experiments and underlines that, for results to be interpretable, we must transplant random samples of the variety of genotypes from each population, a difficult thing to do with, for example, social classes in man.

Correlation of Genetic and Social Inheritance in Man

Now let us turn to man. The sort of environmental differences we were concerned with in these plant examples are to be paralleled in man by the climatic differences in which different human populations grow. But with human populations we have an additional problem: part of the environmental component is cultural, and cultural differences may transplant with the people so that in a human transplant experiment we transplant in part environmental as well as genetic differences. Furthermore, cultural differences are in part socially inherited, so that a transplanted population may preserve some of its environmental differences over the generations.

We should, in this connection, remember that we have no warrant for supposing that cultural inheritance is any more labile than biological inheritance. In fact, the available evidence suggests that sometimes cultural inheritance can be very conservative. Gini (1954) studied five-century-old Albanian settlements in Southern Italy which still preserve "their language, their traditional costumes, and other cultural characteristics," and less ancient Italian settlements in Sardinia. He concluded that "in contrast with the current opinions of students of anthropological sciences and perhaps also of students of social sciences, the cultural tradition may be more tenacious and persistent than the physical heredity." Social inheritance as a cause of persistent differences between peoples has to be taken seriously, and the problem of distinguishing between genetic and social inheritance as causes of any observed difference is very real.

To put the point another way: The social determination and inheritance of cultural differences are correlated with genetic relationship, and therefore environmental and genetic variation are confounded. When we are comparing human groups, therefore, whether social classes or ethnic groups, it is problematic whether we can separate genetic and environmental components completely. This is why, of course, twin studies play so profound a part in human genetics. But twin studies are mostly within-family studies and can only give us within-family estimates of genetic variance and heritability.

When we are interested in social classes, twins reared apart are transplant experiments and can allow us to make some comparisons of within-family genetic variance and between-family environmental variance. Also adoption and orphanage studies are useful transplant studies, provided we know enough about the natural parents as well as the foster parents, and there is some hope also, with the same prerequisite of knowledge of the parents, that social mobility may provide us with revealing transplant experiments that will enable us to assess the relative importance of the environmental and genetic components of class differences. The progeny of artificial insemination would also help, given adequate knowledge of the real father.

When, however, we are comparing races we are stuck completely. You cannot transplant individuals from one race to another without transplanting at least some of the environmental-cultural condition membership of race implies, and racial hybrids are both different genotypes and are presented with a special environment. Hence, it follows that no one can make any meaningful statement about the extent to which a racial difference is cultural or environmental unless they can actually determine the frequencies of relevant genes. And we know nothing about the specific genes that mediate the genetic component of IQ variance or any other component of human ability that shows continuous variance.

The only objectively honest answer anyone can give to such a question as, "Are Negroes innately inferior to whites in IQ?" has therefore two components. First, "Individual Negroes are not inferior to whites in IQ. Their mean as a population is, however, lower." Second, "Having shown that the average of a particular Negro population is lower than that of a particular white population, nobody can say whether the population difference is all genetic, all cultural, or something in between, or even whether the genetic average of the population with the lower phenotypic average may not be the higher of the two."

In these circumstances, we need not find it surprising that the conclusions that workers reach about this question vary not so much according to their experimental data but according to the social factors affecting themselves, as Sherwood and Nataupsky (1968) have recently demonstrated by showing that birth order, whether grandparents were American-born or no, level of parents' education, rural or urban childhood, and undergraduate scholastic standing, are all variables correlated with the conclusion the workers reached. I say that this is not surprising, for it is, I hope, obvious from what I have said that the conclusions cannot have been scientific conclusions since no scientific conclusions can be reached. The conclusions, however much dressed up as if they were scientific conclusions related to experimental data, must in fact be mere opinions, and mere opinions are to be expected to relate to social factors in the individual experimenter's background.

With these considerations in mind, I now want to look at one or two examples of the data about race and IQ so that we can see a little of what it tells us and what it does not tell us.

Firstly, ever since the original findings on recruits to the army in the 1914–1918 war, United States Negro populations have scored on the average lower than United States white populations in IQ tests. This is not at issue: the point at issue is the extent to which the difference is innate or cultural in origin. I have made it clear that at present we cannot know, so that I only wish to make one further comment. This concerns the unfortunate way in which some workers express their findings quantitatively (*see*, for example, McGurk, 1967). These workers express the difference in terms of what they call overlap. By this they mean the proportion of Negroes who are above the average IQ of the white population with which they compare them, a measure which may easily lead us to forget that 50 percent of whites are below the white average! I am reminded of the story of the Member of Parliament who tabled a question for the Minister of Education: "Was the Minister aware that 50 percent of our population are below average intelligence and what was he going to do about it?"

Overlap figures given by McGurk (1967) range from 13 to 30 percent. Let us look at a different way of expressing these findings.

A normal distribution curve includes 34 percent within one and 47½ percent within two standard deviations of the mean on any one side. If two populations differ in mean by one standard deviation, the overlap computed according to McGurk's criteria is about 15 percent. But the real overlap is more like 95 percent. The emotive possibilities of the difference between the two measures of overlap are striking. The proper way of expressing this difference is as follows: If we were to try to classify people into Negroes and whites according to IQ alone, we should only be right 5 percent more often than if we classified them by chance. Such data should not be represented in such an emotively biased way. To represent a difference as 15 percent overlap rather than 95 percent overlap is asking for misinterpretation.

I also wish to refer to the findings of McGurk concerning Negro-white differences of IQ in relation to socioeconomic status. This work comes from Alabama and some people might be inclined to ignore it simply because of this, but I believe it is only proper to look at it as objectively as we can, for it does purport to be a serious-minded attempt to test the culture hypothesis of the racial difference of IQ mean. McGurk reports the following findings, which he regards as failing to support the culture hypothesis as an explanation of Negro-white differences in psychological test performance. They include:

1. When Negroes and whites are paired for similarity of socioeconomic background, the difference in IQ mean persists.
2. The difference was greater among pairs of high socioeconomic status than among those of low socioeconomic status.
3. The difference between Negroes and whites was less for those components of the test battery regarded as having high cultural loading than for those components regarded as having low cultural loading.

Now I am not competent to judge the validity of these findings, for this involves, among other things, assessment of the validity of the division of test questions according to cultural loading, and the validity of the criteria of socioeconomic status. Further, some of the arguments McGurk uses are wide open to question: for example, one of the assumptions which he does not make explicit, related to findings I have not quoted, is that the socioeconomic status of Negroes in the United States has not only improved over the last fifty years but has improved *relative* to that of whites. However, it does seem to me that the kind of comparisons McGurk has made are the right kind of comparison, and should be made on a wider scale by sociologists as well as psychologists.

Supposing, however, that such investigations were made and supposing McGurk's findings were supported. Would they tell us anything about the degree to which Negro-white IQ mean differences are biological or social in origin? I submit they would not. They are open to interpretation on either of two hypotheses or any mixture of the two. The first hypothesis is that the difference is genetic, the second that the difference arises from socially inherited differences in racial culture, so that we are left almost where we started. Not quite, however, for the second hypothesis would now concern differences in racial culture, whether autonomous or involving racial interaction. This type of experiment would, in other words, help to disentangle racial from class differences, whether the race differences and the class differences involve genetic differences or not.

We already have strong evidence that these two must not be confused in the findings that Vandenberg (1968) very briefly showed us at the last symposium (fig. 3 of his published paper). Again, the data cannot distinguish between biological differences and differences of racial social heredity. But that the differences between races in mean IQ profile (as distinct from IQ mean) cannot be explained by differences in the distribution of the races among the social classes is clear, for it is evidently independent of social class. If the differences originate from cultural variation, we must distinguish between class and race differences in culture, an obvious point, perhaps, but one not always given sufficient weight.

I have tried to show you that it is at present logically impossible to determine whether or not a difference in IQ mean (among other continuous variables) distinguishing two ethnic groups has a genetic component or not.

In doing so I may have given the impression that such characters in man are totally intractable. I do not think this is in fact so, and since I consider that variation in human ability factors is the most important topic that we can study, I wish to point out in outline what we can do if we do it carefully enough. This requires that we distinguish carefully between at least four questions:

1. What is the genetic component of the within-sibship variance?
2. What is the genetic component of the between-sibship variance within a population?
3. What is the genetic component of any particular between-population variance?
4. Do any two particular populations differ in genes or gene frequencies affecting the character in question?

I wish to stress that questions 3 and 4 are totally different questions. Answering question 3 positively (if this were possible) would imply a positive answer to question 4. But a positive answer to question 4 would tell us absolutely nothing about the answer to question 3.

Question 1—What is the genetic component of within-sibship variance?—is answerable from comparisons of fraternal and identical twin pairs. Ideally we should have the four-way comparisons, between identical twin pairs thought by their family to be identical, identicals thought to be fraternal, fraternals thought to be identical, and fraternals thought to be fraternal, so that we can assess the effect of family attitude on twin similarity. However, whatever its weaknesses, the equation $\overline{V}_{\mathrm{F}} - \overline{V}_{\mathrm{I}} = \overline{V}_{\mathrm{GS}}$* gives us our only estimate of within-sibship genetic variance, and $\overline{V}_I$ gives us the within-family, sex, and birth date environmental variances (together with the more complex genetic components of variance, which I enumerate to avoid expert criticism as, the variance arising from genetic dominance, gene interaction, and genotype-environment interaction). Given $\overline{V}_{\mathrm{GS}}$, we can then partition the within-sibship phenotypic variance into additive genetic variance and birth order, sex ratio, and other components if we wish to do so, thus providing considerable information on all sibship sizes greater than one. Only children, of course, can only contribute to between-family variance.

Question 2 can now be tackled, but is much more difficult because different families have different cultural and physical environments. Here the environmentalist's approach would be to classify the families according to sociocultural and other environmental factors and see whether the classification is associated with

* The average variance within fraternal twin pairs (of like sex) minus the average variance within identical twin pairs equals the average genetic variance within sib-pairs.

differences in mean IQ, taking due account, of course, of the within-family variables such as birth order. But this is clearly illegitimate unless it be assumed a priori that the sociocultural factors that differentiate the classes of family are in no way a consequence of genetic differences. We cannot assume that if so-called deprived families have lower average IQs that the low IQ is a consequence rather than a cause of the deprivation, for this is to beg the whole question at issue.

The geneticist, on the other hand, tends to take midparent-offspring mean regressions as estimates of population heritability, but this is also illegitimate because the genetic component so isolated will include the environmental parent-offspring correlation.

In these circumstances, our only legitimate approach is through theoretical population genetics. Knowledge of the within-sibship genetic variance allows us to predict, *assuming random mating*, the between sibship within-population genetic variance for in these circumstances, since the theoretical genetic correlation between siblings is one half, half the population genetic variance is between sibships. Departure from random mating at the phenotypic level is in the direction of positive assortative mating for IQ and many other human metrics, and positive assortative mating at the genetic level will *increase* the between, relative to the within, sibship genetic variance. Unless we suppose that the positive phenotypic correlation between mates conceals a negative genetic correlation, then the assumption of random mating will give us a minimum estimate of population genetic variance. The phenotypic variance less this will therefore give us a maximum estimate of environmental variance, and hence a maximum estimate of the cultural component of variation. We may go further, because the midparent-offspring mean regression gives us an estimate of the total phenotypic variance attributable to heredity (biological *and* social), so that we can estimate the heritable variance and subtract the genetic variance, thus estimating the proportion of variance attributable to "heritable" cultural or environmental differences.

Now we must turn to our third question, the heritability of group differences. This is not only the most emotive but also scientifically the least tractable question, for insofar as group differences such as class or race differences have genetic components, these must necessarily be correlated with environmental differences; and we can make no extrapolation of our estimates of within-group heritabilities to between-group differences, for we may not extrapolate from within- to between-population genetic variances.

As I have said, only transplant experiments could solve this problem. The equivalent of transplant experiments for man are twins reared apart, orphanages, or mobility between social classes. Given knowledge of the within-sibship genetic variance obtained as I have described above, such experiments can tell us something of the relationship between the within-sibship and within-group environmental variance. There are, however, well-known difficulties of determining whether the "transplanted" individuals are a random sample of the group from which they came, and of assessing the extent of the differences between the two environments involved, so that at present we can make no satisfactory assessment of between-group genetic differences, and in consequence no satisfactory assessment even between social classes where transplant experiments in principle are possible. We should, however, press on in our attempts to do so.

Question 3 is different when we concern ourselves with mean IQ differences between races or ethnic groups, that is, genetically more or less isolated populations having different environments. I hope I have made it clear that we cannot know and that, unless we can get at gene frequencies, we will never know whether these different ethnic groups differ in mean IQ for genetic or cultural reasons. It is impossible to transplant people between racial groups, and only such transplants could give us estimates of between–ethnic group heritability. We are therefore restricted to *describing* differences between ethnic groups and should not attempt to allocate them to genetic or environmental causes.

Question 4, however, is in principle answerable by a technique which has not, as far as I am aware, been used, that could, even where ethnic groups are involved, in principle provide answers to the question "Do two groups or populations differ genetically?"

Assuming approximately additive gene action, if two populations differ genetically, either in gene frequency or in fixed genes, the mean genetic variance of the progeny of hybrids will be greater than the mean of the two genetic variances within the populations. Now we have a means of assessing within-sibling–pair genetic variance from comparisons of fraternal and identical twins reared together. If therefore we can show whether $\bar{V}_F - \bar{V}_I$ for F_2 hybrids is greater than the average of its value for the two parent populations, then we can obtain evidence that the populations are genetically different that is not confounded with environmental factors.

It must be stressed, however, that valuable though such evidence might be, it will *not* provide an answer to question 3, for evidence that two populations, living in different environments, differ genetically is in no sense evidence that the difference between their means has a genetic component. This can be seen easily if we consider the extreme example of a genetically "superior" population in an "inferior" environment, with a lower mean than a genetically "inferior" population in a "superior" environment. The test could show they were genetically different, but the environment, not the genetic difference, would be the cause of the lower population mean. Thus the test could not give positive evidence that the populations differed genetically even if the genetic difference was of the opposite sign to the phenotypic difference. It could also give positive results even if there were no phenotypic difference, or for that matter, no difference of "genetic mean."

Conclusions

Our question 3 is therefore difficult. So difficult, in fact, that we cannot at present answer the question whether there is any genetic component of social class or race differences in mean IQ. And if we cannot do this, equally we cannot answer the question whether there is any environmental cause of these differences, for the two questions are in fact the same.

In this situation, there is no reason why people should be permitted to assume the answer to either of these questions, especially when those people are supposed to be qualified to pronounce on the matter. Hence the paramount need for sociologists and geneticists to preserve an open mind as far as the nature-nurture problem is concerned. This, of course, in no way precludes the search for environmental factors that will raise IQ, or even environmental conditions that will specifically improve Negro performance (*see* Katz et al., 1968). It is merely that closed minds on the nature-nurture problem are unjustifiable and, whichever kind they are, lend weapons to those who need to rationalize their prejudices.

References

Clausen, J., Keck, D. D., and Heisey, W. M. 1940. *Experimental Studies on the Nature of Species, I. Effects of Varied Environments on Western North American Plants.* Carnegie Institute of Washington Publications, no. 520.

GINI, C. 1954. "The Physical Assimilation of Foreign Settlements in Italy." *Atti del IX Congresso Internazionali de Genetica Caryologia,* vol. 6, suppl., p. 246.

KATZ, I., HENCHY, T., and ALLEN, H. 1968. "Effects of Race of Tester, Approval-Disapproval, and Need on Negro Children Learning." *Journal of Personality and Social Psychology* 8:38.

MCGURK, F. C. J. 1967. "The Culture Hypothesis and Psychological Tests." In *Race and Modern Science,* ed. R. E. Kuttner. New York: Social Science Press.

SHERWOOD, J. J., and NATAUPSKY, M. 1968. "Predicting the Conclusions of Negro-White Intelligence Research from Biographical Characteristics of the Investigator." *Journal of Personality and Social Psychology* 8:53.

VANDENBERG, S. G. 1968. "Primary Mental Abilities or General Intelligence? Evidence from Twin Studies." In *Genetic and Environmental Influences on Behaviour,* ed. J. M. Thoday and A. S. Parkes. Edinburgh: Oliver & Boyd.

Educability and Group Differences

J. M. Thoday

THE RECENTLY PUBLISHED BOOK BY ARTHUR R. JENSEN (1973) IS A very difficult one to review, not so much because the subject is controversial, but because so much of opinion in the area is strongly held, so that critical evaluation of the relation of data to hypothesis is limited. Thus, to review the book properly one would need not only to consider other books that have resulted from the publication of Jensen's (1969) article in *Harvard Educational Review* but also to consider each important statement in each book in relation to the original literature on which the statement is based. Even this would be inadequate, for it is my experience that the original sources seldom give enough information about sample structure or detail about results and statistical techniques to permit satisfying judgment, and that correspondence with the authors of the paper is needed.

No reviewer can do all this, especially in such an interdisciplinary field. If one does any of it, however, one is brought face to face with a most unsatisfactory situation: a scientific area in which, because of the impact of strongly felt ideopolitical points of view, there is wholly inadequate objective criticism of the relation between data and conclusions either by authors or their readers.

Jensen (1969), in his original *Harvard Educational Review* article, raised old questions because he was, I am sure rightly, disturbed that much literature, and a good deal of policy, is based on the assumption that variance in performance of IQ tests, and thus in correlated aspects of educability, arises largely or wholly from environmental reasons, despite the strong evidence for a large within-group genetic component. He also raised the question of between-group differences, in particular the average difference between American white and Negro social groups, maintaining that it was plausible to postulate a genetic component of this difference also. It should be stressed that Jensen only put this forward as a hypothesis, not as proved, but it was for putting forward this hypothesis that he has been most attacked. It should, however, be remembered that to postulate genetic variance in "educability" even in a nonracial context is unacceptable to many.

(The groups mentioned in the previous paragraph are primarily social groups, though they correlate partially with biological ancestry. Leach, reviewing this book in the *Listener*, made this a major criticism of Jensen, holding that he must not make a biological comparison between two groups unless the groups are defined by purely biological criteria. This is illogical. The whole question is whether two socially defined groups differ biologically in a certain variable or whether the difference is all a consequence of social factors.)

Jensen's book is about group differences, but is, as was his original article, based on the evidence concerning within-group differences. Here there are two points which may be regarded as well established.

1. Relative performance in IQ tests correlates with relative performance in aspects of education that are highly relevant to success in and the needs of modern societies. This statement remains true even if we recognize that relative IQ is not to be equated with relative "intelligence," and that the related aspects of educability are not to be equated with the whole of educability.
2. Within-group IQ variance has a large genetic component. There is room for argument whether Jensen's estimate of 80 percent may not be on the high side, but there can be no doubt that heritability of IQ ranks high compared with heritabilities reported in the generality of studies of continuous variables, whether behavioral or physical, in man or other organisms.

The high within-group heritability of this variable does not necessarily imply anything about between-group differences, but such differences need explanation. There are, however, fundamental methodological problems, of which Jensen is well aware, which arise from the necessary fact that different social groups differ in their environmental circumstances and are also partial genetic isolates. Any environmental differences are therefore necessarily correlated with possible gene-frequency differences and vice versa. Peculiarly critical experimental design and cautious interpretation are therefore required in this field.

The two main kinds of between-group difference of mean IQ are those concerning socioeconomic groups (or social "classes") and those concerning "racial" groups, particularly the United States white-"Negro" comparison (I place Negro in quotation marks since the United States Negro population has much white ancestry).

Considering within-race socioeconomic groups (the correlation between IQ and "class" is about 0.5), since (among United States and British whites) within-group heritability is high, social mobility is high, and social mobility is correlated with IQ, it would be very surprising if gene frequency differences were not the explanation of part of the differences of mean IQ. This, however, is a long way from proof, still less estimation of the relative magnitude of the genetically and environmentally caused components of the group differences.

In his book Jensen gives little attention to "social class" differences. His one argument about the evidence (1973, pp. 155–156) is, however, weak, for though he shows certain correlations are incompatible with a zero correlation between IQ genotype and social class, he makes no allowance for error in the estimates, and the correlations would be compatible if, for example, IQ heritability were 70 percent instead of the 80 percent he takes.

Three Questions

Most of his book concerns the "racial" comparison. The essential fact here is that United States "Negroes" score on the average one standard deviation below the United States white mean in IQ tests.

Three questions can be asked about the causation of this difference. Is the difference (in whole or part) a result of gene-frequency differences? Is the difference a result of environmental (including cultural) differences in factors that also vary within the groups? Is the difference a result of some environmental-cultural factor exclusive to the United States "Negro" population?

I shall consider the second question first. It is, of course, clear that there are within-group environmental variables that affect IQ. Otherwise within-group heritability would be 100 percent, and the correlation between identical twins would be 1. That some of this nonheritable variance is associated with socioeconomic group is strongly indicated by Skodak and Skeel's study which showed that adopted children from mothers of low IQ had higher mean IQs than expected on a purely genetic model. Jensen deals with this in a note to chapter 11 (p. 241).

The Negro population is, of course, low in mean socioeconomic status (SES), so that it is reasonable to postulate this as an environmental factor accounting for some of the group difference. Attempts to equate for SES have, however, failed to account for much if any of the difference between white and Negro social groups. Likewise, though extreme malnutrition can lower IQ, attempts to investigate effects of nutrition in the United States show negative results. (Malnutrition at levels found say, in Africa is, of course, irrelevant to an assessment of the United States Negro-white difference.) Likewise, most of the other often postulated environmental factors, motivation, reaction to race of tester, language deprivation, inequalities of schooling, culture loading of tests, seem unable to account for the difference. Indeed, when tests which are regarded as differing on culture loading are used, quite the opposite result than that predicted on the culture-loading hypothesis is obtained. The white-Negro difference is least on the most culture-loaded tests. Furthermore, other "racial" minority groups have higher average IQ than United States Negroes, some higher than United States whites. Having considered the evidence, Jensen regards the evidence relating to my second question as quite inadequate to account for the Negro-white difference in the United States.

This brings me to the third question, which is, as Jensen puts it, to postulate an unknown factor *X*. This has to have certain negative features if it is to fulfill its postulated function, for it must be a factor that does not affect minority groups other than Negroes, for these do not show the low IQ mean that the "Negro" social group does, nor can the factor be one that affects "deprived" whites. It has to be exclusive to Negroes.

It needs, however, to be pointed out that there are cultural factors exclusive to this Negro social group, such as awareness of Negro or slave ancestry and awareness of real or supposed social attitudes to Negroes, and it is not reasonable to discount these simply because they present hypotheses difficult to test, and because some other hypothesis fits the data. Jensen seems to do this.

For example, in arguing for the plausibility of the genetic hypothesis Jensen puts considerable stress on the argument from regression. This argument goes as follows. The "white" social group has a mean IQ of about 100, variance about 15^2 and heritability 80 percent: the "Negro" social group has a mean one standard deviation lower, about 85. Then if we take a sample of whites of IQ about 120, an unbiased sample of their siblings must regress to the mean IQ of 100, that is, genetic theory requires that the sibling mean should be about $100 + 20 \times 0.5 \times 0.8$ or 108. (20 is the deviation from the mean, 0.5 the genetic correlation between siblings, 0.8 the heritability.) This it does, which is not surprising since it is part of the evidence that heritability is 80 percent. By contrast, the siblings of a sample of Negroes of IQ 120 should regress to a mean IQ of 85, that is, assuming that heritability within the Negro social group is also 80 percent then it should be $85+35\times 0.5 \times 0.8$, or 99. By the same argument, the siblings of whites of IQ 80 should have a mean of 92, but those of Negroes of IQ 80, only 83. Jensen claims that these expectations are found in fact; in other words, the data fit genetic predictions. (He does not quote the data, but after correspondence with him, I am satisfied that the results are available and do not seem to involve biases such as might have permitted alternative explanations.) Jensen therefore regards the evidence as supporting the genetic hypothesis.

For some time, I fell into the same trap. But it is a trap, for populations must regress to their own mean whatever the cause, genetic or environmental, of the mean differences between the populations. This evidence is therefore as compatible with explanation in terms of the environmental factor *X* as in terms of the genetic hypothesis. It adds nothing whatsoever to the strength of the genetic hypothesis. All these results reveal is that heritability within the Negro group is of the same order as that within the white group.

I do not think this a culpable error on Jensen's part. So much of the argument concerning environmental factors is open to criticism, and he has been subject to such virulent attack (*see* the documentation in the long preface to his *Genetics and Education*[3]), that it must be well-nigh impossible for him to criticize the evidence relating to genetic hypotheses as closely and objectively as he has that concerning environmental factors.

Summarized Data

But another example is less pardonable. Jensen, Eysenck, and also Shockley, among others, have each proposed that evidence might be obtainable from intergroup hybrids and from studies of correlations, among Negroes, between IQ and amount of white ancestry as estimated from marker genes. Jensen is well aware of the difficulties such studies must present, notably with respect to nonrandomness of those involved in intergroup matings (*see* his Chapter 9), and he deals in a different chapter with the one piece of evidence of this kind that he seems to think good, giving it much less stress than Eysenck (1971) did in his book *Race, Intelligence and Education.* It is, however, a piece of evidence that illustrates well the difficulties and extreme dangers of reliance on summarized data.

This is the evidence of De Lemos (1969) on Australian aborigines, and it concerns Piagetian tests whose results correlate strongly with those of culture-reduced IQ tests. De Lemos gave Piagetian conservation tests to aboriginal children who were divided into two groups, according to Mission records of their ancestry, as full and part aborigines. Most of the part aborigines had, it seems, one white great-grandparent. Jensen reproduces the data direct from page 316 of De Lemos's paper (*see* table 1), and in this form it looks convincing.

But Jensen (1969) had already presented these data before in a different form taken from De Lemos's (1966) Ph.D. thesis as shown in table 2.

Not only are the significance levels now much less impressive, though total sample size is for some reason larger, but one immediately sees that table 1 confounds an age difference with the ancestry difference, whereas table 2 shows age to be, as of course it must, most important in relation to these tests. One is left asking whether age differences within the 8-to-11 and within the 12-to-15-year groups might account for the significant results left in table 2. Examination of De Lemos's original paper shows that age differences are not adequately controlled, and that the data cannot be regarded as demonstrating that the ancestry difference has significant effects. It turns out that this conclusion has already been pub-

Table 1. *Table 17.1 of Jensen, 1973. (Comparison of the Number of Part-Aboriginal and Full-Aboriginal Children showing Conservation*)*

Test	*Full Aboriginal* N = *38*	*Part Aboriginal* N = *34*	χ^2	P
Quantity	4	18	15.21	< 0.001
Weight	16	25	7.23	< 0.01
Volume	2	8	3.50	$0.05 < P < 0.10$
Length	12	20	5.37	< 0.05
Area	3	10	4.23	< 0.05
Number	3	9	3.22	$0.05 < P < 0.10$
Total†	40	90	36.14	< 0.001

* From De Lemos (1969).
† The chi square test for Total (given by De Lemos) is statistically inappropriate here, since pooling more than one observation from the same subject violates the requirement of independence of observations upon which the chi square test depends.

Table 2. *Table 1 from Jensen, 1969. (Numbers of Full-blood and Part-blood Australian Aboriginal Children Passing Piagetian Conservation Tests and the Significance Level* (P) *of the Difference*.)*

	Age 8–11 years			*Age 12–15 years*		
Total N =	*Full*	*Part*	P	*Full*	*Part*	P
	25	*17*		*17*	*21*	
Tests						
Quantity	2	6	< 0.1	2	15	< 0.01
Weight	9	11	< 0.1	7	17	< 0.01
Volume	0	5	< 0.05	2	4	N.S.
Length	10	10	N.S.	3	13	< 0.05
Number	0	4	< 0.05	3	8	N.S.
Area	1	4	N.S.	2	8	N.S.

* Source: De Lemos (1966).

lished by Vetta (1972), who also refers to a replication study by Dasen, which had negative results. These papers came out too late for Jensen to refer to them, for his bibliography contains nothing later than 1971. But he criticizes others in other contexts for failing to equate for age, and he really should have seen the implications of the difference between these two tables reproduced from his own publications.*

I have considered this example in detail because it illustrates the difficulties one is faced with in judging conclusions in this area. Jensen has, often cogently, criticized evidence purporting to support relation between IQ and various environmental factors. Some of his own arguments are open to comparable criticism. The treatment of De Lemos's data and the consideration of regressions suggest that Jensen suffers from the same kind of conscious or unconscious bias as many of his opponents, that is to say, he is prepared to accept evidence that seems to support his hypothesis with less critical examination than he would give to evidence purporting to be against him.

Lessons

One is left, therefore, despite all the literature, with these conclusions. First, there is no evidence which reveals whether the Negro-white IQ difference has any genetic component or any environmental component. Both hypotheses (and any intermediate hypothesis) are equally consistent with the facts. (There is one exception: United States Negro women score higher than United States Negro men, a difference for which no genetic model seems to fit the facts. Jensen (1971) himself has elsewhere put the difference down to environmental factors of the nature of "X," but does not mention the matter in this book. It should account for at least one-tenth of the United States Negro-white difference.)

Second, no statement about causation of IQ variation should be taken at its face value, whoever the authority. Every statement requires most careful consideration of the detailed data on which it is based.

Third, the more we would like to believe some statement about the causation of IQ variations, the more closely should we examine the data and logic behind it.

If we learn these lessons, Jensen may have done us a service. He is perfectly correct in maintaining that the hypothesis that group differences in IQ (or anything else) have a genetic component is

* I thank Dr. J. B. Gibson for drawing my attention to the difference between these tables.

plausible and that there is no evidence against it. This means that no evidence which purports to demonstrate the relevance of some environmental component can be regarded as adequate unless it be demonstrated from the same evidence that the groups being compared could not differ genetically. It follows also that we have no reason to expect that different groups will have the same distribution of attributes and that demonstrations that different groups have different frequencies of success, for example in some educational selection process, cannot be regarded as good evidence that that selection process does not provide equality of opportunity for the individuals concerned. These are important lessons that all—including editors—should bear in mind and try to act on.

But I believe there is a risk that Jensen may have also done us a disservice, for the controversy about group differences is proving a stick to beat him with, to the detriment of rational discussion of individual variation and the importance of the genetic component of individual variation in attributes that correlate with IQ.

There is now a danger that the controversy about group differences may lead the evidence about individual differences to be swept under the carpet. But our relatively recent knowledge concerning the ubiquity of genetic variation is of the utmost scientific, political, and philosophical importance. Variation in IQ, and variation in the related aspects of educability, are but one example. Let us remember that to assume that IQ variation is of no moment, or to assume that it does not have a substantial genetic component, not only flies in the face of the facts, but puts us in a position where it can be held that the state through the educational system may make what it will of anyone. In truth, individuals, however malleable, are different and should not be treated as if they were the same. Controversy about the causation of group differences must not lead us to ignore this, and difficulties that arise because groups differ in the frequency with which they meet certain educational criteria must not lead to our assuming that everyone is able to meet the same educational criteria. It is not true that everyone can reach the same academic standards if provided with adequate opportunity, and the heritability of IQ is a partial measure of that untruth. Equality of educational achievement must prove an unrealizable ideal. Those who raise the hope that it is realizable must bear responsibility for the resulting widespread individual disappointment and all its consequences.

Our societies need to be organized to allow for individual differences, including such differences in aspects of educability. Indeed, if our societies were so organized we might perhaps slowly come to realize that it does not necessarily matter if group differences prove to have a genetic component, for the genetic variation within groups and the overlap between groups are more important and no demonstration that a group difference has a genetic component could justify racism. Perhaps also if we spend less effort considering group differences and more on individual differences and their genetic component we may become more capable of reasonable discussion of the implications of this individuality, the ways in which society should accommodate it, and the extent to which any particular dimension of individuality, such as IQ and its associates, may or may not be overstressed in our present system of status and economic rewards.

References

De Lemos, M. M. 1966. Ph.D. thesis, Australian National University.

———. 1969. "The Development of Conservation in Aboriginal Children." *International Journal of Psychology* 4: 255.

Eysenck, H. J. 1971. *Race, Intelligence and Education.* London: Maurice Temple-Smith.

Jensen, A. R. 1969a. "Reducing the Heredity-Environment Uncertainty." In *Environment, Heredity, and Intelligence. Harvard Educational Review* Reprint no. 2, p. 211.

———. 1969b. "How Much Can We Boost IQ and Scholastic Achievement?" *Harvard Educational Review* 39: 1–123.

———. 1971. "The Race x Sex x Ability Interaction." In *Intelligence,* ed. R. Cancro. New York: Grune & Stratton.

———. 1972. *Genetics and Education.* London: Methuen & Co.

———. 1973. *Educability and Group Differences.* London: Methuen & Co.

Vetta, A. 1972. "Conservation in Aboriginal Children and 'Genetic Hypothesis'." *International Journal of Psychology* 7: 247.

Behavior-Genetic Analysis and Its Biosocial Consequences

Jerry Hirsch

As a psychology student I was taught that a science was founded on the discovery of lawful relations between variables. During my student days at Berkeley, the true psychological scientist was preoccupied with the major learning theories. We read, studied, and designed experiments to test the theories of Thorndike, Guthrie, Hull, and Tolman. Many of their verbally formulated laws of behavior were replaced by the mathematical models that have since come into vogue.

Afterwards I learned empirically the truth of what might be the most general of all behavioral laws, the Harvard law of animal behavior: "Under the most carefully controlled experimental conditions the animals do as they damn please." Still later I discovered the low esteem in which post–World War II psychology was held by two of the best minds this century has seen. In 1947, John Dewey, eighth president of the American Psychological Association, wrote to discourage young Robert V. Daniels from studying psychology at Harvard:

> Psychology . . . is on the whole, in my opinion, the most inept and backwards a tool . . . as there is. It is much of it actually harmful because of wrong basic postulates—maybe not all stated, but actually there when one judges from what they do—the kind of problems attacked and the way they attack them. [1959, p. 570]

On the final page of the last book written before his death in 1951 Ludwig Wittgenstein, perhaps the most influential of the founders of modern philosophical analysis, observed:

> The confusion and barrenness of psychology is not to be explained by calling it a "young science"; its state is not comparable with that of physics, for instance, in its beginning. (Rather with that of certain branches of mathematics. Set theory.) For in psychology there are experimental methods and *conceptual confusion.* (As in the other case conceptual confusion and methods of proof.)
>
> The existence of the experimental method makes us think we have the means of solving the problems which trouble us; though problem and method pass one another by. [1963, p. 232]

Laws of Genetics

It was then while overcome by feelings of disenchantment (obviously without laws behavior study could never be science) that I embraced genetics. There was true science. My passion became even more intense when I realized that, like thermodynamics, genetics had three laws: segregation, independent assortment, and the Hardy-Weinberg law of population equilibria. What a foundation they provided for my beloved individual differences.

Since both my teaching and research involved considerable work with *Drosophila,* I knew and would recount to my classes in somewhat elaborate detail the story of Calvin Bridges's classic experiments on sex determination as a function of a ratio between the sex chromosomes and the autosomes. As the important discoveries in human cytogenetics were made throughout the 1950s and 1960s and "abnormalities" like Klinefelter's, Turner's, and Down's syndromes and the violence-prone males with an extra Y-chromosome became genetically comprehensible, I began to realize that the so-called laws of genetics were no more universal than the so-called laws of behavior. Every one of the above-mentioned clinical conditions involved, at the very least, a violation of Mendel's law of segregation. Of course, so did Bridges's experiments, but it had been too easy to rationalize them as clever laboratory tricks.

Behaviorism

Over the past two decades the case against behaviorist extremism has been spelled out in incontrovertible detail. The behaviorists committed many sins: they accepted the mind at birth as Locke's *tabula rasa,* they advocated an empty-organism psychology, they asserted the uniformity postulate of no prenatal individual differences; in short, they epitomized typological thinking. Many times we have heard quoted the famous boast by the first high priest of behaviorism, John B. Watson:

> Give me a dozen healthy infants, well-formed, and my own specified world to bring them up in, and I'll guarantee to take any one at random and train him to become any type of specialist I might select—doctor, lawyer, artist, merchant-chief and yes, even beggarman and thief, regardless of his talents, penchants, tendencies, abilities, vocations, race of his ancestors.

However, it is only when we read the next sentence, which is rarely, if ever, quoted, that we begin to understand how so many people might have embraced something intellectually so shallow as radical behaviorism. In that all-important next sentence Watson explains:

> I am going beyond my facts and I admit it, but so have the advocates of the contrary and they have been doing it for many thousands of years. [1959, p. 104]

Racism

Who were the advocates of the contrary and what had they been saying? It is difficult to establish the origins of racist thinking, but certainly one of its most influential advocates was Joseph Arthur de Gobineau, who published a four-volume *Essay on the Inequality of the Human Races* in the mid-1850s. De Gobineau preached the superiority of the white race, and among whites, it was the Aryans who carried civilization to its highest point. In fact, they were responsible for civilization wherever it appeared. Unfortunately, de Gobineau's essay proved to be the major seminal work that inspired some of the most perverse developments in the intellectual and political history of our civilization. Later in his life, de Gobineau became an intimate of the celebrated German composer, Richard Wagner. The English-born Houston Stewart Chamberlain, who emigrated to the Continent, became a devoted admirer of both de Gobineau and Wagner. In 1908, after Wagner's death, he married Wagner's daughter, Eva, settled in and supported Germany against England during World War I, becoming a naturalized German citizen in 1916.

In the summer of 1923, an admirer who had read Chamberlain's writings, Adolf Hitler, visited Wahnfried, the Wagner family home in Bayreuth where Chamberlain lived. After their meeting, Chamberlain wrote to Hitler: "My faith in the Germans had never wavered for a moment, but my hope . . . had sunk to a low ebb. At one stroke you have transformed the state of my soul!" (Heiden, 1944, p. 198) We all know the sequel to that unfortunate tale. I find that our modern scientific colleagues, whether they be biological or social scientists, for the most part, do not know the sad parallel that exists for the essentially political tale I have so far recounted. The same theme can be traced down the mainstream of biosocial science.

Today not many people know the complete title of Darwin's most famous book: *On the Origin of Species by Means of Natural Selection or the Preservation of Favoured Races in the Struggle for Life.* I find no evidence that Darwin had the attitudes we now call racist. Unfortunately, many of his admirers, his contemporaries, and his successors were not as circumspect as he. In Paris in 1838, J. E. D. Esquirol first described a form of mental deficiency later to become well known by two inappropriate names unrelated to his work. Unhappily one of these names, through textbook adoption and clinical jargon, puts into wide circulation a term loaded with race prejudice. Somewhat later (1846 and 1866), E. Seguin described the same condition under the name "furfuraceous cretinism" and his account has only recently been recognized as "the most ingenious description of physical characteristics." (Benda, 1962, p. 163)

Unhappily, that most promising scientific beginning was ignored. Instead the following unfortunate events occurred: In 1866, John Langdon Haydon Down published the paper entitled "Observations on an Ethnic Classification of Idiots."

> . . . making a classification of the feeble-minded, by arranging them around various ethnic standards—in other words, framing a natural system to supplement the information to be derived by an inquiry into the history of the case.
>
> I have been able to find among the large number of idiots and imbeciles which comes under my observation, both at Earlswood and the out-patient department of the Hospital, that a considerable portion can be fairly referred to one of the great divisions of the human family other than the class from which they have sprung. Of course, there are numerous representatives of the great Caucasian family. Several well-marked examples of the Ethiopian variety have come

under my notice, presenting the characteristic malar bones, the prominent eyes, the puffy lips, and retreating chin. The woolly hair has also been present, although not always black, nor has the skin acquired pigmentary deposit. They have been specimens of white negroes, although of European descent.

Some arrange themselves around the Malay variety, and present in their soft, black, curly hair, their prominent upper jaws and capacious mouths, types of the family which people the South Sea Islands.

Nor have there been wanting the analogues of the people who with shortened foreheads, prominent cheeks, deep-set eyes, and slightly apish nose, originally inhabited the American Continent.

The great Mongolian family has numerous representatives, and it is to this division, I wish, in this paper, to call special attention. A very large number of congenital idiots are typical Mongols. So marked is this, that when placed side by side, it is difficult to believe that the specimens compared are not children of the same parents. The number of idiots who arrange themselves around the Mongolian type is so great, and they present such a close resemblance to one another in mental power, that I shall describe an idiot member of this racial division, selected from the large number that have fallen under my observation.

The hair is not black, as in the real Mongol, but of a brownish colour, straight and scanty. The face is flat and broad, and destitute of prominence. The cheeks are roundish, and extended laterally. The eyes are obliquely placed, and the internal canthi more than normally distant from one another. The palpebral fissure is very narrow. The forehead is wrinkled transversely from the constant assistance which the levatores palpebrarum derive from the occipito-frontalis muscle in the opening of the eyes. The lips are large and thick with transverse fissures. The tongue is long, thick, and is much roughened. The nose is small. The skin has a slightly dirty yellowish tinge and is deficient in elasticity, giving the appearance of being too large for the body.

The boy's aspect is such that it is difficult to realize that he is the child of Europeans, but so frequently are these characters presented, that there can be no doubt that these ethnic features are the result of degeneration. [Reprinted in McKusick, 1966, p. 432]

And he means degeneration from a higher to a lower race. The foregoing represents a distasteful but excellent example of the racial hierarchy theory and its misleadingly dangerous implications. That was how the widely used terms Mongolism and Mongolian idiocy entered our "technical" vocabulary. For the next century, this pattern of thought is going to persist and occupy an important place in the minds of many leading scientists.

Alleged Jewish Genetic Inferiority

In 1884, Francis Galton, Darwin's half-cousin, founder of the Eugenics movement and respected contributor to many fields of science, wrote to the distinguished Swiss botanist, Alphonse de Candolle: "It strikes me that the Jews are specialized for a parasitical existence upon other nations, and that there is need of evidence that they are capable of fulfilling the varied duties of a civilized nation by themselves" (Pearson, 1924, p. 209). Karl Pearson, Galton's disciple and biographer, echoed this opinion forty years later during his attempt to prove the undesirability of Jewish immigration into Britain: ". . . for such men as religion, social habits, or language keep as a caste apart, there should be no place. They will not be absorbed by, and at the same time strengthen the existing population; they will develop into a parasitic race" (Pearson and Moul, 1925, p. 125).

Beginning in 1908 and continuing at least until 1928, Karl Pearson collected and analyzed data in order to assess "the quality of the racial stock immigrating into Great Britain" (Pastore, 1949, p. 33). He was particularly disturbed by the large numbers of East European Jews, who near the turn of the century began coming from Poland and Russia to escape the pogroms. Pearson's philosophy was quite explicitly spelled out:

> Let us admit . . . that the mind of man is for the most part a congenital product, and the factors which determine it are racial and familial; we are not dealing with a mutable characteristic capable of being moulded by the doctor, the teacher, the parent or the home environment. . . .
>
> The ancestors of the men who pride themselves on being English today were all at one time immigrants; it is not for us to cast the first stone against newcomers, solely because they are newcomers. But the test for immigrants in the old days was a severe one; it was power, physical and mental, to retain their hold on the land they seized. So came Celts, Saxons, Norsemen, Danes and Normans in succession and built up the nation of which we are proud. Nor do we criticize the alien Jewish immigration simply because it is Jewish; we took the alien Jews to study, because they were the chief immigrants of that day and material was readily available. [Pearson and Moul, 1925, pp. 124, 127]

His observations led him to conclude: "Taken *on the average*, and regarding both sexes, this alien Jewish population is somewhat inferior physically and mentally to the native population" (p. 126). Pearson proclaimed this general Jewish inferiority despite his own failure to find any differences between the Jewish and non-Jewish boys when comparisons (reported in the same article) were made for the sexes separately.

Alleged Black Genetic Inferiority

Quite recently there has appeared a series of papers disputing whether or not black Americans are, in fact, genetically inferior to white Americans in intellectual capacity. The claims and counterclaims have been given enormous publicity in the popular press in America. Some of those papers contain most of the fallacies that can conceivably be associated with this widely misunderstood problem.

The steps toward the intellectual cul-de-sac into which this dispute leads and the fallacious assumptions on which such "progress" is based are the following: (1) A trait called intelligence, or anything else, is defined and a testing instrument for the measurement of trait expression is used; (2) the heritability of that trait is estimated; (3) races (populations) are compared with respect to their performance on the test of trait expression; (4) when the races (populations) differ on the test whose heritability has now been measured, the one with the lower score is genetically inferior, Q.E.D.

The foregoing argument can be applied to any single trait or to as many traits as one might choose to consider. Therefore, analysis of this general problem does *not* depend upon the particular definition and test used for this or that trait. For my analysis I shall pretend that an acceptable test exists for some trait, be it height, weight, intelligence, or anything else. (Without an acceptable test, discussion of the "trait" remains unscientific.)

Even to consider comparisons between races, the following concepts must be recognized: (1) the genome as a mosaic, (2) development as the expression of one out of many alternatives in the genotype's norm of reaction, (3) a population as a gene pool, (4) heritability is not instinct, (5) traits as distributions of scores, and (6) distributions as moments.

Since inheritance is particulate and not integral, the genome, genotype, or hereditary endowment of each individual is a unique mosaic—an assemblage of factors many of which are independent. Because of the lottery-like nature of both gamete formation and fertilization, other than monozygotes no two individuals share the same genotypic mosaic.

Norm of Reaction

The ontogeny of an individual's phenotype (observable outcome of development) has a norm or range of reaction not predictable in advance. In most cases the norm of reaction remains largely unknown; but the concept is nevertheless of fundamental importance, because it saves us from being taken in by glib and misleading textbook clichés such as "heredity sets the limits but environment determines the extent of development within those limits." Even in the most favorable materials only an approximate estimate can be obtained for the norm of reaction, when, as in plants and some animals, an individual genotype can be replicated many times and its development studied over a range of environmental conditions. The more varied the conditions, the more diverse might be the phenotypes developed from any one genotype. Of course, different genotypes should not be expected to have the same norm of reaction; unfortunately psychology's attention was diverted from appreciating this basic fact of biology by a half century of misguided environmentalism. Just as we see that, except for monozygotes, no two human faces are alike, so we must expect norms of reaction to show genotypic uniqueness. That is one reason why the heroic but ill-fated attempts of experimental learning psychology to write the "laws of environmental influence" were grasping at shadows. Therefore, those limits set by heredity in the textbook cliché can never be specified. They are plastic within each individual but differ between individuals. Extreme environmentalists were wrong to hope that one law or set of laws described universal features of modifiability. Extreme hereditarians were wrong to ignore the norm of reaction.

Individuals occur in populations and then only as temporary attachments, so to speak, each to particular combinations of genes. The population, on the other hand, can endure indefinitely as a pool of genes, maybe forever recombining to generate new individuals.

Instincts, Genes, and Heritability

What is heritability? How is heritability estimated for intelligence or any other trait? Is heritability related to instinct? In 1872, Douglas Spalding demonstrated that the ontogeny of a bird's ability to fly is simply maturation and not the result of practice, imitation, or any demonstrable kind of learning. He confined immature birds and deprived them of the opportunity either to practice flapping their wings or to observe and imitate the flight of older birds; in spite of this, they developed the ability to fly. For some ethologists this deprivation experiment became the paradigm for proving the innateness or instinctive nature of a behavior by demonstrating that it appears despite the absence of any opportunity for it to be learned. Remember two things about this approach: (1) the observation involves experimental manipulation of the conditions of experience during development, and (2) such observation can be made on the development of one individual. For some people the results of a deprivation experiment now constitute the operational demonstration of the existence (or non-existence) of an instinct (in a particular species).

Are instincts heritable? That is, are they determined by genes? But what is a gene? A gene is an inference from a breeding experiment. It is recognized by the measurement of individual differences —the recognition of the segregation of distinguishable forms of the expression of some trait among the progeny of appropriate matings. For example, when an individual of blood type AA mates with one of type BB, their offspring are uniformly AB. If two of the AB offspring mate, it is found that the A and B gene forms have segregated during reproduction and recombined in their progeny to produce all combinations of A and B: AA, AB, and BB. Note that the only operation involved in such a study is *breeding* of one or more generations and then at an appropriate time of life, observation of the separate individuals born in each generation—controlled breeding with experimental material or pedigree analysis of the appropriate families with human subjects. In principle, only one (usually brief) observation is required. Thus we see that genetics is a science of *differences*, and the breeding experiment is its fundamental operation. The operational definition of the gene, therefore, involves observation in a breeding experiment of the segregation among several individuals of distinguishable differences in the expression of some trait from which the gene can be inferred. Genetics does not work with a single subject, whose development is studied. (The foregoing, the following, and all discussions of genetic analysis presuppose sufficiently adequate control of environmental conditions so that all observed individual differences have developed under the same, homogeneous environmental conditions, conditions never achieved in any human studies.)

How does heritability enter the picture? At the present stage of knowledge, many features (traits) of animals and plants have not yet been related to genes that can be recognized individually. But the role of large numbers of genes, often called polygenes and in most organisms still indistinguishable one from the other, has been demonstrated easily (and often) by selective breeding or by appropriate comparisons between different strains of animals or plants. Selection and strain crossing have provided the basis for many advances in agriculture and among the new generation of research workers are becoming standard tools for the experimental behaviorist. Heritability often summarizes the extent to which a particular population has responded to a regimen of being bred selectively on the basis of the expression of some trait. Heritability values vary between zero and plus one. If the distribution of trait expression among progeny remains the same no matter how their parents might be selected, then heritability has zero value. If parental selection does make a difference, heritability exceeds zero, its exact value reflecting the parent-offspring correlation. Or more generally, as Jensen (1969) says: "The basic data from which . . . heritability coefficients are estimated are correlations among individuals of different degrees of kinship" (p. 48). Though, many of the heritabilities Jensen discusses have been obtained by comparing mono- and dizygotic twins (Jensen, 1967).

A heritability estimate, however, is a far more limited piece of information than most people realize. As was so well stated by Fuller and Thompson (1960): "heritability is a property of populations and not of traits." In its strictest sense, a heritability measure provides for a given population an estimate of the proportion of the variance it shows in trait (phenotype) expression which is correlated with the segregation of the alleles of independently acting genes. There are other more broadly conceived heritability measures, which estimate this correlation and also include the combined effects of genes that are independent and of those that interact. Therefore, heritability estimates the proportion of the total phenotypic variance (individual differences) shown by a trait that can be attributed to genetic variation (narrowly or broadly interpreted) in some particular population at a single generation under one set of conditions.

The foregoing description contains three fundamentally important limitations which have rarely been accorded sufficient attention: (1) The importance of limiting any heritability statement to a specific population is evident when we realize that a gene, which shows variation in one population because it is represented there by two or more segregating alleles, might show no variation in some other population because it is uniformly represented there by only a single allele. Remember that initially such a gene could never have been detected by genetic methods in the second population. Once it has been detected in some population carrying two or more of its segregating alleles, the information thus obtained might permit us to recognize it in populations carrying only a single allele. Note how this is related to heritability: the trait will show a greater-than-zero heritability in the segregating population but zero heritability in the nonsegregating population. This does *not* mean that the trait is determined genetically in the first population and environmentally in the second!

Up to now my discussion has been limited to a single gene. The very same argument applies for every gene of the polygenic complexes involved in continuously varying traits like height, weight, and intelligence. Also, only *genetic* variation has been considered—the presence or absence of segregating alleles at one or more loci in different populations.

(2) Next let us consider the ever-present environmental sources of variation. Usually, from the Mendelian point of view, except for the genes on the segregating chromosomes, everything inside the cell and outside the organism is lumped together and can be called environmental variation: cytoplasmic constituents, the maternal effects now known to be so important, the early experience effects studied in so many psychological laboratories, and so on. None of these can be considered unimportant or trivial. They are ever present. Let us now perform what physicists call a *Gedanken*, or thought, experiment. Imagine Aldous Huxley's *Brave New World* or Skinner's *Walden II* organized in such a way that every individual is exposed to precisely the same environmental conditions. In other words, consider the extreme, but *un*realistic, case of complete environmental homogeneity. Under those circumstances the

heritability value would approach unity, because only genetic variation would be present. Don't forget that even under the most simplifying assumptions, there are over 70 trillion potential human genotypes—no two of us share the same genotype no matter how many ancestors we happen to have in common (Hirsch, 1963). Since mitosis projects our unique genotype into the nucleus, or executive, of every cell in our bodies, the individuality that is so obvious in the human faces we see around us must also characterize the unseen components. Let the same experiment be imagined for any number of environments. In each environment heritability will approximate unity, but each genotype *may* develop a different phenotype in every environment and the distribution (hierarchy) of genotypes (in terms of their phenotypes) must not be expected to remain invariant over environments.

(3) The third limitation refers to the fact that because gene frequencies can and do change from one generation to the next, so will heritability values or the magnitude of the genetic variance.

Now let us shift our focus to the entire genotype or at least to those of its components that might co-vary at least partially with the phenotypic expression of a particular trait. Early in this century Woltereck called to our attention the norm-of-reaction concept: the same genotype can give rise to a wide array of phenotypes depending upon the environment in which it develops (Dunn, 1965). This is most conveniently studied in plants where genotypes are easily replicated. Later Goldschmidt (1955, p. 257) was to show in *Drosophila* that, by careful selection of the environmental conditions at critical periods in development, various phenotypes ordinarily associated with specific gene mutations could be produced from genotypes that did not include the mutant form of those genes. Descriptively, Goldschmidt called these events *phenocopies*—environmentally produced imitations of gene mutants or phenotypic expressions only manifested by the "inappropriate" genotype if unusual environmental influences impinge during critical periods in development, but regularly manifested by the "appropriate" genotype under the usual environmental conditions.

In 1946, the brilliant British geneticist J. B. S. Haldane analyzed the interaction concept and gave quantitative meaning to the foregoing. For the simplest case but one, that of two genotypes in three environments or, for its mathematical equivalent, that of three genotypes in two environments, he showed that there are sixty possible kinds of interaction. Ten genotypes in ten environments generate 7.09×10^{144} possible kinds of interaction. In general, m genotypes in n environments generate $\frac{(mn)!}{m!n!}$ kinds of interaction. Since the characterization of genotype-environment interaction can only be ad hoc and the number of possible interactions is effectively unlimited, it is no wonder that the long search for general laws has been so unfruitful.

For genetically different lines of rats showing the Tyron-type "bright-dull" difference in performance on a learning task, by so simple a change in environmental conditions as replacing massed-practice trials by distributed-practice trials, McGaugh, Jennings, and Thomson (1962) found that the so-called dulls moved right up to the scoring level of the so-called brights. In a recent study of the open-field behavior of mice, Hegmann and DeFries, (1968) found that heritabilities measured repeatedly in the same individuals were unstable over two successive days. In surveying earlier work, they commented (p. 27): "Heritability estimates for repeated measurements of behavioral characters have been found to increase (Broadhurst & Jinks, 1961), decrease (Broadhurst & Jinks, 1966), and fluctuate randomly (Fuller & Thompson, 1960) as a function of repeated testing." Therefore, to the limitations on heritability due to population, situation, and breeding generation, we must now add developmental stage, or, many people might say, just plain unreliability! The late and brilliant Sir Ronald Fisher (1951), whose authority Jensen cites (1969, p. 34), indicated how fully he had appreciated such limitations when he commented: "the so-called coefficient of heritability, which I regard as one of those unfortunate short-cuts which have emerged in biometry for lack of a more thorough analysis of the data" (Fisher, 1951, p. 217). The plain facts are that in the study of man a heritability estimate turns out to be a piece of "knowledge" that is both deceptive and trivial.

The Roots of One Misuse of Statistics

The other two concepts to be taken into account when racial comparisons are considered involve the representation of traits in populations by distributions of scores and the characterization of distributions by moment-derived statistics. Populations should be compared only with respect to one trait at a time, and comparisons should be made in terms of the moment statistics of their trait distributions. Therefore, for any two populations, on each trait of interest, a separate comparison should be made for every moment of their score distributions. If we consider only the first four moments, from which are derived the familiar statistics for mean, variance, skewness, and kurtosis, then there are four ways in which populations or races may differ with respect to any single trait. Since we possess twenty-three independently assorting pairs of chromosomes, certainly there are at least twenty-three uncorrelated traits with respect to which populations can be compared. Since comparisons will be made in terms of four (usually independent) statistics, there are $4 \times 23 = 92$ ways in which races can differ. Since the integrity of chromosomes is *not* preserved over the generations, because they often break apart at meiosis and exchange constituent genes, there are far more than twenty-three independent hereditary units. If instead of twenty-three chromosomes we take the 100,000 genes man is now estimated to possess (McKusick, 1966, p. ix) and we think in terms of their phenotypic trait correlates, then there may be as many as 400,000 comparisons to be made between any two populations or races.

A priori, at this time, we know enough to expect no two populations to be the same with respect to most or all of the constituents of their gene pools. "Mutations and recombinations will occur at different places, at different times, and with differing frequencies. Furthermore, selection pressures will also vary" (Hirsch, 1963, p. 1441). So the number and kinds of differences between populations now waiting to be revealed in "the more thorough analysis" recommended by Fisher literally staggers the imagination. It does not suggest a linear hierarchy of inferior and superior races.

Why has so much stress been placed on comparing distributions only with respect to their central tendencies by testing the significance of mean differences? There is much evidence that many observations are not normally distributed and that the distributions from many populations do not share homogeneity of variance. The source of our difficulty traces back to the very inception of our statistical tradition.

There is an unbroken line of intellectual influence from Quetelet through Galton and Pearson to modern psychometrics and biometrics. Adolphe Quetelet (1796–1874), the Belgian astronomer-statistician, introduced the concept of "the average man"; he also applied the normal distribution, so widely used in astronomy for error variation, to human data, biological and social. The great Francis Galton followed Quetelet's lead, and then Karl Pearson elaborated and perfected their methods. I know of nothing that has contributed more to impose the typological way of thought on, and perpetuates it in, present-day psychology than the feedback from these methods for describing observations in terms of group averages.

There is a technique called composite photography to the perfection of which Sir Francis Galton contributed in an important way. Some of Galton's best work in this field was done by combining—literally averaging—the separate physiognomic features of many different Jewish individuals into his composite photograph of "the Jewish type." Karl Pearson, his disciple and biographer, wrote: "There is little doubt that Galton's Jewish type formed a landmark in composite photography . . ." (1924, p. 293). The part played by typological thinking in the development of modern statistics and the way in which such typological thinking has been feeding back into our conceptual framework through our continued careless use of these statistics is illuminated by Galton's following remarks:

> The word generic presupposes a genus, that is to say, a collection of individuals who have much in common, and among whom medium characteristics are very much more frequent than extreme ones. The same idea is sometimes expressed by the word typical, which was much used by Quetelet, who was the first to give it a rigorous interpretation, and whose idea of a type lies at the basis of his statistical views. No statistician dreams of combining objects into the same generic group that do not cluster towards a common centre; no more can we compose generic portraits out of heterogeneous elements, for if the attempt be made to do so the result is monstrous and meaningless. [Quoted in Pearson, 1924, p. 295]

The basic assumption of a type, or typical individual, is clear and explicit. They used the normal curve and they permitted distributions to be represented by an average because, even though at times they knew better, far too often they tended to think of races as discrete, even homogeneous, groups and individual variation as error.

It is important to realize that these developments began before 1900, when Mendel's work was still unknown. Thus at the inception of biosocial science there was no substantive basis for understanding individual differences. After 1900, when Mendel's work became available, its incorporation into biosocial science was bitterly opposed by the biometricians under Pearson's leadership. Galton had promulgated two "laws": his Law of Ancestral Heredity (1865) and his Law of Regression (1877). When Yule (1902) and Castle (1903) pointed out how the Law of Ancestral Heredity could be explained in Mendelian terms, Pearson (1904) stubbornly denied it. Mendel had chosen for experimental observation seven traits, each of which, in his pea-plant material, turned out to be a phenotypic correlate of a single gene with two segregating alleles. For all seven traits one allele was dominant. Unfortunately, Pearson assumed the universality of dominance and based his disdain for Mendelism on this assumption. Yule (1906) then showed that without the assumption of dominance, Mendelism becomes perfectly consistent with the kind of quantitative data on the basis of which it was being rejected by Pearson. It is sad to realize that Pearson never appreciated the generality of Mendelism and seems to have gone on for the next thirty-two years without doing so.

Two Fallacies

Now we can consider the recent debate about the meaning of comparisons between the "intelligence" of different human races. We are told that intelligence has a high heritability and that one race performs better than another on intelligence tests. In essence we are presented with a racial hierarchy reminiscent of that pernicious "system" which John Haydon Langdon Down used when he misnamed a disease entity "mongolism."

The people who are so committed to answering the nature-nurture pseudoquestion (Is heredity or environment more important in determining intelligence?) make two conceptual blunders. (1) Like Spalding's question about the instinctive nature of bird flight, which introduced the ethologist's deprivation experiment, their question about intelligence is, in fact, being asked about the development of a single individual. Unlike Spalding and the ethologists, however, they do not study development in single individuals. Usually, they test groups of individuals at a single time of life. The proportions being assigned to heredity and to environment refer to the relative amounts of the variance between individuals comprising a population, not how much of whatever enters into the development of the observed expression of a trait in a particular individual has been contributed by heredity and by environment, respectively. They want to know how instinctive is intelligence in the development of a certain individual, but instead they measure differences between large numbers of fully, or partially, developed individuals. If we now take into consideration the norm-of-reaction concept and combine it with the facts of genotypic individuality, then there is no general statement that can be made about the assignment of fixed proportions to the contributions of heredity and environment either to the development of a single individual, because we have not even begun to assess his norm of reaction, or to the differences that might be measured among members of a population, because we have hardly begun to assess the range of environmental conditions under which its constituent members might develop!

(2) Their second mistake, an egregious error, is related to the first one. They assume an inverse relationship between heritability magnitude and improvability by training and teaching. If heritability is high, little room is left for improvement by environmental modification. If heritability is low, much more improvement is possible. Note how this basic fallacy is incorporated directly into the title of Jensen's (1969) article "How Much Can We Boost IQ and Scholastic Achievement?"* That question received a straightforward, but fallacious, answer on his page 59: "The fact that scholastic achievement is considerably less heritable than intelligence . . . . means there is potentially much more we can do to improve school performance through environmental means than we can do to change intelligence. . . ." Commenting on the heritability of intelligence and "the old nature-nurture controversy," one of Jensen's respondents makes the same mistake in his rebuttal: "This is an old estimate which many of us have used, but we have used it to determine what could be done with the variance left for the environment." He then goes on "to further emphasize some of the implications of environmental variance for education and child rearing" (Bloom, 1969, p. 419).

High or low heritability tells us absolutely nothing about how a given individual might have developed under conditions different from those in which he actually did develop. Heritability provides

* This article, what it represents, and the academic-intellectual-political climate in which it appeared have now been analyzed in much greater depth in my "Jensenism: The Bankruptcy of 'Science' without Scholarship," *Educational Theory* 25 (1975):3–27.

no information about norm of reaction. Since the characterization of genotype-environment interaction can only be ad hoc and the number of possible interactions is effectively unlimited, no wonder the search for general laws of behavior has been so unfruitful, and *the* heritability of intelligence or of any other trait must be recognized as still another of those will-o'-the-wisp general laws. And no magic words about an interaction component in a linear analysis-of-variance model will make disappear the reality of each genotype's unique norm of reaction. Such claims by Jensen or anyone else are false. Interaction is an abstraction of mathematics. Norm of reaction is a developmental reality of biology in plants, animals, and people.

In Israel, the descendants of those Jews Pearson feared would contaminate Britain are manifesting some interesting properties of the norm of reaction. Children of European origin have an average IQ of 105 when they are brought up in individual homes. Those brought up in a Kibbutz on the nursery rearing schedule of twenty-two hours per day for four or more years have an average IQ of 115. In contrast, the mid-Eastern Jewish children brought up in individual homes have an average IQ of only 85, Jensen's danger point. However, when brought up in a Kibbutz, they also have an average IQ of 115. That is, they perform the same as the European children with whom they were matched for education, the occupational level of parents, and the Kibbutz group in which they were raised (Bloom, 1969, p. 420). There is no basis for expecting different overall results for any population in our species.

Some Promising Recent Developments

The power of the approach that begins by thinking first in terms of the genetic system and only later in terms of the phenotype (or behavior) to be analyzed is now being demonstrated by an accumulating and impressive body of evidence. The rationale of that approach derives directly from the particulate nature of the gene, the mosaic nature of the genotype, and the manner in which heredity breaks apart and gets reassembled in being passed on from one generation to the next. We now have a well-articulated picture of the way heredity is shared among biological relatives.

That madness runs in families has been known for centuries. The controversy has been over whether it was the heredity or the environment supplied by the family that was responsible for the madness. Franz Kallmann and some others collected large amounts of data in the 1940s and 1950s showing that monozygotic twins were much more concordant than dizygotic twins. Since David Rosenthal of NIMH has provided some of the best criticism of the incompleteness, and therefore inconclusiveness, of the twin-study evidence for the role of heredity in schizophrenia, Rosenthal's own recent findings become especially noteworthy.

He has divided foster-reared children from adoptive homes into two groups: those with a biological parent who is schizophrenic and those without a schizophrenic biological parent. It was found by Rosenthal (1968), and by Heston (1966) in a completely independent but similar study, that the incidence of schizophrenia was much greater among the biological children of schizophrenics. Most significantly, combining the two studies, the risk of schizophrenia in offspring is four to five times greater if a biological parent is schizophrenic. Still other recent studies support the Rosenthal and the Heston findings. Both Karlsson (1966) and Wender (Rose, 1969) found a high incidence of schizophrenia in the foster-reared relatives of schizophrenics.

Thinking genetically first in terms of biological relationship has already paid off in the analytical detail revealed as well as in the mere demonstration of concordance with respect to diagnostic category. Lidz and co-workers (1958) reported marked distortions in communicating among many of the nonhospitalized parents of schizophrenic hospital patients. McConaghy (1939), using an objective test of thought disorder, assessed the parents of ten schizophrenic patients and compared them to a series of control subjects. Sixty percent of the patients' parents, including at least one parent in every pair, registered test scores in the range indicative of thought disturbance. In contrast, less than ten percent of the controls had such scores.

The major features of McConaghy's findings have since been replicated by Lidz and co-workers (1958). More recently, Phillips and co-workers (1965) studied forty-eight relatives of adult schizophrenics and forty-five control subjects using a battery of tests to assess thought disorder. They found cognitive disorders to be much more frequent among the relatives of schizophrenics; seventeen of eighteen parents registered "pathological" scores, even though their social behavior had never been diagnosed as pathological.

In 1962, Anastasopoulos and Photiades assessed susceptibility to LSD-induced "pathological reactions" in the relatives of schizophrenic patients. After studying twenty-one families of patients and nine members of two control families, they reported "it was almost invariable to find reactions to LSD in one of the parents, and often in one or more of the siblings and uncles and aunts, which were neither constant nor even common during the LSD-intoxication of healthy persons" (p. 96).

Analogous work has been done studying the responses of the relatives of patients with depressive disorders using antidepressant drugs like imipramine (Tofranil) or an MAO inhibitor. Relatives tend to show a response pattern similar to that of their hospitalized relations.

Some very interesting human behavior-genetic analyses are currently being done on these affective disorders by George Winokur and his colleagues in St. Louis (Rose, 1969). Out of 1,075 consecutive admissions to a psychiatric hospital, 426 were diagnosed as primary affective disorders. So far, these appear to fall into two subtypes, the first of which shows manic episodes; some first-degree relatives show similar manifestations. The other subtype is characterized by depressive episodes and lack of concordance among close relatives. Furthermore, evidence is now accumulating implicating a dominant factor or factors on the X-chromosome in the manic subtype: (1) the condition is considerably more prevalent in females than in males; (2) the morbid risk among siblings of male probands is the same for males and females, but the morbid risk among siblings of female probands is quite different—sisters of female probands are at a 21 percent risk while their brothers are only at a 7.4 percent risk. More detailed study in several appropriately chosen family pedigrees suggest that there is a dominant gene on the short arm of the X-chromosome. The condition has, so far, shown linkage with color-blindness and the Xg blood groups, both of which are loosely linked on the short arm of the X-chromosome.

To examine the structure of the phenotypic variation in a trait whose development is in no obvious way influenced by environment and which, though ostensibly a simple trait, has been sufficiently well analyzed phenotypically to reveal its interesting complexity, we have chosen to study dermatoglyphics, or fingerprints, in my laboratory. For his doctoral dissertation, R. Peter Johnson is making these observations on both parents and offspring in individual families. His preparatory survey of the previous literature revealed one study which reported data on a cross-sectional sample of 2,000 males (Waite, 1915). Scoring them on all ten fingers with respect to four distinguishable pattern types, the following data reveal the interesting but sobering complexity that exists in such a "simple" trait: The same type of pattern was shown on all ten fingers by 12 percent, on nine of ten fingers by 16 percent, and on eight of ten fingers by 10 percent of the men. In addition, 5 percent of the men showed all four pattern types. This included 1 percent of the individuals who had all four pattern types on a single hand.

While probably everybody has heard that there are some unusual hospitalized males who carry two Y-chromosomes, are rather tall, and prone to commit crimes of violence, few people know that when a comparison was made between the first-order relatives of both the Y-Y-chromosome males and control males hospitalized for similar reasons (but not carrying two Y-chromosomes), there was a far greater incidence of a family history of crime among the controls. In this control group, there were over six times as many individual first-order relatives convicted and many, many times the number of convictions.

In summary, the relationship between heredity and behavior has turned out to be one of neither isomorphism nor independence. Isomorphism might justify an approach like naïve reductionism, independence a naïve behaviorism. Neither one turns out to be adequate. I believe that in order to study behavior, we must understand genetics quite thoroughly. Then, and only then, can we as psychologists forget about it intelligently.

Note

Invited address presented to the Nineteenth International Congress of Psychology, London, England, July 30, 1969, and dedicated to Professor T. Dobzhansky on his seventieth birthday.

This work was prepared with the support of Mental Health Training Grant 1 TO1 10715-04 BLS for Research Training in the Biological Sciences.

References

Anastasopoulos, G., and Photiades, H. 1962. "Effects of LSD-25 on Relatives of Schizophrenic Patients." *Journal of Mental Science* 108:95–98.

Benda, C. E. 1962. " 'Mongolism' or 'Down's Syndrome'." *Lancet* 1:163.

Bloom, B. S. 1969. Letter to the Editor. *Harvard Educational Review* 39:419–421.

CASTLE, W. E. 1903. "The Laws of Heredity of Galton and Mendel, and Some Laws Governing Race Improvement by Selection." *Proceedings of the American Academy of Arts and Sciences* 39:223–242.

DEWEY, J. 1959. "Correspondence with Robert V. Daniels, 15 February, 1947." *Journal of the History of Ideas* 20:570.

DOWN, J. L. H. 1866. "Observations on an Ethnic Classification of Idiots." *London Hospital Reports.* Reprinted in McKusick, V. A. (ed.), 1962, "Medical Genetics, 1961." *Journal of Chronic Diseases* 15:432.

DUNN, L. C. 1965. *A Short History of Genetics.* New York: McGraw-Hill Book Co.

FISHER, R. A. 1951. "Limits to Intensive Production in Animals." *British Agricultural Bulletin* 4:217–218.

FULLER, J. L., and THOMPSON, W. R. 1960. *Behavior Genetics.* New York: John Wiley & Sons.

DE GOBINEAU, J. A. 1967. *Essai sur l' inégalité des races humaines.* Reissue, complete in 1 vol., with a preface by Hubert Juin. Paris: P. Belfond.

GOLDSCHMIDT, R. B. 1955. *Theoretical Genetics.* Berkeley: University of California Press.

HALDANE, J. B. S. 1946. "The Interaction of Nature and Nurture." *Annals of Eugenics* 13:197–205.

HEGMANN, J. P., and DEFRIES, J. C. 1968. "Open-field Behavior in Mice: Genetic Analysis of Repeated Measures." *Psychonomic Science* 13:27–28.

HEIDEN, K. 1944. *Der Führer.* London: Houghton, 1944. Cited in Bullock, A., 1962, *Hitler: A Study in Tyranny,* p. 80. Harmondsworth: Penguin Books.

HESTON, L. L. 1966. "Psychiatric Disorders in Foster Home Reared Children of Schizophrenic Mothers." *British Journal of Psychiatry* 112:819–825.

HIRSCH, J. 1963. "Behavior Genetics and Individuality Understood: Behaviorism's Counterfactual Dogma Blinded the Behavioral Sciences to the Significance of Meiosis." *Science* 142:1436–1442.

JENSEN, A. R. 1967. "Estimation of the Limits of Heritability of Traits by Comparison of Monozygotic and Dizygotic Twins." *Proceedings of the National Academy of Sciences* 58:149–156.

JENSEN, A. R. 1969. "How Much Can We Boost IQ and Scholastic Achievement?" *Harvard Educational Review* 39:1–123.

KARLSSON, J. L. 1966. *The Biologic Basis of Schizophrenia.* Springfield, Ill.: Charles C. Thomas.

LIDZ, T., CORNELISON, A., TERRY, D., and FLECK, S. 1958. "Intrafamilial Environment of the Schizophrenic Patient: VI. The Transmission of Irrationality." AMA *Archives of Neurology and Psychiatry* 79:305–316.

LIDZ, T., WILD, C., SCHAFER, S., ROSMAN, B., and FLECK, S. 1962. "Thought Disorders in the Parents of Schizophrenic Patients: A Study Utilizing the Object Sorting Test." *Journal of Psychiatric Research* 1:193–200.

McConaghy, N. 1959. "The Use of an Object Sorting Test in Elucidating the Hereditary Factor in Schizophrenia." *Journal of Neurology, Neurosurgery, and Psychiatry* 22:243–246.

McGaugh, J. L., Jennings, R. D., and Thomson, C. W. 1962. "Effect of Distribution of Practice on the Maze Learning of Descendants of the Tryon Maze Bright and Maze-Dull Strains." *Psychological Reports* 10:147–150.

McKusick, V. A. 1966. *Mendelian Inheritance in Man: Catalogs of Autosomal Dominant, Recessive, and X-Linked Phenotypes.* Baltimore: Johns Hopkins University Press.

Pastore, N. 1949. *The Nature-Nurture Controversy.* New York: King's Crown Press (Columbia University).

Pearson, K. 1904. "On a Generalized Theory of Alternative Inheritance, with Special Reference to Mendel's Laws." *Philosophical Transactions of the Royal Society of London* A203:53–86.

———. 1924. *The Life, Letters and Labours of Francis Galton,* vol. 2, *Researches of Middle Life.* Cambridge: Cambridge University Press.

Pearson, K., and Moul, M. 1925. "The Problem of Alien Immigration into Great Britain, Illustrated by an Examination of Russian and Polish Jewish Children." *Annals of Eugenics* 1:5–127.

Phillips, J. E., Jacobson, N., and Turner, W. J. 1965. "Conceptual Thinking in Schizophrenics and Their Relatives." *British Journal of Psychiatry* 111: 823–839.

Rose, R. J. 1969. Department of Psychology, University of Indiana, private communication.

Rosenthal, D., Wender, P. H., Kety, S. S., Schulsinger, F., Welner, J., and Østergaard, L. 1968. "Schizophrenics' Offspring Reared in Adoptive Homes." *Journal of Psychiatric Research* 6:377–391.

Waite, H. 1915. "Association of Fingerprints." *Biometrika* 10:421–478.

Watson, J. B. 1959. *Behaviorism.* Chicago: University of Chicago Press.

Wittgenstein, L. 1963. *Philosophical Investigations.* 2nd ed. Trans. G. E. Anscombe. Oxford: Basil Blackwell.

Yule, G. U. 1902. "Mendel's Laws and Their Probable Relation to Intra-racial Heredity." *New Phytologist* 1:193–207, 222–238.

Yule, G. U. 1906. "On the Theory of Inheritance of Quantitative Compound Characters on the Basis of Mendel's Laws—A Preliminary Note." *Report, Third International Conference on Genetics,* pp. 140–142.

The Analysis of Variance and the Analysis of Causes

Richard C. Lewontin

THIS ISSUE OF THE *American Journal of Human Genetics* CONTAINS two articles by Newton Morton and his colleagues (Morton, 1974; Rao et al., 1974) that provide a detailed analytic critique of various estimates of heritability and components of variance for human phenotypes. They make especially illuminating remarks on the problems of partitioning variances and covariances between groups such as social classes and races. The most important point of all, at least from the standpoint of the practical, social, and political applications of human population genetics, occurs at the conclusion of the first paper (Morton, 1974) in which Morton points out explicitly the chief programmatic fallacy committed by those who argue so strongly for the importance of heritability measures for human traits. The fallacy is that a knowledge of the heritability of some trait in a population provides an index of the efficacy of environmental or clinical intervention in altering the trait either in individuals or in the population as a whole. This fallacy, sometimes propagated even by geneticists, who should know better, arises from the confusion between the technical meaning of heritability and the everyday meaning of the word. A trait can have a heritability of 1.0 in a population at some time, yet could be completely altered in the future by a simple environmental change. If this were not the case, "inborn errors of metabolism" would be forever incurable, which is patently untrue. But the misunderstanding about the relationship between heritability and phenotypic plasticity is not simply the result of an ignorance of genetics on the part of psychologists and electronic engineers. It arises from the entire system of analysis of causes through linear models, embodied in the analysis of variance and covariance and in path analysis. It is indeed ironic that while Morton and his colleagues dispute the erroneous programmatic conclusions that are drawn from the analysis of human phenotypic variation, they nevertheless rely heavily for their analytic techniques on the very linear models that are responsible for the confusion.

I would like in what follows to look rather closely at the problem of the analysis of causes in human genetics and to try to understand how the underlying model of this analysis molds our view of the real world. I will begin by saying some very obvious and elementary things about causes, but I will come thereby to some very annoying conclusions.

Discrimination of Causes and Analysis of Causes

We must first separate two quite distinct problems about causation that are discussed by Morton. One is to discriminate which of two alternative and mutually exclusive causes lies at the basis of some observed phenotype. In particular, it is the purpose of *segregation analysis* to attempt to distinguish those individuals who owe their phenotypic deviation to their homozygosity for rare deleterious gene alleles from those whose phenotypic peculiarity arises from the interaction of environment with genotypes that are drawn from the normal array of segregating genes of minor effect. This is the old problem of distinguishing major gene effects from "polygenic" effects. I do not want to take up here the question of whether such a clear distinction can be made or whether the spectrum of gene effects and gene frequencies is such that we cannot find a clear dividing line between the two cases. The evidence at present is ambiguous, but at least *in principle* it may be possible to discriminate two etiological groups, and whether such groups exist for any particular human disorder is a matter for empirical research. It is possible, although not necessary, that the form of clinical or environmental intervention required to correct a disorder that arises from homozygosity for a single, rare recessive allele (the classical "inborn error of metabolism") may be different from that required for the "polygenic" class. Moreover, for the purposes of genetic counseling, the risk of future affected offspring will be different if a family is segregating for a rare recessive than if it is not. Thus, the discrimination between two *alternative* causes of a human disorder is worth making if it can be done.

The second problem of causation is quite different. It is the problem of the *analysis* into separate elements of a number of causes that are interacting to produce a single result. In particular, it is the problem of analyzing into separate components the interaction between environment and genotype in the determination of phenotype. Here, far from trying to discriminate individuals into two distinct and mutually exclusive etiological groups, we recognize that all individuals owe their phenotype to the biochemical activity of their genes in a unique sequence of environments and to developmental events that may occur subsequent to, although dependent upon, the initial action of the genes. The analysis of interacting causes is fundamentally a different concept from the discrimination of alternative causes. The difficulties in the early history of genetics embodied in the pseudoquestion of "nature versus nurture" arose precisely because of the confusion between these two problems in causation. It was supposed that the phenotype of an individual could be the result of *either* environment *or* genotype, whereas we understand the phenotype to be the result of *both*. This confusion has persisted into modern genetics with the concept of the phenocopy, which is supposed to be an environmentally caused phenotypic deviation, as opposed to a mutant which is genetically caused. But, of course, both "mutant" and "phenocopy" result from a unique interaction of gene and environment. If they are etiologically separable, it is not by a line that separates environmental from genetic causation but by a line that separates two kinds of genetic basis: a single gene with major effect or many genes each with small effect. That is the message of the work by Waddington (1953) and Rendel (1959) on canalization.

Quantitative Analysis of Causes

If an event is the result of the joint operation of a number of causative chains and if these causes "interact" in any generally accepted meaning of the word, it becomes conceptually impossible to assign quantitative values to the causes of that *individual event*. Only if the causes are utterly independent could we do so. For example, if two men lay bricks to build a wall, we may quite fairly measure their contributions by counting the number laid by each;

but if one mixes the mortar and the other lays the bricks, it would be absurd to measure their relative quantitative contributions by measuring the volumes of bricks and of mortar. It is obviously even more absurd to say what proportion of a plant's height is owed to the fertilizer it received and what proportion to the water, or to ascribe so many inches of a man's height to his genes and so many to his environment. But this obvious absurdity appears to frustrate the universally acknowledged program of Cartesian science to analyze the complex world of appearances into an articulation of causal mechanisms. In the case of genetics, it appears to prevent our asking about the relative importance of genes and environment in the determination of phenotype. The solution offered to this dilemma, a solution that has been accepted in a great variety of natural and social scientific practice, has been the *analysis of variation.* That is, if we cannot ask how much of an individual's height is the result of his genes and how much a result of his environment, we will ask what proportion of the deviation of his height from the population mean can be ascribed to deviation of his environment from the average environment and how much to the deviation of this genetic value from the mean genetic value. This is the famous linear model of the analysis of variance which can be written as

$$Y - \mu_Y = (G - \mu_Y) + (E - \mu_Y) + (GE) + e, \qquad (1)$$

where μ_Y is the mean score of all individuals in the population; Y is the score of the individual in question; G is the average score of all individuals with the same genotype as the one in question; E is the average score of all individuals with the same environment as the one in question; GE, the genotype-environment interaction, is that part of the average deviation of individuals sharing the same environment and genotype that cannot be ascribed to the simple sum of the separate environmental and genotypic deviations; and *e* takes into account any individual deviation not already consciously accounted for, and assumed to be random over all individuals (measurement error, developmental noise, and so on).

I have written this well-known linear model in a slightly different way than it is usually displayed in order to emphasize two of its properties that are well known to statisticians. First, the environmental and genotypic effects are in units of *phenotype.* We are not actually assessing how much variation in environment or genotype exists, but only how much perturbation of phenotype has been the outcome of average difference in environment or genotype. The analysis in expression (1) is completely *tautological,* since it is framed entirely in terms of phenotype and both sides of the equation must balance by the definitions of GE and *e*. To turn expression (1) into a contingent one relating actual values of environmental variables like temperature to phenotypic score, we would need functions of the form

$$(E - \mu_Y) = f(T - \mu_T) \tag{2}$$

and

$$GE = h[(g - \mu_g),(T - \mu_T)], \tag{3}$$

where g and T are measured on a genetic and a temperature scale rather than on a scale of phenotype. Thus, the linear model [equation (1)] makes it impossible to know whether the environmental deviation $(E - \mu_Y)$ is small because there are no variations in actual environment or because the particular genotype is insensitive to the environmental deviations, which themselves may be quite considerable. From the standpoint of the tautological analysis of expression (1), this distinction is irrelevant, but as we shall see, it is supremely relevant for those questions that are of real importance in our science.

Second, expression (1) contains population means at two levels. One level is the grand mean phenotype μ_Y and the other is the set of so-called "marginal" genotypic and environmental means, E and G. These, it must be remembered, are the *mean* for a given environment averaged over all genotypes in the population and the *mean* for a given genotype averaged over all environments.

But since the analysis is a function of these phenotypic means, it will, in general, give a different result if the means are different. That is, the linear model is a *local analysis.* It gives a result that depends upon the actual distribution of genotypes and environments in the particular population sampled. Therefore, the result of the analysis has a historical (i.e., spatiotemporal) limitation and is not in general a statement about *functional* relations. So, the genetic variance for a character in a population may be very small because the functional relationship between gene action and the character is weak for any conceivable genotype or simply because the population is homozygous for those loci that are of strong functional significance for the trait. The analysis of variation cannot distinguish between these alternatives even though for most purposes in human genetics we wish to do so.

What has happened in attempting to solve the problem of the analysis of causes by using the analysis of variation is that a totally different object has been substituted as the object of investigation, almost without noticing it. The new object of study, the deviation of phenotypic value from the mean, is not the same as the phenotypic value itself; and the tautological analysis of that deviation is not the same as the analysis of causes. In fact, the analysis of variation throws out the baby with the bath water. It is both too specific in that it is spatiotemporally restricted in its outcome and too general in that it confounds different causative schemes in the same outcome. Only in a very special case, to which I shall refer below, can the analysis of variation be placed in a one-to-one correspondence to the analysis of causes.

Norm of Reaction

The real object of study both for programmatic and theoretical purposes is the relation between genotype, environment, and phenotype. This is expressed in the *norm of reaction*, which is a table of correspondence between phenotype, on the one hand, and genotype-environment combinations on the other. The relations between phenotype and genotype and between phenotype and environment are many-many relations, no single phenotype corresponding to a unique genotype or vice versa.

In order to clarify the relation between the two objects of study (i.e., the norm of reaction and the analysis of variance, which analyzes something quite different), let us consider the simplified norms of reaction shown in figures 1*a*–*h*. We assume that there is a single, well-ordered environmental variable E, say, temperature, and a scale of phenotypic measurement P. Each line is the norm of reaction, the relationship of phenotype to environment, for a particular hypothetical genotype (G_1 or G_2).

The first thing to observe is that the phenotype is sensitive to differences in both environment and genotype in every case. That is, there is a reaction of each genotype to changing environment, and in no case are the two genotypes identical in their reactions. Thus in any usual sense of the word, both genotypes and environment are *causes* of phenotypic differences and are necessary objects of our study.

Figure 1*a* is, in one sense, the most general, for if environment extends uniformly over the entire range and if the two genotypes are equally frequent, there is an overall effect of genotype (G_1 being on the average superior to G_2) and an overall effect of environment (phenotype gets smaller on the average with increasing temperature). Nevertheless, the genotypes cross so that neither is always superior.

Figure 1*b* shows an overall effect of environment, since both

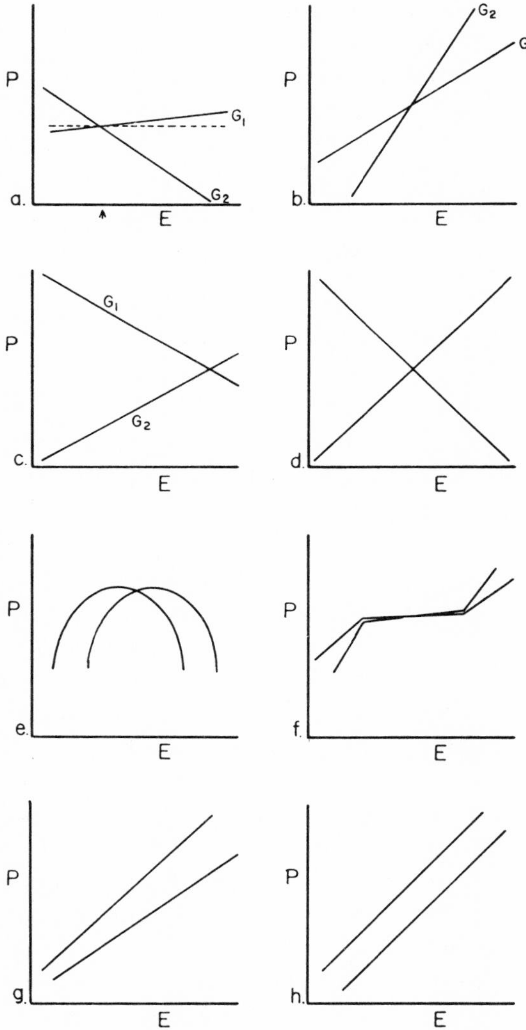

Figure 1, a–h. *Examples of different forms of reaction norms. In each case, the phenotype* (P) *is plotted as a function of environment* (E) *for different genotypes* (G_1, G_2).

genotypes have a positive slope: but there is no overall effect of genotype, since the two genotypes would have exactly the same *mean* phenotype if all environments are considered equally. There is no a priori way from figure 1*b* of ranking the two genotypes. However, if because of particular circumstances the distribution of environments were heavily weighted toward the lower temperatures, then G_1 would be consistently superior to G_2, and an analysis of variance would show a strong effect of genotype as well as of environment, but very little genotype-environment interaction. Thus the analysis of variance would reflect the particular environmental circumstances and give a completely incorrect picture of the general relationship between cause and effect here, where there is overall no effect of genotype but a strong genotype-environment interaction.

Figure 1*c* is the complementary case to that shown in figure 1*b*. In figure 1*c*, there is no overall effect of environment, but G_1 is clearly superior to G_2, overall. In this case, a strong environmental component of variance will appear, however, if either one of the genotypes should be in excess in the population. So the historical events that mold the genotypic distribution of a population will have an effect on the judgment, from the analysis of variance, of the importance of environment.

The overall lack of genetic effect in figure 1*b* and of environmental effect in figure 1*c* can both appear in a trait like that shown in figure 1*a*, which overall has both effects if the distribution of environments or of genotypes is asymmetric. Thus if environments are distributed around the arrow in figure 1*a*, there will *appear* to be no average effect of genotype, while if the population is appropriately weighted toward an excess of G^1, the average phenotype across environments will be constant as shown by the dashed line. Here real overall effects are obscured because of spatiotemporal events, and the analysis of variance fails to reveal significant overall differences.

These last considerations lead to two extremely important points about the analysis of variance. First, although expression (1) appears to isolate distinct causes of variation into separate elements, it does not because the amount of *environmental* variance that appears depends upon the *genotypic* distribution, while the amount of *genetic* variance depends upon the *environmental* distribution. Thus the appearance of the separation of causes is a pure illusion. Second, because the linear model appears as a sum of variation from different causes, it is sometimes erroneously supposed that removing one of the sources of variation will reduce the total variance. So, the meaning of the genetic variance is sometimes given as "the amount of variation that would be left if the environment were held constant," and the environmental variance is described as "the amount of variance that would remain if all the genetic variation were removed," an erroneous explanation offered by Jensen (1969), for example. Suppose that the norms of reaction were as in figure 1*a* and a unimodal distribution of environments were centered near the arrow, with a roughly equal mixture of the two genotypes. Now suppose we fix the environment. What will happen to the total variance? That depends on which environment we fix upon. If we choose an environment about one SD or more to the right of the mean, there would actually be an *increase* in the total variance because the difference between genotypes is much greater in that environment than on the average over the original distribution. Conversely, suppose we fix the genotype. If we chose G_2 to be our pure strain, then, again, we would *increase* the total variance because we had chosen the more environmentally plastic genotype. The apparent absurdity that removing a source of variance actually increases the total variance is a consequence of the fact that the linear model does not really effect a separation of causes of variation and that it is a purely local description with no predictive reliability. Without knowing the norms of reaction, the present distribution of environments, the present distribution of genotypes, and without then specifying which environments and which genotypes are to be eliminated or fixed, it is impossible to predict whether the total variation would be increased, decreased, or remain unchanged by environmental or genetic changes.

In figure 1*d*, there is neither an overall effect of genotype or environment, but both can obviously appear in a particular population in a particular environmental range as discussed above.

Case *e* of Figure 1 has been chosen to illustrate a common situation for enzyme activity, a parabolic relation between phenotype and environment. Here genotypes are displaced horizontally (having different temperature optima). There is no overall superiority of either genotype nor is there any general monotone environmental trend for any genotype. But for any distribution of environments except a perfectly symmetrical one, there will appear a component of variance for genotypic effect. Moreover, if the temperature distribution is largely to either side of the crossover point between these two genotypes, there will be very large components of variance for both genotype and environment and a vanishingly small interaction component; yet over the total range of environments, exactly the opposite is true!

Figure 1*e* also shows a second important phenomenon, that of differential phenotypic sensitivity in different environmental ranges. At intermediate temperatures, there is less difference between genotypes and less difference between the effect of environments than at more extreme temperatures. This is the phenomenon of canalization and is more generally visualized in figure 1*f*. Over a range of intermediate phenotypes there is little effect of either genotype or environment, while outside this zone of canalization phenotype is sensitive to both (Rendel, 1959). The zone of canalization corresponds to that range of environments that have been historically the most common in the species, but in new environments much greater variance appears. Figure 1*f* bears directly on the characteristic of the analysis of variance that all effects are measured in phenotypic units. The transformations [equations (2) and (3)] that express the relationship between the phenotypic deviations ascribable to genotype or environment and the actual values of the genotypes or environmental variables are not simple linear proportionalities. The sensitivity of phenotype to both environment and genotype is a function of the particular range of environments and genotypes. For the programmatic purposes of human genetics, one needs to know more than the components of variation in the historical range of environments.

Figures 1*a–f* are meant to illustrate how the analysis of variance will give a completely erroneous picture of the causative relations between genotype, environment, and phenotype because the particular distribution of genotypes and environments in a given population at a given time picks out relations from the array of reaction norms that are necessarily atypical of the entire spectrum of causative relations. Of course, it may be objected that any sample from nature can never give exactly the same result as examining the universe. But such an objection misses the point. In normal sampling procedures, we take care to get a representative or unbiased sample of the universe of interest and to use unbiased sample estimates of the parameters we care about. But there is no question of sampling here, and the relation of sample to universe in statistical procedures is not the same as the relation of variation in spatiotemporally defined populations to causal and functional variation summed up in the norm of reaction. The relative size of genotypic and environmental components of variance estimated in any natural population reflect in a complex way three underlying relationships: (1) they reflect the actual functional relations embodied in the norm of reaction; (2) they reflect the actual distribution of genotype frequencies, and this distribution, a product of long-time historical forces like natural selection, mutation, migration, and breeding structure, changes over periods much longer than a generation; and (3) they reflect the actual structure of the environments in which the population finds itself, a structure that may change very rapidly indeed, especially for human populations. The effects of historical forces and immediate environment are inextricably bound up in the outcome of variance analysis which thus is not a tool for the elucidation of functional biological relations.

Effect of Additivity

There is one circumstance in which the analysis of variance can, in fact, estimate functional relationships. This is illustrated exactly in figure 1*h* and approximately in figure 1*g*. In these cases there is perfect or nearly perfect additivity between genotypic and environmental effects so that the differences between genotypes are the same in all environments and the differences between environments are the same for all genotypes. Then the historical and immediate circumstances that alter genotypic and environmental distributions are irrelevant. It is not surprising that the assumption of additivity is so often made, since this assumption is necessary to make the analysis of variance anything more than a local description.

The assumption of additivity is imported into analyses by four routes. First, it is thought that in the absence of any evidence, additivity is a priori the simplest hypothesis and additive models are dictated by Occam's razor. The argument comes from a general Cartesian world view that things can be broken down into parts without losing any essential information and that in any complex interaction of causes, main effects will almost always explain most of what we see while interactions will tend to be of a smaller order of importance. But this is a pure a priori prejudice. Dynamic systems in an early stage in their evolution will show rather large main effects of the forces acting to drive them, but as they approach equilibrium, the main effects disappear and interactions predominate. For example, that is what happens to additive genetic variance under selection. Exactly how such considerations apply to genotype and environment is not clear.

Second, it is suggested that additivity is a *first approximation* to a complex situation, and the results obtained with an additive scheme are then a first approximation to the truth. This argument is made by analogy with the expansion of mathematical functions by Taylor's series. But this argument is self-defeating since the justification for expanding a complex system in a power series and considering only the first-order terms is precisely that one is interested in the behavior of the system in the neighborhood of the point of expansion. Such an analysis is a local analysis only, and the analysis of variance is an analysis in the neighborhood of the population mean only. By justifying additivity on this ground, the whole issue of the global application of the result is sidestepped.

Third, it is argued that if an analysis of variance is carried out and the genotype-environment interaction turns out to be small, the assumption of additivity is justified. Like the second argument, there is some circularity. As the discussion of the previous section showed, the usual outcome of an analysis of variance in a particular population in a restricted range of environments is to underestimate severely the amount of interaction between the factors that occur over the whole range.

Finally, additivity or near additivity may be assumed without offering any justification because it suits a predetermined end. Such is the source of figure 1*g*. It is the hypothetical norm of reaction for IQ taken from Jensen (1969). It purports to show the relation between environmental "richness" and IQ for different genotypes. While there is not a scintilla of evidence to support such a picture, it has the convenient properties that superior and inferior genotypes in one environment maintain that relation in all environments, and that as environment is "enriched," the genetic variance (and therefore the heritability) grows greater. This is meant to take care of those foolish egalitarians who think that spending money and energy on schools generally will iron out the inequalities in society.

Evidence on actual norms of reaction is very hard to come by. In man, measurements of reaction norms for complex traits are impossible because the same genotype cannot be tested in a variety of environments. Even in experimental animals and plants where genotypes can be replicated by inbreeding experiments or cloning, very little work has been done to characterize these norms for the genotypes that occur in natural populations and for traits of consequence to the species. The classic work of Clausen et al. (1940) on ecotypes of plants shows very considerable nonadditivity of the types illustrated in figures 1*a–d*.

As an example of what has been done in animals, figure 2 has

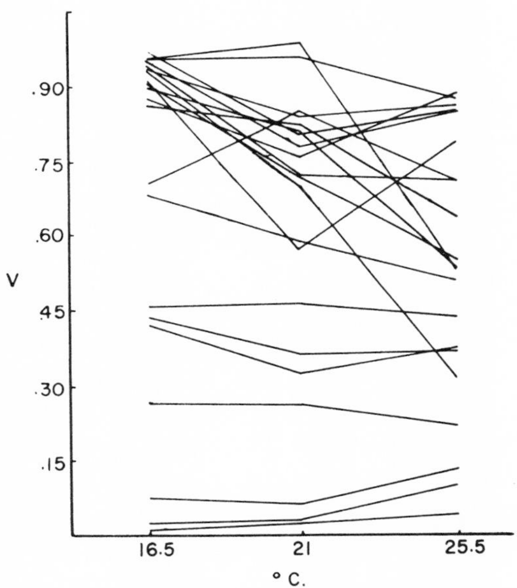

Figure 2. *Actual reaction norms for viability of fourth chromosome homozygotes of* Drosophila pseudoobscura. *Data from Dobzhansky and Spassky (1944).*

been drawn from the data of Dobzhansky and Spassky (1944) on larval viability in *Drosophila pseudoobscura.* Each line is the reaction norm for larval viability at three different temperatures for a fourth chromosome homozygote, where the chromosomes have been sampled from a natural population. As the figure shows, there are a few genotypes that are of uniformly poor viability, probably corresponding to homozygosity for a single deleterious gene of strong effect. However, most genotypes are variable in their expression, and there is a great deal of genotype-environment interaction with curves crossing each other and having quite different environmental sensitivities.

Purpose of Analysis

Just as the objects of analysis are different when we analyze causes and when we analyze variance, so the purposes of these analyses are different. The analysis of causes in human genetics is meant to provide us with the basic knowledge we require for correct schemes of environmental modification and intervention. Together with a knowledge of the relative frequencies of different human genotypes, a knowledge of norms of reaction can also predict the demographic and public health consequences of certain massive environmental changes. Analysis of variance can do neither of these because its results are a unique function of the present distribution of environment and genotypes.

The legitimate purposes of the analysis of variance in human genetics are to predict the rate at which selection may alter the genotypic composition of human populations and to reconstruct, in some cases, the past selective history of the species. Neither of these seems to me a pressing problem since both are academic. Changes in the genotypic composition of the species take place so slowly as compared to the extraordinary rate of human social and cultural evolution that human activity and welfare are unlikely to depend upon such genetic change. The reconstruction of man's genetic past, while fascinating, is an activity of leisure rather than of necessity. At any rate, both these objectives require not simply the analysis into genetic and environmental components of variation, but require absolutely a finer analysis of genetic variance into its additive and nonadditive components. The simple analysis of variance is useless for these purposes and indeed it has no use at all. In view of the terrible mischief that has been done by confusing the spatiotemporally local analysis of variance with the global analysis of causes, I suggest that we stop the endless search for better methods of estimating useless quantities. There are plenty of real problems.

References

Clausen, J., Keck, D. D., and Heisey, W. M. 1940. *Experimental Studies on the Nature of Species. I. Effects of Varied Environments on Western North American Plants.* Carnegie Institute of Washington, Publication no. 520, pp. 1–452.

Dobzhansky, T., and Spassky, B. 1944. "Genetics of Natural Populations. XI.

Manifestation of Genetic Variants in *Drosophila pseudoobscura* in Different Environments." *Genetics* 29:270–290.

JENSEN, A. R. 1969. "How Much Can We Boost IQ and Scholastic Achievement?" *Harvard Educational Review* 39:1–123.

MORTON, N. E. 1974. "Analysis of Family Resemblance. I. Introduction." *American Journal of Human Genetics* 26:318–330.

RAO, D. C., MORTON, N. E., and YEE, S. 1974. "Analysis of Family Resemblance. II. A Linear Model for Familial Correlation." *American Journal of Human Genetics* 26:311–359.

RENDEL, J. M. 1959. "Canalization of the Acute Phenotype of Drosophila." *Evolution* 13:425–439.

WADDINGTON, C. H. 1953. "Genetic Assimilation of an Acquired Character." *Evolution* 7:118–126.

Science or Superstition?

A Physical Scientist Looks at the IQ Controversy

David Layzer

> That valid judgments about the biological significance of differences in tests of mental abilities are impossible has already been stressed in preceding sections. This is a point that cannot be overemphasized in view of the immediacy of the racial problems confronting the United States at this time. Various proposals have been advanced by probably well-intentioned people that suggest how meaningful investigation of genetic rather than phenotypic differences in intelligence and achievement might be carried out. Most of these individuals suffer from the obstinate inability to see the methodological difficulties and inherent biases of their schemes. Some anthropologists even opine that such studies are irrelevant or too vulnerable to misinterpretation and too fraught with political danger to be undertaken. This may or may not be true, but it is a fact that generations of discrimination have made direct comparisons of mental traits between Negroes and whites not biologically meaningful.
>
> —I. M. LERNER (1968, p. 234)

Abstract

JENSEN AND OTHERS HAVE APPLIED THE POLYGENIC THEORY OF INheritance to IQ measurements and have derived a value of about .8 for the "heritability of IQ." From this result, they draw a number of far-reaching conclusions, in particular, that children with low IQs cannot acquire higher cognitive skills (those involved in abstract reasoning and problem-solving) and that ethnic differences in average IQ probably have a significant genetic component.

This paper analyzes the implicit assumptions underlying Jensen's theoretical analysis and demonstrates that they are untenable. Like any other quantitative scientific theory, the theory of polygenic inheritance applies only to measurements that satisfy certain formal requirements. A detailed discussion of IQ measurements shows that they do not satisfy these requirements. Consequently, estimates of the heritability of IQ are not merely unreliable but meaningless.

Next, IQ correlation data and other relevant observations are examined in the light of current ideas concerning cognitive development. It is shown that the data provide no support for the view that children with low IQs, or children of parents with low IQs, have limited *capacity* for acquiring higher cognitive skills.

Finally, the "hypothesis" of genetic differences in intelligence between ethnic groups is shown to be untestable by existing or foreseeable methods. Hence, it should not be regarded as a scientific hypothesis but as a metaphysical speculation.

Introduction

A number of years ago, when high school teachers in North Carolina were being paid a starting salary of $120 per month, I happened to ask a member of that state's legislature whether he considered this to be an adequate salary. "Certainly," he said, "they're not worth any more than that." "How do you know?" I asked. "Why, just look at what they're paid." Circular reasoning? I think not. Our views on salary and status reflect our basic assumptions concerning the individual and his relation to society. One possible assumption is that society should reward each of its members according to his needs and contributions. Another is that society has a fixed hierarchic structure, and each individual gravitates inevitably toward the level where he belongs. My question was based on the first assumption, the legislator's reply on the second.

The idea that, by and large, we get what we deserve—that there is a preordained harmony between what we are and what we achieve—was an essential ingredient in the Calvinist doctrine of New England's Puritan settlers. What really mattered to them was not, of course, how well they did in this world but how well they would do in the next. The first was important only insofar as it provided a clue to the second. Although Calvinism's other-worldly orientation has long since gone out of fashion, its underlying social attitudes persist and continue to play an important part in shaping our social, educational, and political institutions. Because we still tend to interpret wealth and power as tokens of innate worth (and poverty and helplessness as tokens of innate worthlessness), we tend to believe that it is wicked to tamper with "natural" processes of selection and rejection (Thou shalt not monkey with the Market), to erect artificial barriers against economic mobility (downward or upward), or to penalize the deserving rich in order to benefit the undeserving poor.

Not unnaturally, such attitudes have always appealed strongly to the upwardly mobile and those who already inhabit society's upper strata. Besides, they offer a convenient rationalization for our failure to cope with, or even to confront, our most urgent social problem: the emergence of a growing and self-perpetuating lower class, disproportionately Afro- and Latin-American in its ethnic composition, excluded from the mainstream of American life and alienated from its values, isolated in rural areas and urban ghettos, and dependent for the means of bare survival on an increasingly hostile and resentful majority. Faced with this problem, many people find it comforting to believe that human nature, not the System, is responsible for gross inequalities in the human condition. As Richard Nixon has said, "Government could provide health, housing, means, and clothing for all Americans. That would not make us a great country. What we have to remember is that this country is going to be great in the future to the extent that individuals have self-respect, pride, and a determination to do better."

Although such attitudes are deeply ingrained, increasing numbers of Americans are beginning to question their validity. The System may be based on eternal moral truths, but, in practice, it seems to be working less and less well; and one of the eternal moral truths does, after all, assert that practical success is inner virtue's outward aspect. Yet the quality of life in America is deteriorating in many ways, not only for the downwardly mobile lower class (who, according to Mr. Nixon, are not trying hard enough) but also for the upwardly mobile middle class (who are already trying as hard as they can). In these circumstances, any argument that lends support to the old, embattled attitudes is bound to arouse strong emotional responses both among those who recognize a need for basic social reform and among those who oppose it.

Jensenism

This may help to explain the furor generated by the publication, in a previously obscure educational journal, of a long, scholarly article provocatively entitled, "How Much Can We Boost IQ and Scholastic Achievement?" (Jensen, 1969). Very little, concludes the author—because differences in IQ largely reflect innate differences in intelligence. Children with low IQs, he argues, lack the capacity to acquire specific cognitive skills, namely, those involved in abstract reasoning and problem-solving. Such children should be taught mainly by rote and should not be encouraged to aspire to occupations that call for higher cognitive skills.

What is true of individuals could also well be true of groups, continues Jensen: differences between ethnic groups in average performance on IQ tests probably reflect average differences in innate intellectual capacity. Jensen does not shirk the unpleasant duty of pointing out that this conclusion has an important bearing on fundamental questions of educational, social, and political policy:

> Since much of the current thinking behind civil rights, fair employment, and equality of educational opportunity appeals to the fact that there is a disproportionate representaion of different racial groups in the various levels of educational, occupational and socioeconomic hierarchy, we are forced to examine all the possible reasons for the inequality among racial groups in the attainments and rewards generally valued by all groups within our society. To what extent can such inequalities be attributed to unfairness in society's multiple selection processes? . . . And to what extent are these inequalities attributable to really relevant selection criteria which apply equally to all individuals but at the same time select disproportionately between some racial groups because there exist, in fact, real average differences among the groups—differences . . . indisputably relevant to educational and occupational performance?

The contention that IQ is an index of innate cognitive capacity is, of course, not new, but it has not been taken very seriously by most biologists and psychologists. Jensen's article purports to put it on a sound scientific basis. In outline, his argument runs as follows. IQ test scores represent measurements of a human trait which we may call intelligence. It is irrelevant to the argument that we do not know what intelligence "really is." All we need to know is that IQ tests are internally and mutually consistent, and that IQ correlates strongly with scholastic success, income, occupational status, and so forth. We can then treat IQ as if it was a metric character such as height or weight, and use techniques of population genetics to estimate its "heritability." In this way, we can discover the relative importance of genetic and environmental differences as they contribute to differences in IQ. Such studies show, according to Jensen, that IQ differences are mainly genetic in origin.

Jensen's 123-page article is largely devoted to fleshing out this argument and developing its educational implications. Jensenism has also been expounded at a more popular level: in Great Britain by H. J. Eysenck (1971) and in America by R. J. Herrnstein (1971). While Eysenck's main concern is to stress the genetic basis of differences between ethnic groups, Herrnstein is more concerned with the social and political implications of Jensenism. He argues that the more successful we are in our efforts to equalize opportunity and environment, the more closely will the structure of society come to reflect inborn differences in mental ability. Thus "our present social policies" must inevitably give rise to a hereditary caste system based largely on IQ. Indeed, the lowest socioeconomic classes *already* consist of people with the lowest IQs. Since, according to Jensen, IQ is essentially genetically determined, Herrnstein's argument implies that the current inhabitants of urban ghettos and depressed rural areas are destined to become the progenitors of a hereditary caste, its members doomed by their genetic incapacity to do well on IQ tests to remain forever unemployed and unemployable, a perpetual burden and a perpetual threat to the rest of society.

Many of Jensen's and Herrnstein's critics have accused them of social irresponsibility. In reply, Jensen and Herrnstein have invoked the scholar's right to pursue and publish the truth without fear or favor. They point out that we cannot escape the consequences of unpleasant truths either by shutting our eyes to them or by denouncing them on ideological grounds. But how firmly based are these "unpleasant truths"?

The educational, social, and political implications of Jensen's doctrine justify a careful examination of this question. It is easy to react emotionally to Jensenism, but teachers and others who help to shape public attitudes toward education and social policy cannot allow themselves to be guided wholly by their emotional responses to this issue.

There is another reason why Jensen's technical argument repays analysis. It exemplifies—almost to the point of caricature—a research approach that is not uncommon in the social sciences. Taking the physical sciences as their putative model, the practitioners of this approach eschew metaphysical speculation and work exclusively with hard, preferably numerical, data, from which they seek to extract objective and quantitative laws. Thus Jensen deduces from statistical analyses of IQ test scores that 80 percent of the variance in these scores is attributable to genetic differences. By exposing in some detail the logical and methodological fallacies underlying Jensen's analysis, I hope to draw attention to the weaknesses inherent in the "operational" approach that it exemplifies.

The Irrelevance of Heritability

Jensen's central contention, and the basis for his and Herrnstein's doctrines on education, race, and society, is that the heritability of IQ is about .8. This means that about 80 percent of the variation in IQ among (say) Americans of European descent is attributable to genetic factors. Other authors have made other estimates of the heritability of IQ—some higher, some considerably lower than .8. In the following pages, I shall try to explain why all such estimates are unscientific and, indeed, meaningless. But before we embark on a discussion of heritability theory and its applicability to human intelligence, it is worth noticing that, even if Jensen's central contention were meaningful and valid, it would not have the implications that he and others have drawn from it. Suppose, for the sake of the argument, that IQ was a measure of some metric trait like height, and that it had a high heritability. This would mean that under prevailing developmental conditions, variations in IQ are due largely to genetic differences between individuals. It would tell us nothing, however, about what might happen under different developmental conditions. For example, despite the high heritability of height in human populations, systematic dietary changes can produce large systematic changes in height between one generation and the next. The high heritability of height means that relatively tall parents tend to have relatively tall children; it does not imply—nor is it true—that stature is insensitive to environmental influences.

This argument applies with even greater force to cognitive and other behavioral traits. Consider literacy, as measured by performance on standardized tests of reading ability. Within a given population, differences in reading ability result from a complex set of interactions between genetic and experiential factors. Yet recent history contains several instances of populations that have made the transition from illiteracy to literacy in the course of a single generation. This example is highly relevant to the issues raised by Jensen's article. A significant proportion of the United States population is excluded from full participation in the society by functional illiteracy. However, most functionally illiterate people have grown up under precisely the sort of environmental conditions that one would expect to produce this condition. In these circumstances, and in view of the historical evidence, the genetic explanation of functional illiteracy seems far-fetched, to say the least.

The concept of heritability is not only irrelevant to the question of how systematic environmental changes affect the expression of phenotypically plastic traits, it is also of limited relevance to the question: How do genetic differences affect the expression of a trait under fixed environmental conditions? Figure 1 shows the phenotypic responses P of a hypothetical metric trait to a continuous range of environments (y) for three representative genotypes (x_1, x_2, x_3). The responses are characterized by different *thresholds, slopes,* and *ceilings.* Many phenotypically plastic traits, including cognitive skills and other behavioral traits, exhibit these qualitative

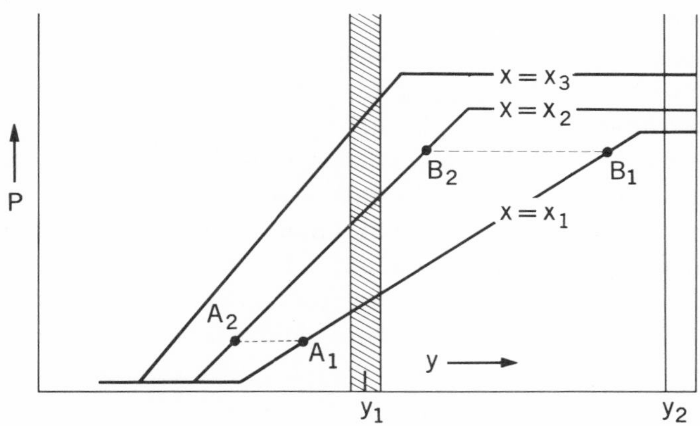

Figure 1. *Phenotypic value (P) of a hypothetical metric trait as a function of an environmental variable (y) for three values of a genotypic variable (x). A_1 and A_2 (also B_1 and B_2) indicate individuals with a common phenotypic value but distinct genotypes, x_1 and x_2, respectively.*

characteristics. For a given environment, the phenotypic differences result entirely from genetic differences—but not always from the same genetic differences. Thus when the environment is optimal for all three genotypes ($y = y_2$, for example), the phenotypic differences are relatively small and result entirely from variations in the developmental ceiling. For suboptimal environments ($y = y_1$ for example), the phenotypic variation is greater and reflects variations in developmental slope and threshold.

Figure 1 does not, however, do justice to the complexity of the situation, because it represents the environmental variation as one-dimensional. Most behavioral traits are sensitive to many environmental factors, and different genotypes demand different environmental profiles for optimal development. Everyone knows that children require differing methods of instruction and modes of interaction with teachers and parents. The heritability of a trait—assuming this to be a meaningful concept—tells us nothing about the variations in that trait that would obtain if the relevant environmental factors were made optimal for each member of the population. But these are precisely the kinds of variations that are—or ought to be—of greatest interest to educators in a democratic society.

What does the alleged high heritability of IQ imply about genetic differences between ethnic groups? The answer to this question is unequivocal: nothing. Geneticists have been pointing out for well over half a century that it is meaningless to try to separate genetic and environmental contributions to measured differences between different strains bred under different developmental conditions.* Between ethnic groups, as between socioeconomic groups, there are systematic differences in developmental conditions (physical, cultural, linguistic, and so on) known to influence performance on IQ tests substantially. Since we have no way of correcting test scores for these differences, the only objectively correct statement that can be made on this subject is the following: The reported differences in average IQ tell us nothing whatever about any average genetic differences that may exist. On the data, black superiority in intelligence (or whatever it is that IQ tests measure) is neither more nor less likely than white superiority.

* A beautiful extended example illustrating this point is given by Lewontin (1970). *See also* Waddington (1957), pp. 92–94, who quotes an exceptionally clear argument by Hogben (1933).

This point is worth emphasizing because it has been widely misunderstood. Professor Herrnstein, for example, writes that the reported differences in mean IQ between United States blacks and whites could be "more genetic, less genetic, or precisely as genetic as implied by a heritability of .8." This seems to imply that, if a genetic difference between blacks and whites exists, it must be, or probably is, in favor of the whites. Such an inference is scientifically indefensible. Even if the heritability of a trait were 100 percent in each of two groups, we could infer nothing about the probable direction of average differences in the relative genetic factors unless we had appropriate information about systematic differences in the relevant environmental factors. Between United States blacks and whites, there are systematic cultural differences and differences in psychological environment, both of which influence the development of cognitive skills in complex ways over long periods of time. Since we have no way of estimating how much these differences affect performance on IQ tests, we can say nothing about genetic differences between the two groups. S. L. Washburn, in a presidential address to the American Anthropological Society, put the matter in these terms:

> I am surprised to hear it stated that if Negroes were given an equal opportunity, their IQ would be the same as the whites'. If one looks at the degree of social discrimination against Negroes and their lack of education, and also takes into account the tremendous amount of overlapping between the observed IQ's of both, one can make an equally good case that, given a comparable chance to that of the whites, their IQ's would test out ahead. Of course, it would be absolutely unimportant in a democratic society if this were to be true, because the vast majority of individuals of both groups would be of comparable intelligence, whatever the mean of these intelligence tests would show.

To sum up, the technical concept of heritability has no direct bearing on the issues of educability or of genetic differences between ethnic groups. Its apparent relevance stems from semantic confusion. In ordinary usage, when we speak of a highly heritable trait, we mean one that is largely inborn. But a trait can have high heritability in the technical sense, either because its expression is insensitive to environmental variation or because the range of relevant environmental variation happens to be small. Jensen and Herrnstein apparently assume that the first of these alternatives is appropriate for IQ. But the available experimental evidence, some of which is cited later in this article, shows that IQ scores are, in fact, highly sensitive to variations in relevant developmental conditions.

Science and Scientism: A Question of Methodology

The theory of heritability, some elementary aspects of which are described below, was developed by geneticists within a well-defined biological context. The theory applies to metric characters of plants and animals—height, weight, and the like. To apply this theory to human intelligence, Jensen and the authors whose work he summarizes must assimilate intelligence to a metric character and IQ to a measurement of that character. Most biologists, I think, would hesitate to take this conceptual leap. Jensen, however, justifies it on the following grounds:

> Disagreements and arguments can perhaps be forestalled if we take an operational stance. First of all, this means that probably the most important fact about intelligence is that we can measure it. Intelligence, like electricity, is easier to measure than to define. And if the measurements bear some systematic relationships to other data, it means we can make meaningful statements about the phenomenon we are measuring. There is no point in arguing the question to which there is no answer, the question of what intelligence *really* is. The best we can do is obtain measurements of certain kinds of behavior and look at their relationships to other phenomena and see if these relationships make any kind of sense and order. It is from these orderly relationships that we gain some understanding of the phenomena.

The "operational stance" recommended by Jensen is thought by many social scientists to be the key ingredient in the "scientific method" as practiced by physical scientists. This belief is mistaken. The first and most crucial step toward an understanding of any natural phenomenon is not measurement. One must begin by deciding which aspects of the phenomenon are worth examining. To do this intelligently, one needs to have, at the very outset, some kind of explanatory or interpretive framework. In the physical sciences, this framework often takes the form of a mathematical theory. The quantities that enter into the theory—mass, electric charge, force, and so on—are always much easier to define than to measure. They are, in fact, completely—if implicitly—defined through the equations that make up the theory.

Once a mathematical theory has been formulated, its predictions can be compared with observation or experiment. This requires appropriate measurements. The aspect of scientific measurements that nonscientists most often fail to appreciate is that they always presuppose a theoretical framewok. Even exploratory measurements, carried out before one has a definite theory to test, always refer to quantities that are precisely defined within a broader theoretical context. (For example, although we do not yet have a theory for the origin of cosmic rays, we know that such a theory must involve the masses, energies, momenta, and charges of cosmic-ray particles. In designing apparatus to measure these quantities, physicists use well-established mathematical theories that describe the behavior of fast particles under a wide variety of conditions.) The theoretical framework for a given set of measurements may be wrong, in which case, the measurements will ultimately lead to inconsistencies, but it must not be vague. In short, significant measurements usually grow from theories, not vice versa. Jensen's views on scientific method derive not from the practice of physical scientists, but from the philosophical doctrine of Francis Bacon (1561–1626), who taught that meaningful generalizations emerge spontaneously from systematic measurements.

These considerations apply equally to biology, where mathematical theories do not yet occupy the commanding position they do in the physical sciences. The following criticism by C. H. Waddington (1957) of *conventional* applications of the heritability theory is illuminating:

> There has been a tendency to regard a refined statistical analysis of incomplete experiments as obviating the necessity to carry the experiments further and to design them in more penetrating fashion. For instance, if one takes some particular phenotypic character such as body weight or milk yield, one of the first steps in an analysis of its genetic basis should be to try to break down the underlying physiological systems into a number of more or less independent factors. Are some genes affecting the milk yield by increasing the quantity of secreting tissue, others by affecting the efficiency of secretion, and others in still other ways?

These views contrast sharply with those of people who believe in the possibility of discovering meaningful relationships between measurable aspects of human behavior without understanding the biological significance of their measurements. The psychometric approach seeks to avoid "metaphysical" speculation, which is an admirable objective. But it is not so easy to operate without a conceptual framework. As we shall see, what Jensen and Herrnstein have done, in fact, is not to dispense with metaphysical assumptions but to dispense with stating them. Such a policy is especially dangerous in the social sciences, where experimental verification of hypotheses is usually difficult or impossible. As Gunnar Myrdal has wisely pointed out, the failure of the social sciences to achieve the same degree of objectivity as the natural sciences can be attributed at least as much to a persistent neglect on the part of social scientists to state and examine their basic assumptions as to the complexity of the phenomena with which they deal.

The operational approach not only spares Jensen the task of trying to understand the nature of intelligence. It also enables him to draw an extremely powerful conclusion from statistical analyses of IQ test scores:

> Regardless of what it is that our tests measure, the heritability tells us how much of the variance in these measurements is due to genetic factors.

Because this assertion holds the key to Jensen's entire argument, we shall analyze it in some detail.

Heritability

In the statement just quoted, Jensen uses the term "heritability" in a specific technical sense that must be elucidated before the statement can be analyzed. Suppose that we have measured an individual character such as height or weight within a given population. The two most fundamental statistical properties of a character are its *mean* and its *variance*. The mean is the average of the measurements; the variance is the average of the squared differences between the individual measurements and the mean. The variance is the most convenient single measure of the spread of individual measurements within a population. Now, this spread results partly from genetic and partly from nongenetic causes. *But this does not mean, nor is it true in general, that a definite fraction of the spread, as measured by the variance, can be attributed to genetic factors and the rest to nongenetic factors.* The variance splits up into separate genetic and nongenetic components only if the variable part of each measurement can be expressed as the sum of *statistically independent* genetic and nongenetic contributions—that is, only if the relevant genetic and nongenetic factors contribute additively and independently to variations of the character in question. (A criterion for statistical independence will be given later.) In this case the genetic fraction or percentage of the variance is called the heritability.

Characters like eye color and blood type, which are entirely genetically determined, have heritability 1. In general, however, the heritability of a character depends on the population considered and on the range of relevant nongenetic factors. Reducing this range always increases the heritability because it increases the relative importance of the genetic contribution to the variance.

It is not easy to find realistic examples of metric characters affected independently by genetic and nongenetic factors. Human height is a possible, though not a proven, example, provided we restrict ourselves to ethnically homogeneous populations. Giraffe height, on the other hand, is a counterexample, since a giraffe's nutritional opportunities may depend strongly on his genetic endowment. Human weight is another counterexample: On a given diet one person may gain weight while another loses weight.

Let us suppose, however, that we have reason to believe that variations of a given character are, in fact, the sum of independent genetic and environmental contributions. To calculate the heritability we need to be able to estimate either the genetic or the environmental contribution to the variance. This can be done if, for example, the population contains a large number of split pairs of one-egg twins. By a split pair, I mean one whose members have been separated since birth and reared in randomly selected, statistically uncorrelated environments. All observable differences between such twins are environmental in origin, and the environmental differences are, by assumption, representative of those between individuals selected at random from the reference population. If, in addition, the genotypes of the twins are representative of those in the population as a whole, then one can use elementary statistical techniques to separate the genetic and environmental contributions to the variance.

Just as the differences between genetically identical individuals reared in randomly selected environments yield an estimate of the magnitude of the environmental contribution to the phenotypic variance, so the differences between genetically unrelated individuals reared in identical environments yield, in principle, a measure of the genetic contribution to the phenotypic variance. For any difference between individuals who have had identical environments must result from differences in genetic endowment. And if both the genotypes and environments in the sample population of unrelated individuals are representative of those in the population at large, the observed phenotypic differences between unrelated individuals reared together yield an estimate of the genetic contribution to the phenotypic variance in the population at large.

If we had suitable data on split, one-egg twins and on unrelated children reared together, and if additivity and statistical independence could be assumed to hold for the trait in question, we would now have independent estimates of the genotypic and environmental contributions to the phenotypic variance. The genotypic contribution to the variance (estimated from phenotypic differences between unrelated individuals reared together) and the environmental contribution (estimated from differences between genetically identical individuals reared apart) should add up to the observed phenotypic variance (estimated from differences between individuals chosen randomly from the population at large). If this sum checks out, it not only provides an estimate of the heritability of the trait; it also vindicates the assumptions underlying the calculation.

This highly idealized example illustrates an important methodological principle. A scientifically respectable data analysis must not only yield estimates of parameters that figure in the underlying theory or "model." It must also provide evidence that the theory actually applies to the data in question. The second requirement is invariably more difficult to satisfy than the first. Anyone versed in statistics can apply techniques analogous to heritability analysis to almost any collection of data, real or imaginary. The results of such an analysis have no scientific value, however, failing a demonstration that the mathematical assumptions underlying the statistical analysis are valid for the data in question. Neglect of this precept is responsible for most of the abuses of statistics and for the low esteem in which statistical arguments are held by the general public.*

The procedures for estimating heritability just sketched depend on certain assumptions about the trait and the population under

* "There are three kinds of lies: plain lies, damned lies, and statistics."—Mark Twain.

consideration. They presuppose that genetic and environmental variations contribute additively to the phenotypic variations. They further presuppose that relevant genetic and environmental differences within the population are statistically unrelated—that is, that "good" genotypes (those that favor the development of a given trait) are as likely to be associated with favorable as with unfavorable environments. Finally, they presuppose that the sample populations of separated identical twins and of genetically unrelated individuals reared together are representative of the parent population. Two questions now arise. Are these assumptions valid for cognitive and other behavioral traits? If not, what sorts of quantitative or semiquantitative inferences can be drawn from available data? Although these questions are obviously crucial for the evaluation of IQ-heritability estimates, they seem to have been totally ignored by Jensen, Herrnstein, Eysenck, and others who emphatically maintain that the high heritability of IQ is a well-established scientific finding. Before considering IQ itself, let us briefly examine these questions in the more general context of animal genetics.

Additivity

By "additivity" I mean the assumption that genetic and nongenetic variations (in a given population) contribute additively to the phenotypic variation. This assumption has been characterized by some writers as a simple and convenient mathematical model for analyzing phenotypic variability. The use of a simplified mathematical model, however, presupposes certain biological assumptions, and it is important to make these assumptions explicit and examine their consequences.

Additively implies that, within a given population, a given environmental variation produces the same phenotypic effect in each member of the population, whatever his genotype or environment. As an example, consider how caloric intake (an environmental factor) affects weight (a phenotypic trait). If the additivity postulate were valid for weight, a given increment in caloric intake —an increase of 200 calories per day, say—would produce the same weight change in each member of the population. For certain populations, this could well be the case; for others, one might expect to observe a wide range of phenotypic responses to the same dietary change. Similarly, the additivity postulate implies that a given genetic variation—something that might, in practice, be quite difficult to define—produces the same phenotypic change in all members of a given population. On general biological grounds, additivity is unlikely to prevail except in circumstances where the ranges of both genetic and environmental variation are severely restricted. It is especially unlikely to be a valid assumption for phenotypically plastic traits. Such traits arise through a complicated developmental process in which genetic and environmental factors are inextricably mingled.

Independence

A second key assumption of conventional heritability analysis is that genetic and environmental variations are statistically unrelated—that "good" environments are not preferentially associated with "good" genotypes. Recognizing the importance of this kind of statistical independence, animal geneticists take great pains to build it into their experimental designs. This involves separating newborns from their mothers at birth and distributing them at random in the population. Since environmental randomization is not feasible for human populations, we must ask whether our inability to reproduce this aspect of the ideal experimental design has serious consequences for the estimation of heritability. I have elsewhere discussed this question in considerable detail (Layzer, 1974). I found that the theoretical consequences of genotype-environment correlation are in fact extremely serious. When such correlations are present, as they undoubtedly are for most phenotypically plastic traits in natural human populations, they make it impossible to estimate heritability—or even to define it in a satisfactory way.

One might seek to avoid this difficulty by studying sample populations consisting of children separated from their biological parents at birth. But this would not resolve the difficulties entirely. In the first place, adoption agencies usually try to match the adoptive environment to their intuitive perceptions of the child's genotype. However unscientific this matching may be, I think it would be unsafe to assume that it introduces no appreciable statistical correlation between genotype and environment. In the second place, important aspects of the environment are selected by the individual himself, and thus, indirectly, reflect his genotype. For example, a child who is genetically predisposed to athletics will tend to devote more time and energy to sports than one who is not so disposed.

Similar considerations may apply in animal genetics. For example, differences in grazing behavior among dairy cattle may be genetically conditioned. One might be tempted to classify all variations of this kind as genetic on the ground that their environmental component has a genetic cause. But this commonsense device would make it impossible to estimate, or even define, heritability in a self-consistent way. Fortunately or unfortunately, the practical impossibility of severing the causal link between genetic and environmental variations greatly limits the applicability of heritability analysis in both animal and human genetics. As Waddington (1957) has remarked in a similar context, "the statistical techniques available [for the analysis of heritability], although imposing and indeed intimidating to most biologists, are in fact very weak and unhandy tools."

Representativeness

Heritability estimates are based on measurements in *sample* populations: separated identical twins, adopted children reared together, and so on. Heritability estimates derived from such measurements are meaningful for the general population only if the samples are representative—that is, if the distribution of genotypes and environments in the sample is the same as in the general population. Suppose that the range of genotypes in a given sample is representative, but the range of environments is less than that of the parent population. Heritability estimates based on the sample will then be too large. This source of systematic bias is especially significant for heritability estimates based on studies of separated identical twins. For such estimates to be meaningful, the relevant environmental differences between separated twins must be statistically indistinguishable from those between individuals selected at random from the general population. None of the major twin studies on which estimates of IQ heritability are based comes close to satisfying this requirement. Even if these studies satisfied other methodological requirements—which they do not—they would substantially overestimate the heritability. What is worse, one cannot reliably estimate the magnitude of the overestimate. Attempts to do so necessarily depend on unverified assumptions that can only be checked by data that do not yet exist.

The foregoing critique of heritability analysis is far from complete. I have not mentioned several important technical issues that further complicate the application of heritability analysis to animal and human data. For a fuller discussion, see the article cited earlier (Layzer, 1974). The present discussion does, however, illustrate a general conclusion of central importance: that heritability is a well-defined concept and susceptible to measurement only for a narrowly defined class of traits under narrowly defined conditions; and it seems highly unlikely on biological grounds that phenotypically plastic traits in human populations ever satisfy these conditions. There are indications that the range of validity of heritability analysis may have been considerably overestimated even in animal genetics. In poultry, for example, the heritabilities of such economically important characters as adult body weight, egg weight, shell thickness, and so on have been repeatedly estimated. Yet for most such characters, the estimates span a considerable range—sometimes as great as 50 percent (Lerner, 1968). Again, estimates of milk yield in dairy cattle range from 25 percent to 90 percent.

IQ as a Measure of Intelligence

We are now ready to analyze the key assertion quoted earlier: "Regardless of what it is that our tests measure, the heritability tells us how much of the variance of these measurements is due to genetic factors." Implicit in this statement are two distinct assumptions: that IQ is a phenotypic character having the mathematical structure (additivity and independence of the genetic and environmental contributions) presupposed by the theory of heritability; and that—assuming this condition to be fulfilled—the heritability of IQ can be estimated from existing data. Now, the IQ data that Jensen and others have analyzed were gathered in eight countries and four continents over a period of fifty years by investigators using a wide variety of mental tests and testing procedures. Geneticists and other natural scientists who make conventional scientific measurements under controlled conditions know from bitter experience how wayward and recalcitrant, how insensitive to the needs and wishes of theoreticians, such measurements can be. Their experience hardly leads one to anticipate that the results of mental tests constructed in accordance with unformulated, subjective, and largely arbitrary criteria possess the special mathematical structure needed to define heritability. It is difficult to imagine how this happy result could have been achieved except through the operation of collective serendipity on a scale unprecedented in the annals of science. Nevertheless, let us examine the case on its merits.

At the very outset we have to ask, is IQ a valid measure of intelligence? Jensen and Herrnstein assure us that it is. "The most important fact about intelligence is that we can measure it," says Jensen, while Herrnstein remarks that the "objective measurement of intelligence" is psychology's "most telling accomplishment." I find these claims difficult to understand. To begin with, the "objective measurement" does not belong to the same *logical category* as what it purports to measure. IQ does not measure an individual phenotypic character like height or weight; it is a measure of the rank order or relative standing of test scores in a given population. Thus the statement, "A has an IQ of 100" means that half the members of a certain reference population scored lower than A on a certain set of tests and half scored higher. "B has an IQ of 115" means that 68 percent of the reference population scored lower than B and 32 percent higher, and so on. (IQ tests are so constructed that the frequency distribution of test scores in the reference population conforms as closely as possible to a normal distribution—the familiar bell-shaped curve—centered on the value of 100 and having a half-width or standard deviation [the square root of the variance] of 15 points.) To call IQ a measure of intelligence conforms neither to ordinary educated usage nor to elementary logic.

One might perhaps be tempted to dismiss this objection as a mere logical quibble. If IQ itself belongs to the wrong logical category to be a measure of intelligence, why not use actual test scores? One difficulty with this proposal is the multiplicity and diversity of mental tests, all with equally valid claims. (This is part of the price that must be paid for a strictly operational definition of intelligence.) Even if one were to decide quite arbitrarily to subscribe to a particular brand of mental test, one would still need to administer different versions of it to different age groups. An appearance of uniformity is secured only by forcing the results of each test to fit the same Procrustean bed (the normal distribution). But this mathematical operation cannot convert an index of rank order on tests having an unspecified and largely arbitrary content into an objective measure of intelligence.

TQ and Tentacle Length

The preceding considerations show that, for logical and methodological reasons, IQ scores cannot be regarded as direct measurements of a phenotypic character. It is conceivable, however, that rank order on mental tests like the Stanford-Binet or Wechsler provide *indirect* measurements of some phenotypic character. This interpretation seems to have been held by Sir Cyril Burt, a convinced hereditarian whose work forms a mainstay of Jensen's technical argument. "Differences in this hypothetical ability [intelligence]," Burt wrote (1966), "cannot be directly measured. We can, however, systematically observe relevant aspects of the child's behavior and record his performances on standardized tests; and in this way we can usually arrive at a reasonably reliable and valid estimate of his 'intelligence' in the sense defined." (Earlier in the same paper, Burt defines intelligence as "an innate, general, cognitive factor.") Burt's conviction that intelligence cannot be directly or objectively measured—a conviction bred by over half a century of active observation—profoundly influenced his practical approach to the problem of assessment. Burt and his assistants usually used group tests, but also relied heavily on subjective impressions of teachers. When a discrepancy arose between a teacher's assessment and the results of group tests, the child was retested individually, sometimes repeatedly. Whether or not Burt's final assessments are "reasonably reliable and valid," they are certainly not objective measurements, nor did he consider them to be so.

The following example illustrates how subjective assessments of rank order could, under appropriate circumstances, yield valid quantitative estimates of a trait that cannot be measured directly. Suppose that members of a superintelligent race of octopuses, unable to construct rigid measuring rods but versed in statistical techniques, wished to measure tentacle length. Through appropriate tests of performance, they might be able to establish rank order of tentacle length in individual age groups. By forcing the frequency distribution of rank order in each group to fit a normal distribution with mean 100 and standard deviation 15, they would arrive at a TQ (tentacle quotient) for each octopus. In all probability, differences in TQ would turn out to be closely proportional to differences in actual tentacle length within a given age group, though the factor of proportionality would vary in an unknown way from one group to another. Thus, our hypothetical race of octopuses would be able to *infer* relative tentacle length within an age group from information about rank order. This inference evidently hinges on the assumption that tentacle length, which the octopuses cannot measure directly, is in reality normally distributed within each age group.

Some Tacit Assumptions Unmasked and Analyzed

Similarly, the *inference* that IQ is a measure of intelligence depends on certain assumptions, namely: (*a*) that there exists an underlying one-dimensional, metric character related to IQ in a one-to-one way, as tentacle length is related to TQ, and (*b*) that the values assumed by this character in a suitable reference population are normally distributed.

If these assumptions do not in themselves constitute a theory of human intelligence, they severely restrict the range of possible theories. Once again we see that the operational stance, though motivated by a laudable desire to avoid theoretical judgments, cannot, in fact, dispense with them. The choice between a theoretical approach and an empirical one is illusory; we can only choose between explicit theory and implicit theory. But let us examine the assumptions on their own merits.

The first assumption is pure metaphysics. Assertions about the existence of unobservable properties cannot be proved or disproved; their acceptance demands an act of faith. Let us perform this act, however—at least provisionally—so that we can examine the second assumption, which asserts that the underlying metric character postulated in the first assumption is normally distributed in suitably chosen reference populations. Why normally distributed? A possible answer to this question is suggested by a remark quoted by the great French mathematician Henri Poincaré: "Everybody believes in the [normal distribution]: the experimenters because they think it can be proved by mathematics, the mathematicians because it has been established by observation." Nowadays, both experimenters and mathematicians know better. Generally speaking, we should expect to find a normal frequency distribution when the variable part of the measurements in question can be expressed as the sum of many individually small, mutually independent, variable contributions. This is thought to be the case for a number of metric characters of animals such as birth weight in cattle, staple length of wool, and (perhaps) tentacle length in octopuses. It is not the case, on the other hand, for measurements of most kinds of skill or proficiency. Typing speed, for example, is not likely to be normally distributed because proficiency in typing does not result from the combined action of a large number of individually small and mutually independent factors.

What about mental ability? Jensen and Herrnstein believe that insight into its nature can be gained by studying the ways in which people have tried to measure it. Jensen argues that because different mental tests agree moderately well among themselves, they must be probing a common factor (Spearman's *g*). Some tests, says Jensen, are "heavily loaded with *g*," others, not so heavily loaded. Thus *g* is something like the pork in cans labeled "pork and beans."

Herrnstein takes a less metaphysical line. Since intelligence is what intelligence tests measure, he argues, what needs to be decided is what we *want* intelligence tests to measure. This is to be decided by "subjective judgment" based on "common expectations" as to the "instrument." "In the case of intelligence, common expectations center around the common purposes of intelligence testing—predicting success in school, suitability for various occupations, intellectual achievement in life." Thus Herrnstein defines intelligence "instrumentally" as the attribute that successfully predicts success in enterprises whose success is commonly believed to depend strongly on—intelligence. That is, intelligence is what is measured by tests that successfully predict success in enterprises whose success is commonly believed to depend strongly on what is measured by tests that successfully predict success in enterprises whose success is commonly believed to depend strongly on . . .

Whatever the philosophical merits of the definitions offered by Jensen and Herrnstein, they afford little insight into the question at hand: Does intelligence depend on genetic and environmental factors in the manner required by heritability theory? In other words, is the heritability of intelligence a meaningful concept? To pursue this question, we must go outside the theoretical framework of Jensen's discussion.

Intelligence Defined; Cognitive Development

Many modern workers believe that intelligence can usefully be defined as information-processing ability. As a physical scientist, I find this definition irresistible. To begin with, it permits us to distinguish as many qualitatively different kinds of information and, hence, as many different kinds or aspects of intelligence, as we wish. Moreover, because information is a precisely defined mathematical concept, there is no obvious reason why it should not be possible to devise practical methods for reliably measuring the ability to process it. (In its broadest sense, information-processing involves problem-solving as well as the extraction and rearrangement of data.) Whether or not such tests would be accurate predictors of "success," I do not know. They could, however, be usefully employed in assessing the effectiveness of teachers, educational procedures, and curricula.

Information-processing skills, like other skills, are not innate, but develop over the course of time. What is the nature of this development? Consider such complex skills as skiing or playing the piano. In order to acquire an advanced technique, one must acquire, in succession, a number of intermediate techniques. Each of these enables one to perform competently at a certain level of difficulty, and each must be thoroughly mastered before one can pass to the next level. The passage to a higher level always involves the mastery of qualitatively new techniques. Through systematic observations carried out over half a century with the help of numerous collaborators, Jean Piaget (1952) has demonstrated that basic cognitive structures also develop in this way, and he has traced the development of a great many of these structures in meticulous detail. Each new structure is always more highly organized and more differentiated than its predecessor. At the same time, it is more adequate to a specific environmental challenge. The intermediate stages in the development of a given structure are not rigidly predetermined (there are many different ways of learning to read, or ski, or play the piano), nor is the rate at which an individual passes through them, but in every case, cognitive development follows two basic rules (Piaget, 1967): "Every genesis emanates from a structure and culminates in another structure. Conversely, every structure has a genesis."

Cognitive development may be compared to the building of a house. Logic and the laws of physics demand that the various stages be completed in a definite order: the foundations before the frame, the frame before the walls, the walls before the roof. The finished product will depend, no doubt, on the skill of the builder and on the available materials, but it will also depend on the builder's intentions and on the nature of the environmental challenge. Similarly, although cognitive development is undoubtedly strongly influenced by genetic factors, it represents an adaptation of the human organism to its environment and must, therefore, be strongly influenced by the nature of the environmental challenge. Thus, we may expect cultural factors to play an important part in shaping all the higher cognitive skills, for the environmental challenges that are relevant to these skills are largely determined by cultural context.

Genotype-environment Interaction

If intelligence, or at least its potentially measurable aspects, can be identified with information-processing skills, and if the preceding very rough account of how these skills develop is substantially correct, then it seems highly unlikely that scores achieved on mental tests can have the mathematical properties that we have been discussing—properties needed to make "heritability of IQ" a meaningful concept. The information-processing skills assessed by mental tests result from developmental processes in which genetic and nongenetic factors interact continuously. The more relevant a given task is to an individual's specific environmental challenges, the more important are the effects of this interaction. A child growing up in circumstances that provide motivation, reward, and opportunity for the acquisition of verbal skills will achieve a higher level of verbal proficiency than his twin reared in an environment hostile to this kind of development. Even if two genetically unlike individuals grow up in the same circumstances—for example, two-egg twins reared together—we cannot assume (as Jensen, Herrnstein, and other hereditarians usually do) that the relevant nongenetic factors are the same for both. If one twin has greater verbal aptitude or is more strongly motivated to acquire verbal skills (usually the two factors go together), he will devote more time and effort to this kind of learning than his twin. Thus, differences between scores on tests of verbal proficiency will not reflect genetic differences only, but also—perhaps predominantly—differences between the ways in which the genetic endowments of the twins have interacted with their common environment.

One might be tempted to classify these interactive contributions to developed skills as genetic, on the grounds that they are not purely environmental and that the genetic factor in the interaction plays the active role. In technical discussions, however, common sense must accommodate itself to definitions and conventions laid down at the outset. If we redraw the line that separates genetic and nongenetic factors, we must formulate a new theory of inheritance; if we wish to use the existing theory, we must stick to the definitions that it presupposes.

Do the IQ Data Fit the Theory of Heritability?

Until now, we have concerned ourselves with the first of the two implicit assumptions underlying Jensen's key assertion about the interpretation of heritability estimates, namely, the assumption that intelligence is a phenotypic character to which the theory of heritability can be applied. We have found no plausible grounds for supposing that genetic and environmental factors contribute to the development of intelligence in the simple way required by heritability theory. To this objection, Jensen and Herrnstein might reply as follows: "Discussions of 'meaning,' 'mathematical structure,' and 'logical categories' are irrelevant. IQ test scores and the statistical quantities that can be derived from them (means, variances and correlations) are hard data. There is nothing to prevent us from applying heritability theory to these data and seeing whether or not it fits. If the theory does fit, we may reasonably assume that it applies to these data, whatever their provenance."

This argument has some validity. If the heritability theory *did* apply to IQ test scores and the statistics derived from them, these statistics would simultaneously satisfy a large number of numerical relations, like the steel girders composing a complex rigid structure. Conversely, if all these relations were indeed accurately satisfied by the statistical data, we would have good reason to suppose that the theory applied to them in spite of a priori arguments to the contrary. Yet such arguments do serve an important purpose: they help us to decide how good the evidence must be to convince us that the theory really does fit the data. Suppose that an astronomer, having made several observations of a newly discovered planet, tries to determine its orbit using Newton's theory of gravitation. If he applies the theory correctly, he can be virtually certain that any discrepancy between the theory and his observations results from observational error: previous experience has firmly established Newton's theory and its applicability to the motion of planets. When the validity or applicability of a theory is not so well established, however, it may not be easy to decide how much of the discrepancy between theory and observation to attribute to experimental error and how much to error or incompleteness in the theoretical description. In such cases, a competent scientist bases his judgment on all the relevant theoretical and observational information available to him. If that information strongly suggests that a given theory does not apply to given data, he will demand highly convincing evidence of internal consistency before taking seriously claims based on a theoretical analysis of the data.

Strangely enough, Jensen and Herrnstein offer little evidence of this kind, and what evidence they do offer is, as we shall see, specious. The data that Jensen, Burt, and other hereditarians have analyzed consist mainly of *correlations* between the measured IQs of more or less closely related persons living in more or less similar environments. Before examining these data, let us recall what statisticians mean by the term correlation.

Suppose we have measured the heights and weights of twelve-year-olds in a certain school. Height and weight are said to be strongly correlated if their *rank orders* are very similar. A perfect correlation is assigned the value +1, a null correlation—between two statistically independent sets of measurements—the value 0. (This property *defines* statistically independent measurements.) Two paired sets of numbers have a large *negative* correlation if the rank order of one closely resembles the *reversed* rank order of the second.* IQ correlations between genetically related individuals are measured in exactly the same way as the correlation between height and weight in our example. To measure the correlation between the IQs of grandparents and grandchildren in a given sample, one can follow the above recipe, substituting the grandchild's IQ for, say, height and the grandparent's IQ for weight.

* One calculates the numerical value of the correlation between height and weight from the following recipe. (1) Calculate the mean height and the variance in height, also the mean weight and the variance in weight. (2) Subtract from each child's height the mean height, and divide the result by the square root of the variance in height. The resulting number may be called the reduced deviation from the mean. Do the same thing for the weight. (3) Multiply corresponding reduced deviations in height and weight together, add up all the products and divide the result by the number of children in the sample. The resulting number is the height-weight correlation for this sample. Notice that it does not depend on what units (inches or centimeters, pounds or kilograms) we use to measure height and weight.

Such studies have usually shown that the IQs of more closely related people tend to be more highly correlated than those between less closely related people; also, that the IQs of children growing up in similar circumstances tend to be more highly correlated than those of children growing up in dissimilar circumstances. For pairs of children reared together, the measured correlations increase systematically with increasing genetic similarity. Thus, the IQs of one-egg (identical) twins tend to be more highly correlated than those of two-egg (fraternal) twins, which, in turn, tend to be more highly correlated than the IQs of unrelated children reared together. These findings show that IQ is strongly influenced by both genetic and environmental factors. Can we disentangle these factors? "By evaluating the total evidence," writes Herrnstein, "and by a procedure too technical to explain here, Jensen concluded (as have most of the other experts in the field) that the genetic factor is worth about 80% and that only 20% is left to everything else. . . ."

As summarized by Herrnstein, the evidence on which this conclusion rests seems quite impressive. Herrnstein compares the "actual" values of IQ correlations between relatives with "theoretical" correlations calculated on the assumption that nongenetic effects on IQ are negligible. In every case the agreement seems to be very close:

> Uncle's (or aunt's) IQ should, by the genes alone, correlate with nephew's (or niece's) by a value of 31%; the actual value is 34%. The correlation between grandparent and grandchild should, on genetic grounds alone, also be 31%, whereas the actual correlation is 27%, again a small discrepancy. And finally for this brief survey, the predicted correlation between parent and child, by genes alone, is 49%, whereas the actual correlation is 50% using the parents' adult IQ and 56% using the parents' childhood IQ's—in either case too small a difference to quibble about.

But let us take a closer look. What does Herrnstein mean by the word "actual" in the passage just quoted? "The foregoing figures," he writes, "are lifted directly out of Jensen's famous article, figures that he himself culled from the literature on intelligence testing." Referring to Jensen's article, we do indeed find the figures quoted by Herrnstein, in a column headed "obtained median correlation." How is the "median" correlation related to the "actual" correlation? Can we assert that the actual value of a quantity lies close to the median of several measurements of that quantity? As every working scientist knows, the answer to this question is, No, not in general. All that can be said is that, in the absence of systematic errors, the actual value is likely to lie within a range of values comparable to the range spanned by the actual measurements (if there are enough of them).

What ranges do the IQ correlations span? Jensen's paper does not supply this important information. His table, however, is adapted from one given by Burt, who, in order to compare his own correlation measurements with those of previous investigators, tabulated the medians of correlation measurements collected by Erlenmeyer-Kimling and Jarvik (1964). Burt did not display the actual ranges of the measured correlations, but he did mention that several of them were large and gave one example: For siblings reared together, the correlations obtained in 55 studies range from .3 to .8 and are spread almost uniformly over the entire range. The correlations reported between parent and child in 11 studies—to give a second example—range from about .2 to about .8.

But these figures still do not tell the whole story. What do the reported correlations actually mean? Each reported correlation refers to a particular sample and to a particular test or set of tests. How homogeneous are the samples with respect to nongenetic variables? How meaningful is it to combine correlations referring to population samples and tests differing in unspecified and unknown ways? The answers to these hard but important questions are to be found, if they are to be found at all, in the primary sources. As we move further and further from these sources, the errors and uncertainties in the data become less and less noticeable, until at last, in the pages of *The Atlantic Monthly*, only the "actual values" remain. Like the reputations of saints, scientific data often improve with transmission.

The "theoretical correlations" quoted by Herrnstein have undergone a similar transformation. The theory in question is a mathematical model especially devised by Burt and Howard (1956) to represent their correlation data—data based, let us recall, on Burt's semiobjective assessments of intelligence. This kind of agreement does little to enhance one's confidence in either the theory or the data.

Suppose, for the sake of the argument, that IQ scores could be assimilated to valid measurements of a phenotypic character. What theoretical inferences could legitimately be drawn from them? As I mentioned earlier, a recent study (Layzer, 1974) has shown that heritability analysis of a phenotypically plastic trait is impossible in principle unless genotype and environment are statistically uncorrelated. This condition is certainly never satisfied for important behavioral traits in natural human populations. Studies of unrelated adopted children reared together and half-siblings reared separately could conceivably yield useful information about heritability, but even in such studies, the problems introduced by genotype-environment correlation present serious, possibly insurmountable, obstacles to meaningful heritability estimation.

Many people I have talked to find it difficult to accept the conclusion that no meaningful inference about IQ heritability can be extracted from the reams of published data. Granted that the data are not as good as one might wish, and the theories not as powerful, still—it is said—it ought to be possible to say *something* meaningful about IQ heritability: for example, that it is probably greater than zero. In fact, in the physical sciences, one encounters many situations in which no quantitative or semiquantitative conclusion can be drawn in spite of an abundance of data and the existence of an underlying mathematical theory. For this reason, a physical scientist tends to bring rather modest expectations to a problem like the one under discussion. In the paper cited earlier (Layzer, 1974), I have suggested that from data on unrelated adopted children reared together one can tentatively conclude that the nominal value of IQ heritability may lie between zero and 50 percent. But I must emphasize that I attach little weight to this estimate. Apart from the difficulties with IQ scores as such, the studies on which the estimates are based are beset by serious methodological difficulties.

What, then, can we infer from the data on IQ correlations? If genetic factors did not appreciably influence IQ, we would expect to find no appreciable differences between the IQ correlations of one-egg twins, two-egg twins, and unrelated children reared in the same home. In fact, the measured correlations tend to be greater for one-egg twins than for two-egg twins, and greater for two-egg twins and siblings than for unrelated children reared in the same home—although the ranges overlap considerably. These findings suggest that genetic factors presumably do influence IQ significantly—but not probably in the manner presupposed by heritability analysis. The internal consistency of the reported data is far too low to lend credence to claims that IQ measurements have the mathematical structure required for that analysis to be meaningful.

Jencks's Analysis

The most elaborate heritability analysis of the IQ data has been made by Jencks (Jencks et al., 1972). Jencks estimated the heritability (h^2) separately for four different kinds of data. His findings may be summarized as follows:

Table 1.

$0.29 < h^2 < 0.76$	from parent-child data
$0.45 < h^2 < 0.60$	from comparisons of identical and fraternal twins
$0.29 < h^2 < 0.50$	from identical twins reared apart
$0 < h^2 < 0.25$	from comparisons of siblings and unrelated children reared together

Within each category, Jencks found inconsistencies among data from different sources, some of them systematic. The most pronounced systematic differences were between English and American studies, the English studies yielding consistently higher estimates of h^2 than the American. As the above tabulation shows, the results of estimates based on comparisons of siblings and unrelated children are inconsistent with all other estimates. Notice, too, that the estimates derived by Jencks from studies of identical twins reared apart are substantially lower than those of Burt, Jensen, and other hereditarians.

Jencks summarized his conclusions in the following terms:

> . . . we think the chances are about two in three that the heritability of IQ scores, as we have defined the term, is between 0.35 and 0.55, and . . . the chances are about 19 out of 20 that [it] is between 0.25 and 0.65.

Yet Jensen, using essentially the sama data, concluded that h^2 lies between .70 and .80, and in most of his discussions, he uses the higher figure. Now, Jencks's analysis indicates that the value .80 is about as likely to be correct as the value 0.10. Indeed, a heritability of 0.10 would at least be consistent with estimates based on comparisons between siblings and unrelated children reared together.

How are we to interpret this discrepancy between inferences drawn from the same data by means of essentially the same theory? A nonscientist might be tempted to suspect that Jensen, a hereditarian, had injected hereditarian bias into his data analysis, and that Jencks, who is more sympathetic to the environmentalist position, had allowed an opposite bias to color *his* analysis. But there is more to it than that. A glance at table 1 shows that appropriately selected data could be found to support any value of h^2 between 0 and .75. But this does not mean that one is free to choose data to support the value one finds most agreeable. Not all kinds of data are equally relevant to a given theory, and not all data of the same kind are equally reliable. It is up to the scientist who analyzes data to assess their relevance and reliability and to weight them accordingly. For example, the four kinds of data used by Jencks are susceptible to different kinds of systematic errors. Those resulting from genotype-environment correlation, for example, are likely to be smallest in studies of unrelated children reared together. Unfortunately, as has already been mentioned, these studies suffer from other defects, so that while heritability estimates based on them are comparatively reliable, they are not reliable in an absolute sense.

How one interprets the results of Jencks's analysis (table 1) depends on one's prior theoretical expectations. If there were good reasons for believing that IQ test scores represent genuine measurements of intelligence and that individual variations in intelligence have the mathematical structure required by heritability theory, then Jencks's estimate of h^2 could be interpreted in the conventional way—as an estimate of the fraction of IQ variance "explained by" genetic variation. If, on the other hand, there are strong a priori reasons to doubt whether IQ test scores represent measurements and whether the heritability theory would apply to them if they did—and we have seen that such reasons do indeed exist—then we must begin by asking whether the results of the analysis serve to allay these theoretical doubts. In my opinion, they do not. On the contrary, it seems to me that they strongly confirm them. The internal inconsistencies and systematic discrepancies revealed by the analysis are just what one would expect to find if the kind of behavior sampled by IQ tests had a strong genetic component that could not be disentangled from environmental and experiential influences in the simple manner presupposed by heritabilty theory.

The philosopher of science, Karl Popper, has emphasized that experimental and observational findings can never *confirm* a theory; they can only falsify it. The results of Jencks's analysis constitute prima facie evidence that heritability analysis cannot meaningfully be applied to IQ test scores, and this interpretation of the evidence accords fully with our earlier considerations of IQ and the limitations of heritability analysis. Burt, Jensen, Herrnstein, and other hereditarians make the argument that IQ is a highly heritable phenotypic trait by pointing to data that harmonize with this thesis. This is good debating strategy but poor science.*

IQ and Culture

How do cultural factors affect IQ correlations? Jensen's views on this question are instructive. They are summarized in a passage that begins with the sentence we have already analyzed in some detail:

> Regardless of what it is that our tests measure, the heritability tells us how much of the variance in these measurements is due to genetic factors. If the test scores get at nothing genetic, the result will simply be that estimates of their heritability will not differ significantly from zero. The fact that heritability estimates based on IQ's differ significantly from zero is proof that genetic factors play a part in individual differences in IQ. To the extent that a test is not "culture-free" or "culture-fair" it will result in a lower heritability measurement.

Jensen is here making two further assertions: only genetic factors can *raise* heritability estimates, and cultural bias in a test always *lowers* the heritability of the results. The term "bias" is misleading, however. Its use presupposes the possibility of constructing an unbiased intelligence test. But if intelligence is a result of adaptation, this is impossible, except under entirely unrealistic assumptions about cultural homogeneity. An adequate test of intelligence for one cultural context must be inadequate for another. The idea that each of us is born with a certain abstract "capacity" for, say, pattern

* Leon Kamin has analyzed the IQ-correlation data from an environmentalist standpoint. He concludes that the data are consistent with vanishingly small IQ heritability—that is, with the thesis that differences in performance on IQ tests are wholly attributable to nongenetic factors.

recognition, which can be measured by an appropriate culture-free test (if only it could be found), is pure metaphysics.

Do cultural factors systematically affect performance on IQ tests? Hereditarians usually argue that such effects are probably small, because the skills assessed by IQ tests are not systematically taught in school. Hence, they argue, all children have the same opportunity to learn them. It is (unfortunately) true that many of these skills are not taught effectively in schools, but no one who has actually visited classrooms in suburban and in ghetto schools would deny that they are usually taught more effectively in the former than in the latter. Moreover, they are deliberately and effectively taught in homes where learning is valued for its own sake. Children who grow up in such homes learn to speak grammatically, to use words with precision, to get information and entertainment from books, to argue consequentially, to solve abstract problems, and to set a high value on these and similar activities. Such home environments tend strongly to run in families and to occur more frequently in some ethnic groups than in others.

In her classic study of some leading American scientists, Anne Roe (1953) discovered that an unexpectedly large proportion of her subjects were the sons of professional men. She suggested as the most likely explanation for this finding the fact that her subjects grew up in homes where

> for one reason or another learning was valued for its own sake. The social and economic advantages associated with it were not scorned, but they were not the important factor. The interest of many of these men took an intellectual form at quite an early age. This would not be possible if they were not in contact of some sort with such interests and if these did not have value for them. This can be true even in homes where it is not taken for granted that the sons will go to college.

We do not choose our cultural background but are born into it. For this reason, people related by blood are more likely to share a common cultural background and common cultural values than unrelated persons. Since parental values help to shape children's intellectual development, it is clear that cultural factors affect IQ correlations between relatives in roughly the same way as genetic factors. The correspondence between cultural and genetic effects is not, of course, perfect. For example, if cultural factors alone were important, we would expect to find nearly the same correlations between the IQs of one-egg and two-egg twins. On the other hand, heritability estimates based on IQ correlations between separated one-egg twins would probably be quite high, because in those cases where the separated twins are not actually reared by relatives, adoption agencies usually strive to match the cultural backgrounds of natural and adoptive parents. In short, cultural factors may be expected to increase some heritability estimates and to decrease others. The available observational evidence is at least consistent with the hypothesis that cultural factors contribute heavily to measured IQ correlations.

Culture is, of course, not the only mock-genetic environmental factor that systematically influences IQ. Because cultural values, wealth, social status, occupational and educational levels are all more or less strongly correlated with one another as well as with IQ, and because they are all transmitted from generation to generation in roughly the same way as genetic information, it would obviously be very difficult under the best of circumstances to disentangle the genetic factors affecting IQ from nongenetic factors. We have seen, however, that even if all these systematic nongenetic factors were absent, such a separation could not be accomplished through the application of heritability theory, because IQ measurements do not satisfy that theory's requirements.

Jensen and other hereditarians not only fail to take cultural and other environmental factors adequately into account, they also ignore *genetic but noncognitive* factors which interact in complex and, as yet, little understood ways with each other, with cognitive factors, and with the environment. Among such factors are sex, color, temperament, and physical appearance. Thus, in societies where mathematical ability is considered unfeminine and feminity is prized, women tend not to develop mathematical ability. Again, skepticism and curiosity may help a middle-class child to develop scientific ability but only cause trouble for a ghetto child. And in societies like our own, where selection and advancement mechanisms often employ positive feedback (nothing succeeds like success), small, "initial" differences in genetic or environmental endowment often become greatly amplified. This is another way of saying that genetic-environmental interaction may well dominate the development of the skills that IQ tests assess.

Inferences from the Twin Studies

Studies of separated one-egg twins provide the most direct evidence bearing on genetic factors in intelligence. Four such studies have been published. The sample sizes range from nineteen to fifty-three pairs, and the studies are very diverse with respect to methodology, selection of subjects, and scientific objectives. Nevertheless, they do permit certain broad conclusions to be drawn. The most striking result, common to all four, is that the tested IQs of separated one-egg twins tend to be very similar if the environmental differences are not too great. In these circumstances, the measured correlations are usually higher than those between two-egg twins reared together. This result clearly indicates that genetic factors can play an important role in the development of cognitive skills.

Among the genetic factors that influence the development of cognitive skills we must include, for example, sex, color, temperament, and various physical characteristics. So far as I know, there is no theoretical or observational justification for supposing that the effects of such noncognitive genetic factors are small compared with those of the specifically cognitive factors. We must, therefore, beware of drawing conclusions about the importance of cognitive genetic factors from studies of one-egg twins.

But why bother to distinguish between cognitive and noncognitive genetic factors? The answer is that the ways in which some noncognitive factors influence the development of intelligence depend strongly on social and cultural attitudes that could change rapidly and drastically. Prevailing attitudes toward women and Afro-Americans are currently among the most attractive candidates for such a change.

Although there appears to be little doubt that one-egg twins reared separately tend to have similar IQs, the measured IQ correlations have no *quantitative* significance, because in none of the studies are the environmental differences between separated twins representative of those between randomly paired children of the same age. The environmental differences that matter in this connection are, of course, those between factors especially relevant to intellectual growth. What are these factors? Or, to put the same question in slightly different terms, what kinds of experience at what ages contribute most effectively to a child's mental development? This question is still the subject of intensive research. Most students of cognitive development agree, however, that the usual indices of socioeconomic status (based chiefly on the occupational status of the head of the family) are not reliable indices of the environmental factors most relevant to cognitive development. Since crucial stages of this development occur during the first years of a child's life, the mother's child-rearing practices, her intelligence, and the quality of the mother-child relationship, are perhaps more relevant than the father's occupation.

Relevant environmental differences between separated twins are most carefully assessed in the classic study of Newman, Freeman, and Holzinger (1937). Summarizing their findings on the environmental differences between the separated twins in their study, these authors write: "The majority of the ratings [of relevant environmental differences] were relatively low. For only a few cases were the environmental differences of the twin pairs judged to be large." These comments apply equally to Shields's study (1962), in which, besides, a large fraction of the subjects not reared in their own homes were reared by relatives. In the third study (Juel-Nielsen and Mogensen, 1957), the sample—twelve pairs—is too small to permit statistical inferences to be drawn from it. The fourth study, that of Burt, is the largest (fifty-three pairs) and the only one in which the subjects were children. Moreover, only five of the fifty-three pairs were reared by relatives. Finally—and this is a point on which Burt, Jensen, and Herrnstein lay great stress—there was no significant correlation in occupational status between the families in which the separated twins were reared. Can we infer from this that the relevant environmental differences between separated twins in Burt's study are likely to be representative of differences between randomly selected children in the same population?

To begin with, it is worth noting that the vast majority of the children in Burt's study attended school in London. Thus the range of educational experiences in this study was considerably narrower than it would have been for a representative sample of English schoolchildren. But the environmental differences between separated twins in Burt's study are unrepresentative in other respects also. As Burt points out, the absence of significant correlation between the occupational categories of foster parents and natural parents does not reflect a deliberate policy on the part of adoption agencies to randomize the placement of foster children. On the contrary, efforts are normally made to match the social classes of foster and natural parents, though such efforts frequently fail. Thus, half the foster parents in Burt's sample belong to occupational category V (semi-skilled), but this category supplied only 28 percent of the children reared in foster homes or residences. For category VI (unskilled), the imbalance is in the reverse direction. These and similar systematic asymmetries, which no doubt reflect systematic differences between the social and economic pressures experienced by families in different occupational categories, are presumably responsible for the failure of efforts to match the social classes of adoptive and natural parents. But is there any reason to suppose that the agencies failed to place children in families that they judged to be suitable in other respects? Or to suppose that their placement criteria, which presumably included emotional and intellectual qualities of the foster mother, were less relevant to the foster child's intellectual development than the occupational status of the foster father? For Herrnstein (if I interpret his remarks on the subject correctly), the "breadwinner's occupation" is the most relevant environmental factor, but I think it is possible that not all psychologists share this view.

Three other points should be mentioned in connection with Burt's study. First, although his sample is relatively large, it is by no means what statisticians would call a "fair sample." For example, among the children reared in their hown homes, the frequency distribution of IQ is not normal (i.e., bell-shaped) as it would be in a fair sample, but strongly skewed: Although the mean and the variance are nearly equal to their values in the standard reference population, only 38 percent of the children in the sample have IQs greater than 100. Second, Burt's assigned IQs are not measurements in the Jensen-Herrnstein sense, but semiobjective assessments. Burt's published figures show that the subjective assessments for separated twins correlate better and differ less than do group-test scores.

Finally, serious questions have recently been raised concerning the reliability of Burt's published IQ data. Leon Kamin (1974, 1976) has pointed out a number of serious internal discrepancies and inconsistencies in Burt's published data. For example, the reported IQ correlation between separated one-egg twins remained constant to three decimal places as the size of his twin sample increased, over a period of twenty-three years, from fifteen to fifty-three pairs. (*See also* Jensen [1974] where many other coincidences of this kind are pointed out. Jensen remarks that all of Burt's data files have been lost or destroyed. This unfortunate circumstance renders Burt's work on kinship correlations—previously held to be the most comprehensive and homogeneous collection of data bearing on IQ heritability—unusable for scientific purposes.)

To sum up, in none of the published twin studies can we consider the environmental differences between separated twins to be representative of those between children of the same age drawn at random from a standard reference population. Because the samples are small and unrepresentative, and because data are wholly lacking on relevant environmental differences that might enable one to correct partially for the unrepresentative character of a sample, we cannot even guess what the IQ correlation between genetically identical children reared in randomly selected, statistically independent environments might turn out to be. This difficulty cannot, of course, be overcome by combining correlations measured in four methodologically disparate studies carried out in three different countries; four sieves will not hold more water than one.

Environmental Differences and IQ Differences in the Twin Studies

Although the twin studies do not provide usable information about theoretically significant correlations, they do tell us something about the effects of environment on IQ. For example, we can ask: Do large environmental differences tend to produce large IQ differences? In his incisive critique of Jensen's article, Martin Deutsch (1969) has summarized several independent studies that throw light on this question, of which the following is representative:

> Using the Newman *et al.* ratings of educational and social differences between pairs of twins, Stone and Church (1968) classified 10 pairs of twins as having "larger differences in educational and social advantages" (DSEA), and 9 pairs of twins as having "smaller DSEA." They found that 7 pairs of the twins in the larger DSEA group had IQ differences of 10 or more points, while only 3 pairs of twins in this group had IQ differences of less than 10 points. In the group with the smaller DSEA, all pairs of twins showed IQ differences of less than 10 points. In the larger DSEA group, 4 pairs of twins showed differences of 15, 17, 19 and 24 IQ points. . . . These analyses of twin data indicate greater differences in intelligence test scores between identical twins reared apart than Jensen acknowledges in his discussion; implied is a greater environmental contribution to the performance of even the most genetically similar individuals.

Interestingly enough, Burt's own data (unpublished) reveal a similar statistical connection between large IQ differences and large differences in occupational status. For example, in every case where the IQ difference between separated twins is at least 10 points and the two families differ by at least two grades in occupational status—five cases in all—the IQ difference is in the same direction as the difference in occupational status. In view of the limited relevance of occupational class to cognitive development in this study (illustrated by the fact that the second largest IQ difference in the study, 15 points, occurs for twins reared in families assigned to the same occupational category), we could hardly expect to find a stronger connection between differences in IQ and socioeconomic status. Burt and his colleagues have themselves examined correlations between the IQ differences of the separated twins and differences in indices relating to "cultural conditions" and "material conditions" in their environments. They report a positive correlation of 43 percent between differences in cultural conditions and differences in IQ *as measured by group tests* and a correlation of 74 percent between differences in cultural conditions and differences in scholastic attainment. No significant correlation was found between IQ differences and differences in material conditions, but differences in scholastic attainments did correlate with differences in material conditions (37 percent). Thus Burt's results, like those of other investigators, do support the thesis that cultural factors play an important part in the growth of cognitive skills.

The Hypothesis of Fixed Mental Capacity

So far we have been chiefly concerned with the arguments by which Jensen and other hereditarians have sought to establish the high heritability of IQ. We have seen that these arguments do not hold water. In the first place, the "heritability of IQ" is a pseudoconcept like "the sexuality of fractions" or "the analyticity of the ocean." Assigning a numerical value to the heritability of IQ does not, of course, make the concept more meaningful, any more than assigning a numerical value to the sexuality of fractions would make *that* concept more meaningful. In the second place, even if we had a theory of inheritance that could be applied to IQ test scores, we could not apply it to the correlation data employed by Jensen. A scientific theory, like a racing car, needs the right grade of fuel.

Jensen's data are to scientific data as unrefined petroleum is to high-test gasoline. Jensen and Herrnstein would have us believe that we can gain important insights into human intelligence and its inheritance by subjecting measurements that we do not understand to a mathematical analysis that we cannot justify. Unfortunately, many people appear to be susceptible to such beliefs, which have their roots in a widespread tendency to attribute magical efficacy to mathematics in almost any context. The perennial popularity of astrology is probably an expression of this tendency. Astrology is based, after all, on hard numerical data, and the success and internal consistency of its predictions are customarily offered as evidence for its validity. The most important difference between astrology and the Jensen-Herrnstein brand of intellectual Calvinism is not methodological but philosophical; one school believes that man's fate is written in the stars, the other that it runs in his genes.

Jensen's and Herrnstein's central thesis is that certain cognitive skills—those involving abstract reasoning and problem-solving—cannot be taught effectively to children with low IQs. From this thesis, and from it alone, flows all the disturbing educational, social, and political inferences drawn by these authors. If social and educational reforms could raise the general level of mental abilities to the point where people with IQs of 85 were able to solve calculus problems and read French, rank order on mental tests would no longer seem very important. It is precisely this possibility that Jensen's argument seeks to rule out. For if only a small fraction of the difference in average IQ between children living in Scarsdale and in Bedford-Stuyvesant can be attributed to environmental differences, it seems unrealistic to expect environmental improvements to bring about substantial increases in the general level of intelligence.

Even if Jensen's theoretical considerations and his analysis of data were beyond reproach, they would afford a singularly indirect means of testing his key thesis. The question to be answered is whether appropriate forms of intervention can substantially raise (*a*) the rate at which children acquire the abilities tested by IQ tests or (*b*) final levels of achievement. This question can be answered experimentally, and it has been. Since we do not yet know precisely what forms of intervention are most effective for different children, negative results (such as the alleged failure of compensatory education) carry little weight. On the other hand, all positive results are relevant. For if IQ can be substantially and consistently raised—by no matter what means—it obviously cannot reflect a fixed mental capacity.

At the very outset, we need to know something about how IQ varies in the course of development under "normal" conditions. This question has recently been answered by McCall, Appelbaum, and Hogarty (1973), who conducted a comprehensive longitudinal study of middle-class children. The study showed that "normal home-reared middle-class children change in IQ performance during childhood, some a substantial amount. . . . The average individual's range of IQ between 2½ and 17 years of age was 28.5 IQ points, one of every three children displayed a progressive change of more than 30 points, and one in seven shifted more than 40 points. Rare individuals may alter their performance as much as 74 points" (p. 70). According to this study, then, the average range of variation for middle-class children is almost two standard deviations. In other words, the normal developmental variability of IQ for a single individual actually exceeds the variability that prevails within the population at any given moment. These experimental findings flatly contradict the hereditarian thesis that IQ tests measure an innate trait, modified by relatively small fluctuations of nongenetic origin.

The study just quoted refers to "normal, home-reared, middle-class children." Do children born into less favorable environments exhibit comparable plasticity? The available evidence strongly suggests that they do. Skodak and Skeels (1949) found a 20-point mean difference between a certain group of adopted children and their biological mothers, whose environments were systematically poorer than those of the adoptive mothers. Another study by Skeels (1966) dealt with thirteen subnormal children who were "adopted" by the inmates of a state institution for mentally retarded females. The mean IQ of the children rose from 64 to 92, and the IQs of eleven of the children who were legally adopted rose subsequently to a mean of 101.

The Milwaukee Project of Rick Heber and his associates (1973) provides a dramatic demonstration of the efficacy of appropriate environmental modifications in accelerating the cognitive development of children with poverty backgrounds. In this study, now in its seventh year, a comprehensive family intervention program produced a sustained 30-point difference in IQ between an experimental group and a control group, each composed of twenty randomly selected children of mothers with tested IQs under 75.

Over a five-year period, the average IQ of the experimental group remained close to 125. The performance of children in the experimental group was evaluated through tests administered by psychologists not connected with the study.

Bronfenbrenner (1974, 1975) has reviewed a number of recent studies of early intervention. He finds that while disadvantaged children enrolled in highly structured, cognitively oriented, preschool programs consistently register substantial gains in IQ, these gains gradually disappear after the children enter school. He concludes that "even the best curriculum cannot immunize the child against developmental decline, once he is cast back into a consistently impoverished environment." On the other hand, home-based intervention programs, in which mothers are trained to interact with their children in ways expected to promote cognitive development, have been found to produce substantial and enduring gains in cognitive performance. "The earlier and more intensely mother and child were stimulated to engage in communication around a common activity, the greater and more enduring was the gain in IQ achieved by the child."

Teachers and therapists who work with children suffering neuropsychiatric disorders (including emotional and perceptual disturbances) regularly report large increases in their tested IQs. One remedial reading teacher of my acquaintance works exclusively with "ineducable" children. So far, she has not had a single failure; every one of her pupils has learned to read. And reading, of course, provides the indispensable basis for acquiring most of the higher cognitive skills.

IQ and Race

Jensen and others have argued that reported differences in average IQ between black and white children are probably attributable, in part, to systematic genetic differences. Jensen (1973) explicitly states the "rule of inference" used to draw this conclusion: "The probability that a phenotypic mean difference between two groups is in the same direction as a genotypic mean difference is greater than the probability that the phenotypic and genotypic mean differences are in opposite directions." This rule fails, however, when systematic effects, whose magnitude cannot be estimated, are known to contribute to the phenotypic mean differences. (Unfortunately, these are the only circumstances in which the rule might be useful.) In order to estimate the probabilities mentioned in the rule, one would need to estimate the probability that the observed phenotypic mean difference exceeds the contribution of systematic effects—which, by assumption, is impossible.

Among the relevant systematic differences between blacks and whites are cultural differences and differences in psychological environment. Both influence the development of cognitive skills in complex ways, and no one has succeeded in either estimating or eliminating their effects. "Culture-free" tests deal with his problem only on the most superficial level, for culture-free and "culture-bound" aspects of cognitive development are inseparable. The difficulties cannot be overcome by refined statistical analyses. As long as systematic differences remain and their effects cannot be reliably estimated, no valid inference can be drawn concerning genetic differences among races.

Many tests have shown that blacks living in the urban North score systematically higher on IQ tests than those living in the rural South. For many years hereditarians and environmentalists debated the interpretation of this finding. The environmentalists attributed the systematic IQ difference to environmental differences, the hereditarians to selective migration (they argued that the migrants could be expected to be more energetic and intelligent than the stay-at-homes). The environmental interpretation was decisively vindicated by O. Klineberg (1935), who showed that the IQs of migrant children increased systematically and substantially with length of residence in the North. In New York (in the early 1930s), migrant black children with eight years of schooling had approximately the same average IQ as whites. These important findings were fully confirmed by E. S. Lee (1951), who, fifteen years later, repeated Klineberg's experiment in Philadelphia. Additional studies bearing on IQ differences between ethnic groups are reviewed and analyzed by L. Plotkin (1971).

The Hypothesis of Unlimited Educability

That the growth of intelligence is controlled, in part, by genetic factors seems beyond doubt. The significant questions are, "What are these factors?" "How do they operate?" "How do they interact with noncognitive and environmental factors?" Experience suggests that children differ in the ease with which they acquire specific kinds of cognitive skills as well as in the intensity of their cognitive drives or appetites. But cognitive appetites, like other appetites, can be whetted or dulled. Nor are aptitude and appetite the only relevant factors. Everyone can cite case histories in which motivation has more than compensated for a deficit in aptitude. There are excellent skiers, violinists, and scientists who have little natural aptitude for any of these activities. None of them will win international acclaim, but few of them will mind. I know of no theoretical or experimental evidence to contradict the assumption that everyone in the normal range of intelligence could, if sufficiently motivated, and given sufficient time, acquire the basic cognitive abilities demanded by such professions as law, medicine, and business administration.

Once we stop thinking of human intelligence as static and predetermined, and instead focus our attention on the growth of cognitive skills and on how the interaction between cognitive, noncognitive, and environmental factors affects this growth, the systematic differences in test performance among ethnic groups appear in a new light. Because cognitive development is a cumulative process, it is strongly influenced by small systematic effects acting over an extended period. Information-processing ability grows roughly in the same way as money in a savings account: the rate of growth is proportional to the accumulated capital. Hence, a small increase or decrease in the interest rate will ultimately make a very large difference in the amount accumulated. Now, the "cognitive interest rate" reflects genetic, cultural, and social factors—all interacting in a complicated way. Membership in the Afro-American ethnic group is a social factor (based in part on noncognitive genetic factors) that, in the prevailing social context, contributes negatively to the cognitive interest rate. The amount of the negative contribution varies from person to person, being generally greatest for the most disadvantaged. But there is no doubt that it is always present to some extent. In these circumstances, we should expect to find exactly the kind of group differences that we do find. I think it is important to take note of these differences. They are valuable indices of our society's persistent failure to eradicate the blight of racism.

It may be that the assumption of unlimited educability will one day be shown to be false. But until then, it could usefully be adopted as a working hypothesis by educators, social scientists, and politicians. We have seen that the widely held belief in fixed mental capacity as measured by IQ has no valid scientific basis. As a device for predicting scholastic success (and thereby for helping to form the expectations of teachers, parents, and students), as a criterion for deciding that certain children should be excluded from certain kinds of education, and as a lever for shifting the burden of scholastic failure from schools and teachers to students, the IQ test has indeed been, in Herrnstein's words, "a potent instrument"—potent and exceedingly mischievous.

Admirers of IQ tests usually lay great stress on their predictive power. They marvel that a one-hour test administered to a child at the age of eight can predict with considerable accuracy whether he will finish college. But, as Burt and his associates have clearly demonstrated, teachers' subjective assessments afford even more reliable predictors. This is almost a truism. If scholastic success is to be predictable, it must be reasonably consistent at different age levels (otherwise, there is nothing to predict). But if it is consistent, then it is its own best predictor. Johnny's second-grade teacher can do at least as well as the man from ETS. This does not mean that mental tests are useless. On the contrary, sound methods for measuring information-processing ability and the growth of specific cognitive skills could be extremely useful to psychologists and educators—not as instruments for predicting scholastic success, but as tools for studying how children learn and as standards for assessing the effectiveness of teaching methods.

Conclusions

To what extent are differences in human intelligence caused by differences in environment, and to what extent by differences in genetic endowment? Are there systematic differences in native intelligence among races or ethnic groups? Jensen, Herrnstein, Eysenck, Shockley, and others assure us that these questions are legitimate subjects for scientific investigation; that intelligence tests and statistical analyses of test results have already gone a long way toward answering them; that the same techniques can be used to reduce still further the remaining uncertainties; that the results so far obtained clearly establish that differences in genetic endowment are chiefly responsible for differences in performance on intelligence tests; that reported differences in mean IQ between Afro- and Europo-Americans may well be genetically based; and that educational, social, and political policy decisions should take these "scientific findings" into account. We have seen, however, that the arguments put forward to support these claims are unsound. IQ scores and correlations are not measurements in any sense known to the natural sciences, and "heritability estimates" based on them have as much scientific validity as horoscopes. Perhaps the single most important fact about human intelligence is its enormous and, as yet, ungauged capacity for growth and adaptation. The more insight we gain into cognitive development, the less meaningful seems any attempt to isolate and measure differences in genetic endowment—and the less important. In every natural science, there are certain questions that can profitably be asked at a given stage in the development of that science, and certain questions that cannot. Chemistry and astronomy grew out of attempts to answer the questions "How can base metals be transmuted into gold?" "How do the heavenly bodies control human destiny?" Chemistry and astronomy never answered these questions; they outgrew them. Similarly, the development of psychology during the present century has made the questions posed at the beginning of this paragraph seem increasingly sterile and artificial. Why, then, are they now being revived? Earlier in this article I suggested that a combination of cultural, historical, and political factors tempts us to seek easy "scientific" solutions to hard social problems. But this explanation is incomplete. It leaves out a crucial psychological factor: once we have acquired a skill, we find it hard to believe that it was not always "there," a latent image waiting to be developed by time and experience. The complex muscular responses of an expert skier to a difficult trail are, to him, as instinctive as a baby's reaction to an unexpected loud noise. For this reason, the doctrine of innate mental capacity exercises an intuitive appeal that developmental accounts can never quite match. This however, makes it all the more important to scrutinize critically the logical, methodological, and psychological underpinnings of that doctrine.

Note

This essay was originally published in *Cognition* 1 (1972):265–300. The present revision incorporates material from subsequent articles in *Cognition* 1 (1972):453–473, and *Science* 183 (1974):1259–1266, and has benefited greatly from criticisms (of earlier versions) and suggestions by Professor James F. Crow, Professor Irven De Vore, Professor Everett Dempster, Professor I. Michael Lerner, Professor Lawrence Plotkin, Professor Sandra Scarr-Salapatek, Professor Robert L. Trivers, Dr. Jean Carew Watts, and many other kind friends and critics. I acknowledge with gratitude their help and encouragement.

References

BRONFENBRENNER, U. 1974. *Is Early Intervention Effective?* Washington, D.C.: Department of Health, Education and Welfare, Office of Child Development.

———. 1975. "Is Early Intervention Effective?" Condensed version reprinted in *Race and IQ*, ed. Ashley Montagu, New York: Oxford University Press.

BURT, C. 1966. "The Genetic Determination of Differences in Intelligence: A Study of Monozygotic Twins Reared Together and Apart." *British Journal of Psychology* 57, nos. 1–2:137–153.

BURT, C., and HOWARD, M. 1956. "The Multifactorial Theory of Inheritance and Its Application to Intelligence." *British Journal of Statistical Psychology* 9:95–131.

DEUTSCH, M. 1969. "Happenings on the Way Back to the Forum." *Harvard Educational Review* 39:423–557.

ERLENMEYER-KIMLING, L., and JARVIK, L. F. 1964. "Genetics and Intelligence." *Science* 142:1477–1479.

EYSENCK, H. J. 1971. *The IQ Argument: Race, Intelligence and Education*. New York: Library Press.

HEBER, R., GARBER, H., HARRINGTON, S., and HOFFMAN, C. "Rehabilitation of Families at Risk for Mental Retardation." Unpublished research report.

HERRNSTEIN, R. J. 1971. "IQ." *Atlantic Monthly*, September 1971, pp. 43–64.

HOGBEN, L. T. 1933. *Nature and Nurture*. New York: W. W. Norton & Co.

JENCKS, C., SMITH, M., ACLAND, H., BANE, M. J., COHEN, D., GINTIS, H., HEYNS, B., and MICHELSON, S. 1972. *Inequality: A Reassessment of the Effect of Family and Schooling in America*. New York: Basic Books.

JENSEN, A. R. 1969. "How Much Can We Boost IQ and Scholastic Achievement?" *Harvard Educational Review* 39:1–123.

———. 1973. "The IQ Controversy: A Reply to Layzer." *Cognition* 1:427–452.

———. 1974. "Kinship Correlations Reported by Sir Cyril Burt." *Behavior Genetics* 4:1–28.

JUEL-NIELSEN, N. 1965. "Individual and Environment: A Psychiatric-psychological Investigation of Monozygous Twins Reared Apart." *Acta Psychiatrica et Neurologica Scandinavica* Monograph suppl. 183. Copenhagen: Munksgaard.

KAMIN, L. J. 1974. *The Science and Politics of I.Q.* Potomac, Md.: Erlbaum Associates.

KAMIN, L. 1976. "Heredity, Intelligence, Politics and Psychology: I," published in this volume, pp. 242–264.

KLINEBERG, O. 1935. *Negro Intelligence and Selective Migration.* New York: Columbia University Press.

LAYZER, D. 1974. "Heritability Analyses of IQ Scores: Science or Numerology?" *Science* 183:1259–1266. Reprinted in Montagu, *Race and IQ.*

LEE, E. S. 1951. "Negro Intelligence and Selective Migration: A Philadelphia Test of the Klineberg Hypothesis." *American Sociological Review* 16:227–233.

LERNER, I. M. 1968. *Heredity, Evolution and Society.* San Francisco: W. H. Freeman & Co.

LEWONTIN, R. C. 1970. "Race and Intelligence." *Bulletin of the Atomic Scientists* 26, no. 3:2–8. Reprinted in Montagu, *Race and IQ,* and in this volume, pp. 78–92.

MCCALL, R. B., APPELBAUM, M. I., and HOGARTY, P. S. 1973. Developmental Changes in Mental Performance. *Child Development Monographs* 38, no. 3:1–84.

NEWMAN, H. H., FREEMAN, F. N., and HOLZINGER, K. J. 1937. *Twins: A Study of Heredity and Environment.* Chicago: University of Chicago Press.

PIAGET, J. 1952. *The Origins of Intelligence in Children.* New York: W. W. Norton & Co.

PIAGET, J. 1967. *Six Psychological Studies.* New York: Random House.

PLOTKIN, L. 1971. "Negro Intelligence and the Jensen Hypothesis." *New York Statistician* 22:3–7.

ROE, A. 1953. *The Making of a Scientist.* New York: Dodd, Mead & Co.

SHIELDS, J. 1962. *Monozygotic Twins Brought Up Apart and Brought Up Together.* London: Oxford University Press.

SKEELS, H. M. 1966. "Adult Status of Children with Contrasting Early Life Experiences: A Follow-up Study." *Child Development Monographs* 31, no. 3 (serial no. 105).

SKODAK, M., and SKEELS, H. M. 1949. "A Final Follow-up Study of One Hundred Adopted Children." *Journal of Genetic Psychology* 75:85–125.

WADDINGTON, C. H. 1957. *The Strategy of the Genes.* London: George Allen & Unwin.

Heredity, Intelligence, Politics, and Psychology: I

Leon J. Kamin

IN THIS SECTION I AM GOING TO DISCUSS WHAT EVIDENCE THERE IS FOR the nonzero heritability of IQ. The relevant literature is very large, and I shall be concerned here with only two classes of studies, those involving separated identical twins, and those involving adopted children. These two types of studies, by general agreement, provide the most powerful evidence for the heritability of IQ. They involve relatively few assumptions, and are conceptually simple to understand. The IQ correlation between separated MZ twins, e.g., itself provides an unbiased estimate of the heritability of IQ under the assumptions that such twins are representative of the population, and that the environments in which they have been reared are uncorrelated.

There have been four major studies of separated MZ twins. The essential data of these studies are presented in table 1. Though the table indicates some discrepancy in the magnitudes of the reported correlations, the major conclusion is clear. The separated twins in all four studies do resemble one another in IQ very substantially, thereby suggesting a predominant role for heredity.

The most important of the four twin studies is that of the late Sir Cyril Burt, recipient, in 1971, of the American Psychological Association's Edward Lee Thorndike Award. Burt's study involved the largest number of twin pairs, and reported the largest correlation; more important, it is the only study which purports to provide quantitative data on the socioeconomic status of the homes in which separated twins were reared. For Burt's twins, at least, there was no detectable correlation between the statuses of the homes in which members of a separated pair were raised.

Table 1. *IQ Correlations for Separated MZ Twins*

Study	*Test*	*Correlation*
Burt, 1966	"Individual Test"	.86 (N = 53)
Shields, 1962	Dominoes + [2 × Mill Hill]	.77 (N = 37)
Newman et al., 1937	Stanford-Binet	.67 (N = 19)
Juel-Nielsen, 1965	Wechsler	.62 (N = 12)

There are, it must be reported, a number of unresolved procedural ambiguities in Burt's published papers. The 1943 review of his work, e.g., presents a large number of relevant correlation coefficients, but virtually nothing is said of when, or to whom, tests were administered, or of what tests were employed. The reader is told only: "Some of the inquiries have been published in L.C.C. reports or elsewhere; but the majority remain buried in typed memoranda or degree theses" (Burt, 1943a).

This lack of procedural detail is unfortunate, since a cross-check of several of Burt's papers, as table 2 indicates, reveals a number of puzzling inconsistencies—as well as a number of astonishing consistencies. The figure indicates that, in 1943, Burt reported a correlation for a large sample between "intelligence and economic status." There was no clear indication of how either intelligence or economic status had been measured, but Burt took pains to indicate that the correlation involved economic, as opposed to cultural, status. The same survey was referred to again in 1956 (Burt and Howard, 1956); but now, it becomes clear that the correlation—involving "socioeconomic status"—was based upon "adjusted assessments" of intelligence, rather than upon "crude test-results." Then, one year later, the *same* correlation from the *same* survey was de-

Table 2. *Correlations from Burt*

Burt, 1943a	Intelligence and economic status ("Economic" differentiated from "cultural")	$r = .32$
Burt and Howard, 1956	Intelligence and socioeconomic status	
	a) adjusted assessments	$r = .315$
	b) crude test results	$r = .453$
Burt and Howard, 1957b	Adjusted assessments of IQ and economic and cultural conditions	
	a) with "cultural [i.e. educational and motivational] background"	$r = .315$
	b) with "material [i.e. financial and hygenic] conditions"	$r = .226$

scribed as involving *"cultural"* status, and an entirely *different* correlation was presented for "economic" status (Burt and Howard, 1957b). From where did this latter correlation emerge? Why was it not utilized in 1943, when Burt wished to differentiate economic from cultural status? How did the measurement of one magically transmute to a measurement of the other over fourteen years? We are not told.

Burt collected correlations for various categories of kinship throughout a lengthy research career, continually cumulating cases. The correlations, based upon increasingly larger Ns, were sporadically reported in a number of his papers. The correlations often exhibited an extraordinary stability. Table 3 presents correlations for the category "siblings reared apart," taken from Burt's 1955 and 1966 papers. The addition of twenty new cases over that eleven-year period, it will be noted, did not change to the third decimal place three separate correlations for school attainments, or the correlations for height and weight. The correlations were similarly robust in the face of unexplained *decreases* in sample size, as table 4 makes clear. The disappearance of forty-five DZ twin pairs over the same period during which twenty separated sibling pairs were located again failed to affect correlations for school attainments and physical traits, though intelligence correlations were affected.

Table 3. *Correlations from Burt*
Siblings Reared Apart

	1955 (N = 131)	*1966 (N = 151)*
Intelligence		
Group Test	.441	.412
Individual Test	.463	.423
Final Assessment	.517	.438
School Attainment		
Reading, Spelling	.490	.490
Arithmetic	.563	.563
General	.526	.526
Physical		
Height	.536	.536
Weight	.427	.427

Table 4. *Correlations from Burt*

DZ Twins Reared Together		
	1955 (N = 172)	*1966 (N = 127)*
Intelligence		
Group Test	.542	.552
Individual Test	.526	.527
Final Assessment	.551	.453
School Attainment		
Reading, Spelling	.915	.919
Arithmetic	.748	.748
General	.831	.831
Physical		
Height	.472	.472
Weight	.586	.586

This type of stability also characterized the unspecified "group test" of intelligence used by Burt in his studies of separated twins, which apparently formed a basis for his "final assessments" of the twins' IQs. Table 5 adds to the two papers already cited two additional "interim reports" (Burt, 1958a; Conway, 1958). With the exception of a minor perturbation in late 1958, which simultane-

Table 5. *Correlations from Burt*
"Group Test" of Intelligence

	MZ Twins Reared Apart	*MZ Twins Reared Together*
Burt, 1955	.771 (N = 21)	.944 (N = 83)
Burt, 1958a	.771 (N = "over 30")	.944 (N = ?)
Conway, 1958	.778 (N = 42)	.936 (N = ?)
Burt, 1966	.771 (N = 53)	.944 (N =95)

ously afflicted the correlations both for twins reared apart and for twins reared together, the IQ correlations for both categories remained identical to three decimal places over the entire series of Burt's cumulated researches.

There is considerable confusion in Burt's reports concerning the relation between "tests" and "assessments" of intelligence. Table 6

Table 6. *How Burt Assessed the Intelligence of Adults*

Burt and Howard, 1957b	"But in each of our surveys, assessments were individually obtained for a representative sample of parents, checked, for purposes of standardization, by tests of the usual type." They refer the reader to a 1955 paper, p. 172.
Burt and Howard, 1956	They report correlations for 963 parent-child pairs and 321 grandparent-grandchild pairs. "The procedures employed, and the results obtained have already been described in previous publications (Burt, 1955)."
Burt, 1955, p. 172, footnote	"For the assessments of the parents we relied chiefly on personal interviews; but in doubtful and borderline cases an open or a camouflaged test was employed."

presents three quotations, all referring to the same survey involving adult "IQs." When the assiduous reader at length tracks down the 1955 footnote, it becomes clear that the IQ correlations discussed by Burt in 1956 and 1957 were based primarily upon "personal interviews" of adults. The spectacle of Professor Burt administering a "camouflaged test" of intelligence to a London grandparent is an amusing one, but it does not inspire scientific confidence. Nor is that confidence bolstered by the observation that, over two years, Professor Burt's memory transformed "doubtful and borderline cases" to "a representative sample of parents," and "an open or a camouflaged test" to "tests of the usual type."

We have a somewhat clearer picture of how the "final assessments" of children's IQs—including the twins'—were arrived at. Burt wrote: "The final assessments for the children were obtained by submitting the marks from the group tests to the judgment of the teachers . . . where the teacher disagreed with the verdict of the marks, the child was interviewed personally, and subjected to further tests, often on several successive occasions" (Burt, 1958a).

The rationale for depending upon such adjusted assessments, rather than raw test scores, was presented in some detail by Burt and Howard (1957a). They argued straightforwardly that, in testing the goodness of fit of a multifactorial model of inheritance, the best possible estimate of genotypic intelligence should be employed; and that the teacher's judgment was a more accurate estimate than that provided by an intelligence test score. This confidence in the teacher as a genotype-detector, however, was not always adhered to by Burt, as table 7 indicates. The contradictory 1943 quotation (Burt, 1943b) reflects Burt's concern that, in postwar Britain, the genetically most gifted should be selected for university preparation and training. For such an important practical purpose, only the science of mental testing—and not the teacher—could be entrusted to sniff out the superior genotypes.

For the twin studies, however, Burt reported correlations based not only upon group test and final assessment, but also upon individual test. The most extended discussion of the tests given to the twins was provided in 1966, and is reproduced in table 8. The description is shrouded in ambiguities; and unfortunately, it is impossible to know what group or individual tests were administered to which twins. The references supplied by Burt are of no help. One contains *no* group intelligence tests, while the other contains

Table 7.

Burt and Howard, 1957a	"We . . . are perfectly willing to admit that, as a means of estimating genotypic differences, even the most carefully constructed tests are highly fallible instruments, and that their verdicts are *far less trustworthy* than the judgments of the pupil's own teachers . . ."
Burt, 1943b	"But in regard to innate general ability there can be no question: the unaided judgments even of the most experienced teachers, shrewd as they are in many cases, are nevertheless *far less trustworthy* in the long run than the results obtained with properly applied intelligence tests."

Emphasis the author's.

Table 8. *What "Individual Test" Did Burt Give to Separated MZ Twins*

Jensen, 1970	"Their IQ's were obtained from an individual test, the English adaptation of the Stanford-Binet. . . ."
Burt, 1966	"The tests employed have been fully described elsewhere (Burt, 1921; 1933). . . . They consisted of (i) a group test of intelligence containing both non-verbal and verbal items, (ii) an individual test (the London Revision of the Terman-Binet Scale) used primarily for standardization, and for doubtful cases (iii) a set of performance tests . . . standardized by Miss Gaw (1925). The test results . . . were submitted to the teachers for comment or criticism; and wherever any question arose, the child was reexamined."
Burt, 1958b	"[A critic's] main point is that the correlations we obtained with our individual tests (.843 for identical twins reared apart . . .) display a wider divergence than those reported by Newman . . . obtained with the old Stanford-Binet . . . the figures he quotes from our own research were based on non-verbal tests of the performance type."

no fewer than seven; but each of these contains exclusively verbal items. With regard to these seven group tests, Burt wrote: "complete tables of age-norms would be unnecessary or even misleading . . . I give only rough averages calculated regardless of sex. . . . I have not thought it worth the necessary time and space to elaborate and print a set of standardised instructions as to procedure or marking" (Burt, 1921).

Presumably, it was the 1966 description which led Jensen (1970) to conclude that the twin "IQ scores" given to him by Burt were raw scores from a Binet test. But as the figure indicates, Burt (1958b) had earlier asserted—in apparent contradiction to his 1966 description—that the figures reported in his tables under the heading "Individual test" were "based on non-verbal tests of the performance type." There is no way out of this morass, except possibly to examine Burt's raw data. That is what Professor Jensen has now attempted to do, but he has recently reported: ". . . alas, nothing remained of Burt's possessions . . . unfortunately, the original data are lost, and all that remains are the results of the statistical analyses. . . ."

There is somewhat more information available on the set of performance tests standardized by Miss Gaw (1925). The standardization sample consisted of "100 pupils in London schools," none of them "of scholarship or central school ability." There were "striking" differences between the sexes in test scores. The reliability of the tests was .76 for boys and .54 for girls.

The use in twin studies of tests which have not been adequately standardized for age and for sex is a *very* serious matter—particularly when, as in Burt's case, no information whatever is provided about the sexes or ages of the twins. Recall that identical twins are necessarily of the same age and same sex. Thus, if the measure on which members of a twin pair are compared varies with either sex or age, and if twin pairs of both sexes and of varying ages are included, the IQ correlation between twins is utterly confounded with sex and age. The close resemblance of twins in IQ under such circumstances is scarcely an unambiguous testament to the genetic determination of IQ variation.

We should comment finally on the unique virtue of Burt's study—the provision of quantitative socioeconomic class data. Here too there are problems. In 1959, when only forty-two separated twin pairs were available, it was clearly indicated that *at least four* children of "professional" parents had been reared in "orphanages"

(Conway, 1959); but, in 1966, with the sample size increased to fifty-three pairs, it was reported that *precisely two* children of such parents had been reared in "residential institutions." Furthermore, a comparison of the marginal totals in Burt's 1966 table with the individual socioeconomic data for the *same* twins which Burt later gave to Jensen and Shockley, indicates that, in *at least* six cases, the classification of a twin was changed *after* 1966. There is also clear evidence that, at least in the case of two twin pairs, the IQs reported were changed after 1966. The twin data were collected by Burt over a period of some fifty years, and it seems quite possible that assessments both of intelligence and of socioeconomic class were subjected to a continuing process of revision and refinement.

The conclusion seems to me inescapable, and I can only regret that time does not permit fuller documentation—which exists in abundance. The numbers left behind by Professor Burt are simply not worthy of serious scientific attention.

The separated twin study by Shields (1962), unlike Burt's, is replete with procedural details. The text (and especially the Appendix), however, indicate some problems. Shields's correlations are not based upon IQs as such, but upon a "total intelligence score" which pooled the raw scores on two separate tests. The raw score on the Mill Hill vocabulary test was multiplied by two before being added to the raw Dominoes score, since the standard deviation of the Mill Hill was about half that of the Dominoes. The Dominoes had been standardized upon a British Army population, and not upon civilians or upon women. The majority of Shields's separated twins were older females. The standardization of the Mill Hill indicated large age effects on raw score, but these were ignored by Shields. There is evidence within Shields's data for significant sex differences in Dominoes score, and for significant differences between separated and nonseparated twins not only in mean test scores, but also in variances.

The Appendix provides a considerably more detailed glimpse than does Burt of what kinds of cases constitute the category "separated twins." These examples, if a bit extreme, are not wholly atypical. Benjamin and Ronald, separated at nine months: "Both brought up in the same fruit-growing village, Ben by the parents, Ron by the grandmother. . . . They were at school together. . . . They have continued to live in the same village." Jessie and Winifred, separated at three months: "Brought up within a few hundred yards of one another. . . . Told they were twins after girls discovered it for themselves, having gravitated to one another at school at the age of 5. . . . They play together quite a lot . . . Jessie often goes to tea with Winifred. . . . They were never apart, wanted to sit at the same desk. . . ." Bertram and Christopher, separated at birth: "The paternal aunts decided to take one twin each and they have brought them up amicably, living next-door to one another in the same Midlands colliery village. . . . They are constantly in and out of each other's houses." That is not, I fancy, the sort of separation conjured up by readers of secondary sources which present the twin data.

The Shields Appendix makes possible a number of calculations which seem to me to have theoretical significance. Some examples are indicated in table 9. Shields presents intelligence scores for

Table 9. *From Shields*

Mean co-twin score difference	35 pairs tested by Shields	8.5*
Mean co-twin score difference	5 pairs *not* tested by Shields	22.4
Intelligence correlation	27 pairs reared in related families	.83*
Intelligence correlation	13 pairs reared in unrelated families	.51

* $p < .05$.

thirty-five separated pairs in which he had personally tested each member; for five remaining pairs, the twins had been examined by two different psychologists (one usually Shields). The twin pairs tested by the same examiner resembled one another more closely in intelligence to a statistically significant degree. It is of some interest to note that, following Jensen's arbitrary procedure for converting Shields's scores to IQs, the mean IQ difference between twins tested by different examiners exceeds 17 points. That is, the mean difference theoretically expected when pairing individuals entirely at random. Thus, it seems very likely that all separated twin studies are afflicted by an unconscious experimenter bias introduced during the administration and/or scoring of the tests. (The same tendency may be at work in the data of Churchill (1965), who reported raw Wechsler IQ scores for thirteen pairs of nonseparated monochorionic twins. The correlation between twins can be computed as

.995. When, again following Jensen, one corrects for unreliability of the test (assuming a .95 reliability), the true correlation between monochorionics stands revealed as 1.05.)

The Shields Appendix also indicates that, in twenty-seven cases, the two separated twins were reared in related branches of the parents' families; only in thirteen cases were the twins reared in unrelated families. The twins reared in related families resembled one another more closely to a statistically significant degree. That is scarcely evidence for an overwhelming genetic determination of IQ scores. Furthermore, the relatively modest correlation of .51 observed in twins reared in unrelated families must in no sense be taken as an estimate of what might be observed if twins were assigned to families at random. The typical case of "unrelated families" was one in which the mother kept one twin, and gave the other to "friends of the family." For his study as a whole, Shields reported, "Large differences in social class do not occur often. . . ." This source of bias, of course, cumulates with biases introduced by imperfectly standardized tests and by experimenter expectation.

The only American study in this series, by Newman, Freeman, and Holzinger (1928), was also the first to be reported. The study involved only nineteen twin pairs, gathered largely by mail response to newspaper and radio appeals for volunteers. We shall pass over the bias introduced by exclusion from the study of volunteer pairs who, by mail, indicated that they were not "very much alike." The detailed analysis of the Newman et al. data affords a clear illustration of the way in which the confounding of IQ score with age may affect the interpretation of twin data.

For one hundred *nonseparated* MZ individuals located in the Chicago schools, Newman et al. reported a correlation of −.49 between Stanford-Binet IQ and age. Their *separated* twins, however, were mostly adults, for whom an uncorrected intraclass IQ correlation of .67 was reported. This correlation was doubly "corrected" by McNemar (1938)—both for age and for restriction of range among the separated twins. The double correction raised the correlation from .67 to .77, with the latter figure by far more commonly cited in textbooks. (The .77 figure has recently received a *third* correction from Jensen [1969]; when corrected for test unreliability, it rises to .81.)

Table 10 indicates, however, some of the problems associated with "correcting for age." The correlations between age and IQ are vastly different—significantly so—for the male and the female sepa-

Table 10. *From Newman, Freeman, and Holzinger*

Correlations of IQ with Age:	
For 14 males, aged 13½–27	$r = -.78^{**}$
For 24 females, aged 11½–59	$r = -.11$
For 6 females, aged 15–27	$r = -.60$
For 14 females, aged 30–59	$r = -.27$
For 20 (mixed-sex) individuals, aged 13½–27	$r = -.58^{**}$

** $p < .01$.

rated twins. This apparent sex difference is, in turn, confounded with the fact that the ages of most of the female twins lie outside the range of the seven male pairs. The figure indicates that, within the age range included by the male twins, substantial age-IQ correlations can be detected for both males and females; but for the older subjects, all of whom happen to be female, the age-IQ correlation appears to drop substantially. Partial correlation corrective techniques are simply not justified in such a situation. There is no elegant way of removing the biasing effect of age in these data from the estimation of twin resemblance produced by identical genes—or by correlated environments.

Table 11 indicates an extraordinarily inelegant procedure which may serve at least some illustrative function. The ordinary intraclass correlational procedure used with twin data of course pairs the IQ scores of each set of twins. The "pseudopairing" procedure illustrated first sets down the scores of all twin pairs, with the order of entry corresponding exactly to the ages of the twins. The twins are then broken up into clusters, each cluster consisting of two twin sets immediately adjacent in age. (When the number of twin sets in the sample is odd, the set midmost in age is discarded.) Then, within each cluster, four pseudopairings are made. Each score is paired with each other score in the cluster—*except that* pairings of the scores of actual twins are *omitted*. Finally, the intraclass correlation is computed for the pseudopairings.

The column of pseudopairings, it will be noted, contains precisely the same numbers as does the column of true pairings, but each number appears twice among the pseudopairings. The virtue of the pseudopairing procedure is obvious. The pseudopairings remove all genetic effects from the correlation, and the only sys-

Table 11. *Pseudopairing Procedure*

	Intraclass		*Pseudopairings*	
	IQ	*IQ*	*IQ*	*IQ*
Pair A, age 27	96	77	96	66
			77	78
Pair B, age 26.7	66	78	96	78
			77	66
Pair C, age 26	91	90	91	99
			90	101
Pair D, age 23	99	101	91	101
			90	99
	etc.		etc.	

Intraclass, 7 male NFH pairs:	$r = .58$
Pseudopaired intraclass, same group:	$r = .67$
Intraclass, 10 NFH pairs aged 13.5–27:	$r = .65$
Pseudopaired intraclass, same group:	$r = .47$

tematic bias introduced is that all pseudopairings involve individuals quite close together in age. Thus, although the correlation computed from the pseudopairings cannot be assessed for statistical significance, it does provide—without the assumptions underlying partial correlation—an estimate of the resemblance between twins to be expected solely on the basis of their shared age. The table indicates that, for the Newman et al. male sample, that estimate is in fact *higher* than the actual IQ correlation. For the mixed-sex sample consisting of pairs being within the same age range as the males, the estimate, though substantial, is lower than the actually observed correlation.

The final separated twin study, conducted by Juel-Nielsen (1965) in Denmark, involved only twelve pairs. The test employed was the Wechsler scale for adults. That test, as the author indicated, had never been standardized on a Danish population. Perhaps some of the peculiarities in her data, present in table 12, reflect that unfortunate fact. Her male subjects had significantly higher IQs than her female subjects. There were significant correlations between age and IQ for each sex—but the two significant correlations

Table 12. *From Juel-Nielsen*

Mean IQ, 18 females = 103; mean IQ, 6 males = 113*	
Correlation, IQ × age, 18 females	$r =$ .61**
Correlation, IQ × age, 6 males	$r = -$ .82*
Intraclass IQ correlations, 9 female pairs	$r =$ .59
Pseudopaired intraclass, same group	$r =$ .59

* $p < .05$.
** $p < .01$.

were opposite in sign! (This, in turn, is confounded with the fact that two of her three male pairs fell outside the age range of her nine female pairs.) The pseudopairing technique is applicable only to the relatively "large" female sample. For that sample, the estimate of the correlation produced by age resemblance alone is identical to the actually observed IQ correlation.

The confounding of age with IQ score, it should be noted, is not confined to very old IQ tests employed in exotic foreign studies. Table 13 presents correlations computed from raw data presented

Table 13. *Recent Data on the Age-IQ Confound in Twins*

	Babson et al., 1964 (Stanford-Binet)	*Willerman and Churchill, 1967 (WISC Verbal)*
For Monozygotics:		
Intraclass IQ correlation	.83** (9 pairs)	.82** (14 pairs)
Age × IQ correlation	.59** (N = 18)	.59** (N = 28)
Pseudopaired intraclass	.28	.77
For Dizygotics:		
Intraclass IQ correlation	.65* (7 pairs)	—
Age × IQ correlation	.75** (N = 14)	—
Pseudopaired intraclass	.44	
Pooling All MZ's:		
IQ correlation	.87** (23 pairs)	
Pseudopaired correlation	.67	

* $p < .05$.
** $p < .01$.

in two relatively recent American studies (Babson et al., 1964; Willerman and Churchill, 1967). The studies employed the latest versions of the Stanford-Binet and of the WISC, with samples of nonseparated MZ twins. The agreement between the studies in the IQ correlation for twins is remarkable; but the agreement between them in the highly significant correlation between age and IQ is even more remarkable. The age-IQ correlation is also observed in the single sample of DZ twins. The pseudopairing technique again indicates that a substantial proportion of the IQ correlation between twins may be simply attributable to age. For the pooled MZ sample, the observed IQ correlation was .87. The estimate derived by the pseudopairing technique was .67.

From the four separated twin studies reviewed above, Jensen (1970) concluded: "The overall intraclass correlation . . . .824 . . . may be interpreted as an upperbound estimate of the heritability of IQ in the English, Danish, and North American populations sampled in these studies." My own conclusion is different. I see no unambiguous evidence whatever in these studies for *any* heritability of IQ test scores. The studies do contain clear data which demonstrate the importance of correlated environments in determining the IQ resemblance of so-called separated twins, and there are strong suggestions that unconscious experimenter expectations inflate the reported correlations. There is also very strong evidence in these studies that either our leading IQ tests are incredibly badly standardized or that general population norms do not apply to twins, or that the twin samples studied by psychologists are bizarre —or all three.

The twin studies, however, are often said to be cross-validated by other types of evidence suggesting a high heritability of IQ. Particularly, the study of adopted children has been asserted by Vandenberg (1971) to provide "the strongest evidence possible for hereditary factors in intelligence." We turn now to that strong evidence.

There have been four major IQ studies of adoptive children— Freeman, Holzinger, and Mitchell (1928), Burks (1928), Leahy (1935), and Skodak and Skeels (1949). The four studies agree in one major particular. The IQ correlation observed between adoptive child and adoptive parent was invariably much lower than that normally observed between biological parent and child. That has been presumed to reflect the fact that biological parent-child pairs share not only a common environment, but also, and much more importantly, common genes.

The Freeman et al. study observed only adoptive families; there were, within the study itself, no natural families to which the adoptive families could be compared. This deficiency was corrected in both the Burks and Leahy studies, which included "matched" control groups of natural families. The Burks study, indeed, according to Jensen (1969), contained "a perfectly matched control group of parents rearing their own children."

With reflection, however, it may seem doubtful whether adoptive and natural families could be "perfectly" matched for very many variables. The Burks study (like Leahy's), in fact, matched families, imperfectly, only with respect to the following: father's occupation and "type of neighborhood;" subject child's age and sex; nonseparation of living parents; and exclusion of blacks, Jews, and southern Europeans. That matching was not sufficient to control for a number of theoretically relevant variables, as the comparisons in table 14 make clear.

Table 14. *Comparisons Between "Matched" Adoptive and Control Homes*

	Adoptive	*Control*
From Leahy, 1935:		
Home's "environmental status score"	137.9	118.7 ($p < .001$)
Correlation, husband and wife IQ	.57	.41*
Correlation, husband and wife education	.59	.71*
From Burks, 1928:		
Father's mean age	45.9	41.0 (sig.)
Mother's mean age	41.0	36.1 (sig.)
Number children in home	1.5	2.3 (sig.)
Mean family income	$6,200	$4,100 (sig.)
Mean value of home	$13,200	$9,500 (sig.)

* $p < .05$.

Though the Leahy report does not include equivalent data, Burks's data demonstrate the obvious—foster parents are older and have fewer children in their homes than do natural parents. They are also, in the Burks study, some 50 percent wealthier, and live in much more expensive homes—though they have been matched to the control parents for "type of neighborhood." Although Burks's adoptive parents lived in more expensive homes and with more money than did her control parents, she could detect no difference in the cultural environments of the homes. However, Leahy, who reported no financial data, did report a significant difference in the "environmental status scores" of the homes.

The most interesting differences between the two types of families are those reported without significance tests by Leahy, involving correlations *between husband and wife.* The resemblance between spouses is significantly different in the two types of families—both for IQ and for amount of education. For IQ, spouses in adoptive families resemble one another more than do spouses in control families; but for education, they resemble one another less! That is to say, the interrelations among variables which may themselves relate to the child's IQ differ in the two types of families. The foster families are a highly selected group and evidently provide a special kind of family environment. The entire nexus of environmental IQ determinants appears to differ across the two types of families. To the extent that this is true, a comparison of parent-child correlations across the two types of families is simply inappropriate.

The studies do contain, however—without comment—some nuggets of very relevant data. There are some adoptive parents who have, in addition to their adopted child, a *biological child* of their own. The relevant question to ask is, what is the IQ correlation between *such* children and their parents? How does it compare to the correlation between the *same* parents and their adopted children, and how does it compare to the correlation between "control" children and parents?

The entirety of the available data is presented in table 15. The correlations in the table are for child and midparent, since that is the only form in which data for the biological children of adoptive parents are presented in these studies. The midparent-child correlation is typically considerably larger than the correlation between child and a single parent.

The first point to be made about table 15 is that, within each column, there is no significant heterogeneity among the correla-

Table 15. *IQ Correlations, Adoption Studies*

	Control Child × True Midparent	*Adoptive Child × Adoptive Midparent*	*Own True Child × Adoptive Midparent*
Freeman et al., 1928	—	.39** (N = 169)	.35* (N = 28)
Burks, 1928	.52** (N = 100)	.20* (N = 174)	—
Leahy, 1935	.60** (N = 173)	.18* (N = 177)	.36 (N = 20)
Three Studies Pooled	.57** (N = 273)	.26** (N = 520)	.35** (N = 48)

* $p < .05$.
** $p < .01$.

tions. Thus, the correlations may reasonably be pooled. For the pooled data, it is clear that the correlation between adoptive child and adoptive parent is significantly lower than that between control child and biological parent. However, the correlation between biological child of an adoptive parent and its parent is *also* significantly lower than that observed in control families. Furthermore, the correlations do *not* differ significantly between adoptive parent and, on the one hand, adoptive child, and on the other hand, biological child. The latter two correlations do not differ significantly in either of the two studies which present data for each. The data, of course, confirm that the parent-child correlation is higher in control than in adoptive families. They also indicate, however, that *within* adoptive families, it makes no difference whether the child is adoptive or biological; in either case, the correlation with parent is low. There is no evidence for heritability in these data; quite the opposite.

The final study, by Skodak and Skeels (1949), involved only a single group of one hundred adoptive children. The same children were compared both to their true mothers and to their foster mothers. The comparison, however, did not directly involve the parents' IQs, since IQs were not available for the foster parents. The child's IQ was correlated with the *number of years of education* of the true

Table 16. *From Skodak and Skeels*

True mother's education × child's IQ	$r =$ .32** (N = 92)	
Foster mother's education × child's IQ	$r =$ .02 (N = 100)	
Proportion of true mothers attending college	8.7%	
Proportion of foster mothers attending college	51.0%	($p < .0001$)
For 8 true mothers > grade 12, *foster* mother's education	13.9 years*	
For 12 true mothers < grade 8, *foster* mother's education	11.8 years	
For 12 "college" foster homes, *true* mother's education	11.3 years ($p < .005$)	
For 22 "grade school" foster homes, *true* mother's education	9.1 years	

* $p < .05$.

and the adoptive parent. The correlations were .32 with true mother's education, and a mere .02 with foster mother's. The contrast appears to suggest a powerful role for heredity; but as table 16 indicates, there are complications.

There were enormous differences in the amount of education of the two types of mothers. More than half of the foster mothers had attended college, while very few of the true (illegitimate) mothers had. It seems doubtful whether *any* measure could be significantly correlated with number of years of college attended. The foster mothers, in short, were a very much more homgeneous group than the true mothers. The logic of a "scale" of educational attainment which equates the difference between completing six and seven years of school with that between completing twelve and thirteen is, in any event, questionable; one suspects that the decision to proceed from high school to college might be correlated with sociocultural variables not reflected between grade 6 and grade 7 dropouts. There is, therefore, little mystery in the fact that the adopted children's IQs did not correlate with their foster parents' education. The significant fact is that the child's IQ *did* correlate with its true mother's education, although the child had never lived with the true mother. What could this fact reflect other than a powerful genetic determination of IQ?

The most obvious fact which it might reflect is a policy of selective placement on the part of adoption agencies. The raw data in the Skodak and Skeels Appendix can, in fact, be used to demonstrate clearly such selective placement. There were eight true mothers who had attended college, and twelve true mothers who had failed to complete grade school. We can ask, were there differences in the types of homes into which illegitimate children of these two groups of mothers were placed? There were indeed; the foster mothers who received children of highly educated true mothers were significantly more highly educated than were the foster mothers who received children of poorly educated true mothers. The same kind of selective placement can be demonstrated in mirror image. There were twelve foster homes in which each parent had *completed* college, and twenty-two in which neither parent had completed high school. The true mothers of children placed into these two classes of homes differed significantly, in the expected direction, in the amount of their own education. The illegitimate child of a college-educated mother, it appears, is very likely to be placed in a college foster home; the probability is much less that the illegitimate child of a grade-school dropout will be so placed.

Perhaps, it is instructive to note that Skodak and Skeels were unable to demonstrate selective placement in terms of the true mother's *IQ*. That suggests that adoption agencies, unlike psychologists, are guided in the real world by significant social facts such as amount of education, more than by such fanciful psychological constructs as the IQ.

The evidence indicates, in any event, that children were placed into foster homes on the basis of their true mothers' education, or of variables correlated with it. We can assume that the children of highly educated true mothers were placed into "good" foster homes, homes conducive to the development of high IQ. The observed correlation between child's IQ and true mother's education follows directly from this fact. Further, on the assumption that foster parent's education correlated only moderately with "goodness" of the foster home, there would be little or no correlation between child's IQ and foster parent's education. The latter assumption is especially likely to be valid, as Skodak and Skeels themselves pointed out, in a farming state such as Iowa in the 1930s. They indicated in their report that many of the foster homes with the greatest cultural and environmental amenities belonged to successful farmers who had relatively little formal education.

The adopted child studies, like the separated twin studies, seem to me to offer no evidence sufficient to reject the hypothesis of zero heritability of IQ scores. The policy recommendations made by contemporary psychologists on the basis of these data seem no more firmly grounded in science than those made during the earlier eugenic and immigration debates. They seem to me to reflect the same elitism and the same ethnocentric narrowness of vision and concern which characterized our psychological ancestors during their anti-immigrant hysteria.

The assertion is abroad that compensatory education has been tried, and has failed; and that that failure is an inevitable consequence of demonstrated genetic truths. Perhaps, however, it is we psychologists who have failed; perhaps again, it is the society in which we live that has failed. Those who care have a double task. We had better build a better psychology; and we had better help to build, quickly, a better society.

Note

This is the text of part of an address given by Professor Kamin at the Eastern Psychological Association meetings in 1973, and at a number of universities in 1972–1973. Kamin's criticisms were greatly expanded and published in 1974 as *The Science and Politics of IQ* (Potomac, Md.: Erlbaum Associates).

The second part of this essay appears on pp. 374–382 of this book.

References

BABSON, S. G., KANGAS, J., YOUNG, N., and BRAMHALL, J. L. 1964. "Growth and Development of Twins of Dissimilar Size at Birth." *Pediatrics* 30:327–333.

BURKS, B. S. 1928. "The Relative Influence of Nature and Nurture upon Mental Development: A Comparative Study of Foster Parent–Foster Child Resemblance and True Parent-Child Resemblance." In *Twenty-seventh Yearbook: Nature and Nurture, Part I, Their Influence on Intelligence,* pp. 219–316. Bloomington, Ind.: National Society for the Study of Education.

BURT, C. 1921. *Mental and Scholastic Tests.* London: King & Son.

———. 1943a. "Ability and Income." *British Journal of Educational Psychology* 13:83–98.

———. 1943b. "The Education of the Young Adolescent: The Psychological Implications of the Norwood Report." *British Journal of Educational Psychology* 13: 126–131.

———. 1955. "The Evidence for the Concept of Intelligence." *British Journal of Educational Psychology* 25: 158–177.

———. 1958a. "The Inheritance of Mental Ability." *American Psychologist* 13:1–15.

———. 1958b. "A Note on the Theory of Intelligence." *British Journal of Educational Psychology* 28:281–288.

———. 1966. "The Genetic Determination of Differences in Intelligence: A Study of Monozygotic Twins Reared Together and Apart." *British Journal of Psychology* 57, nos. 1–2:137–153.

Burt, C., and Howard, M. 1956. "The Multifactorial Theory of Inheritance and Its Application to Intelligence." *British Journal of Statistical Psychology* 9:95–131.

———. 1957a. "Heredity and Intelligence: A Reply to Criticism." *British Journal of Statistical Psychology* 10:33–63.

———. 1957b. "The Relative Influence of Heredity and Environment on Assessments of Intelligence." *British Journal of Statistical Psychology* 10:99–104.

Churchill, J. A. 1965. "The Relationship Between Intelligence and Birth Weight in Twins." *Neurology* 15:341–347.

Conway, J. 1958. "The Inheritance of Intelligence and Its Social Implications." *British Journal of Statistical Psychology* 11:171–190.

———. 1959. "Class Differences in General Intelligence." *British Journal of Statistical Psychology* 12:5–14.

Freeman, F. N., Holzinger, K. J., and Mitchell, B. C. 1928. "The Influence of Environment on the Intelligence, School Achievement, and Conduct of Foster Children." In National Society for the Study of Education, *Twenty-seventh Yearbook,* part 1.

Gaw, F. 1925. "A Study of Performance Tests." *British Journal of Psychology* 15:374–392.

Jensen, A. R. 1969. "How Much Can We Boost IQ and Scholastic Achievement?" *Harvard Educational Review* 39:1–123.

———. 1970. "IQ's of Identical Twins Reared Apart." *Behavior Genetics* 1: 133–146.

Juel-Nielsen, N. 1965. "Individual and Environment: A Psychiatric-Psychological Investigation of Monozygous Twins Reared Apart." *Acta Psychiatrica et Neurologica Scandinavica,* monograph suppl. 183. Copenhagen: Munksgaard.

Leahy, A. M. 1935. "Nature-Nurture and Intelligence." *Genetic Psychology Monographs* 17, whole no. 4.

McNemar, Q., Newman, H. H., and Freeman, F. N. 1938. "Holzinger's Twins: A Study of Heredity and Environment." *Psychological Bulletin,* pp. 237–249.

Newman, H. H., Freeman, F. N., and Holzinger, K. J. 1937. *Twins: A Study of Heredity and Environment.* Chicago: University of Chicago Press.

Shields, J. 1962. *Monozygotic Twins Brought Up Apart and Brought Up Together.* London: Oxford University Press.

Skodak, M., and Skeels, H. M. 1949. "A Final Follow-up Study of One Hundred Adopted Children." *Journal of Genetic Psychology* 75:85–125.

VANDENBERG, S. G. 1971. "What Do We Know Today About the Inheritance of Intelligence and How Do We Know It?" In *Intelligence: Genetic and Environmental Influences,* ed. R. Cancro. New York: Grune & Stratton.

WILLERMAN, L., and CHURCHILL, J. A. 1967. "Intelligence and Birth Weight in Identical Twins." *Child Development* 38:623–629.

Mysteries of the Meritocracy

Arthur S. Goldberger

Glendower: *I can call spirits from the vasty deep.*
Hotspur: *Why, so can I, or so can any man; but will they come when you do call for them?*
—*Henry IV*, First Part

1. Introduction

IN HIS BOOK *I.Q. in the Meritocracy*, RICHARD J. HERRNSTEIN (1973) calls on a classic article by Barbara S. Burks (1928) to support his position that a large part of the variation in intelligence can be accounted for by variation in heredity, as distinguished from variation in environment and from covariation of heredity and environment.

But Herrnstein's report of the Burks study is thoroughly misleading.

2. Herrnstein's Report

In Chapter 4 of his book, after reviewing other empirical evidence on heritability, Herrnstein turns to the Burks study. His presentation (1973, pp. 182–184) is reproduced below in its entirety. For ease of reference, I have italicized and numbered selected passages:

> A still more persuasive case for the relative unimportance of home-genetic covariance can be found in a study published in 1928 by Barbara Burks of Stanford University. The study compared 214 foster children and their adoptive parents (the "experimental" group) to a carefully matched collection of 105 children being raised by their own parents (the "control" group). The control group was chosen to mimic the experimental group for the age and sex of the

children, and for the locality, type of neighborhood, occupation, and ethnic characteristics of the family. Moreover, enough was known about the true fathers of the adopted children to show that there was little if any selective placement as regards fathers. There was no correlation between the occupational level or the cultural rating of the foster fathers and the occupational level of the true fathers. All the children in the experimental group were adopted before the age of twelve months and more than 60 per cent of them did not know they were adopted at the time of the study. The distribution of intelligence-test scores covered about the same range, with close to the same average, for the foster parents and the control parents raising
(1) their own children. *To the extent possible for naturalistic studies of human beings, Burks succeeded in crossing a broad range of genetic endowments with a broad range of home environments.* If covariance were crucial, the study would have shown it.

First, in keeping with the studies of foster children summarized
(2) earlier, *the foster children's I.Q.'s correlated with their natural parents' I.Q.'s more than with their foster parents'.* Even though the natural parents and the foster parents were uncorrelated as regards
(3) cultural or social-class characteristics, *the true father-child or true mother-child correlations were in the .5 range.* In contrast, the foster father–child correlation was essentially zero, while the foster
(4) mother–child correlation was about .2. *The control-group correlations, for parents raising their natural children, were only slightly higher than the true parent-child correlations in the experimental group, comprising adopted children.* The study clearly and unequivocally showed that the home environment, when disentangled from the genetic connection between ordinary parents and their children, accounts for relatively little of the variation in children's I.Q.'s.

(5) *Subjecting Burks's results to the statistical procedures of quantitative genetics yields an estimate of heritability in the familiar .8 range,* even though the design of her study carefully eliminated the covariance of genetic endowment and home environment. From this, and from other comparable results, it can be concluded that covariance of this variety accounts for little concerning the I.Q. in most circumstances. That, of course, is again not to say that in a radically different world, covariance might not be highly significant, or even that in certain limited instances in our own society, it is un-
(6) important. *From her analysis, Burks could properly say that "nearly 70 per cent of schoolchildren have an actual I.Q. within 6 to 9 points of that represented by their innate intelligence." But unusually good or unusually poor environments, so rare as to affect something less than 1 per cent of the total population, might be promoting or*

retarding the I.Q.'s of the people encountering them as much as 20 points.

(7) *Burks's sample was drawn from the white, primarily native American or western European population living around San Francisco*
(8) *and Los Angeles. The families spanned all social classes, from those of unskilled laborers to successful professionals and businessmen.*
(9) *Nevertheless, in racial, ethnic, linguistic, and, no doubt, cultural terms, the study omitted significant parts of the vastly diversified American population.* Since heritability measures a population trait,
(10) it is quite possible that the estimates are off somewhat. *It could well be that there are sources of environmental variation left out of Burks's study,* or, for that matter, the other studies in the literature reviewed here. Including them might reduce the heritability estimate. Much of Jinks and Fulker's analysis is, for example, based on data collected in England by Cyril Burt. If one assumes that the intellectual environment in England is more homogeneous than in America, then Burt's data will set too high a value on heritability for the American population.

These uncertainties inhere in any population statistic—birth and mortality rates, crime rates, and so on—not just in the estimation of the genetic contribution to tested intelligence. Population statistics are not like the timeless constants of physical science, fixed by properties somehow inherent in nature. They are, rather, more like the actuarial data of the insurance business—more or less approximate, contingent, and, above all, changeable features of populations. Both insurance companies and quantitative geneticists are well advised to keep taking soundings.

The fact that a number of independent studies point to a particular narrow range of values for the heritability of I.Q. suggests a robustness to the estimate that should not be overlooked. However, it should not be overinterpreted either, for, while independent, the studies may nevertheless share common methodological weaknesses.
(11) *For example, the poorest, most culturally deprived sectors of the population tend to be omitted, or at least underrepresented, in most assessments of heritability.* If those unfortunate people happen to show most fully the impact of environment, their omission from the population sampled raises the heritability.

3. A Major Mystery

In passages (2), (3), and (4), Herrnstein refers to the correlation between the IQs of the foster (= adopted) children and the IQs of their natural (= true) parents. In (2), he says that it is larger than the correlation between the IQs of the foster children and the IQs of their foster parents. In (4), he says that it is less than the correlation between the IQs of own (= nonadopted) children and the IQs of their parents. And in (3), he says that it (or rather each component of it) is approximately one half.

What makes these statements mysterious is the fact that the Burks study contains no information on the IQs of the natural parents of the foster children. Burks's research group did not meet these parents, did not test them, nor was their intelligence tested by anyone else.

Where then did Herrnstein's figures come from? Burks, on pages 314 and 316, gives .45 and .46 as the correlations of the IQ of "true child" with the IQ (strictly speaking, "mental age") of father and of mother, respectively. These numbers are, indeed, in the .5 range and larger than the .07 and .19 reported for the foster-child–foster-father and the foster-child–foster-mother correlations on pages 313 and 315. However, the .45 and .46 clearly refer to the own, non-foster, children, that is, to the *control* group. It seems that Herrnstein mistook these control group figures for foster-group figures. Furthermore, on page 285, Burks gives .55 and .57 as the control-group correlations, corrected for attenuation, of the IQ of child with the IQ of father and of mother, respectively. These numbers are indeed slightly higher than .45 and .46, but, of course, they refer to the control group. It seems that Herrnstein also mistook a difference between corrected and uncorrected correlations in one group for a difference between the two groups.

It is obvious that Herrnstein's report of Burks's findings on the resemblance between the intelligence of adopted children and the intelligence of their natural parents is untrue; Burks had no such findings.

4. Another Mystery

In items (1) and (8), Herrnstein suggests that Burks's sample was fairly representative of the United States population; this is qualified to some extent in (7), (9), (10), and (11).

Turning to Burks (1928, p. 236) we find that the foster sample was confined to white, non-Jewish, English-speaking, adoptive couples, who were American, British, or north-European born, with husband and wife both alive and living together, resident in the San Francisco, Los Angeles, and San Diego areas.

Proceeding in Burks's article, on page 267, we find that over half of the adoptive fathers were professionals, business owners, or managers (while 2 percent of them were unskilled laborers); on page 268, we learn that 83 percent of them were homeowners; and on page 270, we find that one-third of the boys and one half of the girls had private tutoring outside of school in "music, dancing, drawing, etc."

On the 25-point "Whittier Index" of home quality, the foster families' average score was 23.3 points (p. 269).[1] In intelligence, the foster parents averaged a full standard deviation above the population at large (p. 305). As for "the total complex of environment," Burks's own conservative estimate was that the foster homes averaged somewhere between one half and one standard deviation above the general population (p. 306).

Surely, Burks's families were not representative of the population at large. It should come as no surprise that children were placed for adoption with families located in the upper socioeconomic brackets. What is mysterious is Herrnstein's decision to regard this sample as though it covered a broad range of environments, being merely limited with respect to racial, ethnic, linguistic, and cultural characteristics. His concession that some environmental variation may have been omitted, and his hint that the very poorest groups were underrepresented, hardly do justice to the facts.

Since Burks's sample was so highly selective, the variation in environment will have been much less than in the population at large. If so, the explanatory power of environment in the sample will also have been limited as compared to that in the population.[2]

5. Quantitative Genetics

Next, we turn to Herrnstein's (5), which says that the statistical procedures of quantitative genetics applied to Burks's data yield an estimate of about .8 for heritability (= proportion of variation in intelligence accounted for by variation in heredity). He cites no source, but two possibilities suggest themselves.

The first is the Burks article itself. For the foster group, a multiple correlation of $R = .42$ was obtained when child's IQ was regressed on father's IQ, father's vocabulary score, mother's vocabulary score, and income (p. 287). Interpreting those explanatory variables as environmental measures, Burks takes the multiple R^2, namely .17 ($= .42^2$), to be the proportion of variance in child's IQ that is due to home environment. She then arbitrarily adds .05 or .10 to this to allow for "the possible 'random somatic effects of environment,' " and subtracting the total from 1, produces the conclusion that "probably, then, close to 75 or 80 percent of IQ variance is due to innate and heritable causes" (pp. 303–304).

It is not clear why Herrnstein would feel that a regression of foster child's IQ on three test scores and income involves the statistical procedures of quantitative genetics. Be that as it may, it is not clear why we must accept that particular combination of variables as the relevant measure of home environment.[3]

The second possible source of Herrnstein's heritability estimate is a detailed analysis of Burks's data by the distinguished geneticist, Sewall Wright. Wright (1931) works with five correlations drawn from Burks:

Foster group:	$r_{CP} = .23,$	$r_{CE} = .29$	
Control group:	$r_{CP} = .61,$	$r_{CE} = .49,$	$r_{EP} = .86$

Here C denotes the child's IQ, P denotes the midparental IQ, and E denotes "environment." For environment, Wright uses Burks's "Culture Index."[4]

Wright introduces a variable H to represent the child's heredity, which is of course not directly observed. He develops a simple model in which the basic equation is a regression of C on H and E. He supposes that H is correlated with P and E in the control group, but uncorrelated with them in the foster group. In this model, the set of observed correlations given above produces an estimate of .90 as the "path coefficient" (= standardized regression coefficient) running from H to C. And the square of this, namely .81, measures the proportion of the variation in IQ that is attributable to variation in heredity.

This calculation of Wright, then, may provide a basis for Herrnstein's statement that a heritability estimate in the .8 range results when Burks's data are subjected to "the statistical procedures of quantitative genetics." As such, it merits our attention.

Wright clearly indicates that his model attributes to heredity H, which is not directly measured, all effects which cannot be attributed to *measured* environment E. If so, the heritability estimate may be sensitive to the choice of a measure for E. To see this, let us subject Wright's (1931, p. 160) formulas to the quantitative procedures of elementary algebra. We find that his estimate of the path coefficient, say p, running from H to C, is calculated as

$$p = \sqrt{1 - q^2}\,(-\,qr + \sqrt{q^2r^2 + 1 - 2q^2})/(1 - 2q^2),$$

where q and r are, respectively, the foster-group and control-group correlations of child's IQ with environment. Thus, the coefficient p is completely determined by the two r_{CE}'s, and is quite independent of the r_{CP}'s and r_{EP}. The environmental measure used by Wright is the Culture Index, a single variable reflecting certain aspects of the parents' speech, education, interests, home library, and artistic taste: with that measure of E, we have $q = .29$, $r = .49$, and the formula above indeed gives $p = .90$.

But there is nothing sacred about the Culture Index as a measure of the environmental influences on intelligence. Indeed, we have already seen that Burks found a foster-group multiple correlation of .42 between C and a set of four environmental variables. As it happens, she found a control-group multiple correlation of .61 between C and a set of four environmental variables.[5] For illustrative purposes, we can take $q = .42$ and $r = .61$, instead of $q = .29$ and $r = .49$ as values for the correlations of child's IQ with environment. When the new values are inserted in the formula above, we find $p = .82$ instead of $p = .90$. That is, we get $p^2 = .68$ rather than $p^2 = .81$ as our estimate of heritability.

It is hardly surprising to find in Wright's model that a more refined measure of environment leads to a lower estimate of heritability. After all, that model attributes to heredity all effects which are not attributable to *measured* environment. What is mysterious is that Herrnstein chose to cite only "the .8 range" for heritability.

What deepens the mystery is the fact that, in the same brief article, Wright (1931) himself obtains a different estimate of heritability from Burks's data. The alternative comes from a different model. In this second, more elaborate, model, environment is still measured by the Culture Index alone, but the effects not attributable to measured environment are now allocated between G (additive genotype) and M (a residual which includes *unmeasured* environment, along with any nonadditive genotype and interaction effects). The path coefficient running from G to C is estimated as .71; squaring this yields Wright's second estimate of heritability, namely .49.

To some extent, the reduced value arises because of the shift from broad to narrow heritability. But Wright does not explain it away in that manner. Rather (p. 162), he clearly states that the first estimate is intended as an upper bound, the second as a lower bound. On at least two subsequent occasions, in reviewing his analysis of Burks's data, he emphasized the same point:

> [The first model is] doubtless too simple since heredity is represented as the only factor apart from the measured environment. Any estimates of the importance of hereditary variation will thus be maximum.... [In the second model, we] attempt at obtaining a minimum estimate of heredity.... The path coefficient for influence of hereditary variation lies between the limits + .71 (if dominance and epistasis are lacking) and + .90. [Wright, 1934, pp. 185, 187, 188]
>
> The results are reasonable [for the first model] except that H undoubtedly includes more than heredity. . . . [Wright, 1954, p. 23]

With all this in mind, it seems fair to conclude that Herrnstein's item (5) is not an accurate statement.

6. Environmental Effects

Now let us turn to Herrnstein's item (6), which purports to give the effects of environmental change upon intelligence, measured in IQ points. Here, Herrnstein is accurately reporting these items from Burks's (pp. 308–309) summary of her conclusions:

> 3. Measurable environment one standard deviation above or below the mean of the population does not shift the IQ by more than 6 to 9 points above or below the value it would have had under normal environmental conditions. In other words, nearly 70 percent of school children have an actual IQ within 6 to 9 points of that represented by their innate intelligence.
>
> 4. The maximal contribution of the *best* home environment to intelligence is apparently about 20 IQ points, or less, and almost surely lies between 10 and 30 points. Conversely, the least cultured, least stimulating kind of American home environment may depress the IQ as much as 20 IQ points. But situations as extreme as either of these probably occur only once or twice in a thousand times in American communities.

Burks, in turn, was summarizing her calculations on pages 306–308. Her basic estimate is that a standard-deviation change in environment will change IQ by 6 points. It was obtained as follows. The correlation of foster child's IQ with "environment," namely the previously reported multiple R of .42, was viewed as a standardized regression coefficient: A standard-deviation change in environment produces a .42 standard-deviation change in IQ.

Multiplying this by the standard deviation of IQ, namely 15 points, yielded 6 points.

This 6-point figure was then *tripled* to give "about 20 points" as the change in IQ produced by a *three*-standard-deviation change in environment, that is, by a movement from an average environment to the "best" environment. Her higher estimates for the effects of one- and three-standard-deviation changes in environment, namely 9 and 30 points, respectively, were calculated in the same manner except that .62 was arbitrarily used instead of .42 for the IQ-environment correlation. Finally, "nearly 70 percent" and "once or twice in a thousand" are simply Burks's descriptions of the respective probabilities with which a normally distributed variable lies within one standard deviation, and beyond three standard deviations, of its mean.

Let us focus on Burks' basic estimate, namely that a standard-deviation improvement in environment would raise IQ by 6 points. In constructing this estimate, she uses the environmental standard deviation in the sample, but her conclusion refers to the environmental standard deviation in the population. Her logic is invalid when the sample is systematically different from the population. In particular, if environmental variation is substantially less in the sample than it is in the population, Burks's method will lead to a substantial underestimate of the effect of environmental change upon IQ.

Recall the various respects in which the foster group is nonrepresentative of the population at large, having been drawn from the upper reaches of the socioeconomic scale. The variation of environment within those upper brackets is no doubt less than the variation of environment over the population at large. Consequently, the sample standard deviation of environment is no doubt less than the population standard deviation of environment. A given environmental difference which is large when measured in sample standard deviations, and rare in the sample, may well be small when measured in population standard deviations, and common in the population.

To suggest orders of magnitude, we apply some formulas given by Kelley (1947, pp. 295–298). If we select the top 38 percent of a normally distributed population, we get a group whose mean is one standard deviation above the population mean; the standard deviation within this group is 54 percent as large as that in the population. Recall Burks's guess that on "the total complex of environment," her sample may have averaged one standard deviation above the population average. Consequently, a fair guess is that the standard deviation of environment in her sample was about half as large as it was in the population at large. If so, we are free to *double* her estimates of the effects of environmental change.

To sum up this section: In item (6), Herrnstein accurately reports Burks's conclusions; what is puzzling is that he believes that her conclusions were properly drawn from her data.

7. Remarks

What lessons are to be drawn from our critical reading of Herrnstein? First, his report cannot be taken at face value: to find out what the Burks study contains, it is necessary to read Burks, not Herrnstein. Second, the Burks study cannot support strong conclusions about the relative contributions of heredity and environment to the determination of intelligence.

Throughout the IQ controversy, the advocates of high heritability have, to a considerable extent, developed their case by reporting on several studies of adopted children along with several studies of separated identical twins: see Herrnstein (1973, chap. 4) and Jensen (1972, pp. 121–130, 307–326; 1973, chaps. 7 and 8). My own assessment is that those reports cannot be taken at face value and, moreover, that the studies themselves cannot support strong conclusions.

But a thoughtful reader will hardly take *my* word for it, and is advised to consult the original studies.[6] To assist in this detective work, Bronfenbrenner (1975) and Kamin (1974, chaps. 3–5) provide many helpful clues.

Notes

The research reported here was supported by funds granted to the Institute for Research on Poverty at the University of Wisconsin by the Office of Economic Opportunity pursuant to the Economic Opportunity Act of 1964, and by Grant GS–39995 of the National Science Foundation. I am deeply indebted to my colleague Glen Cain for many instructive discussions. I am also grateful to Lee Bawden, Dudley Duncan, Irwin Garfinkel, Richard Herrnstein, Leon Kamin, David Layzer, and Sewall Wright for helpful comments on an earlier draft. But the opinions expressed here are mine and should not be attributed to the institutions and individuals named above.

1. This "Whittier Scale for Home Grading" is the sum of scores on five 5-point items: necessities, neatness, size, parental conditions, and parental supervision. The respective mean scores (with standard deviations in parentheses) in the foster group were 4.7 (0.4), 4.5 (0.6), 4.7 (0.5), 4.8 (0.4), 4.7 (0.4). To convey the meaning of such averages, we give the verbatim descriptions of the prototypic conditions for scores of 4 and 5 on each item:

Necessities

4 = Income, salary, and tips of head waiter in a large hotel. Clothing neat, well-kept, apparently made to last. Good table set. Half modern bungalow. Furniture good quality, plentiful. Wicker and reed chairs, piano, rugs, good pictures. Rather poor lighting from windows, but modern electric fixtures. Running water, modern sanitary conveniences. Rear porch bedroom, couch in living room.

5 = Architect, well-to-do. Well-dressed. Tableware indicates abundant food. Large modern bungalow, frame construction, well-finished. Furniture fine quality, plentiful. Fine carpets, rugs, and pictures. Modern conveniences, built-in cupboards, electric fixtures, plumbing.

Neatness

4 = Rooms clean, but dark, closed, and stuffy most of the time. Furniture neatly arranged and kept in good order. Exterior cleanliness good. House somewhat in need of paint. Lawn well-kept. Considerable attention given to home when possible.

5 = Interior clean and sanitary. Furniture neatly arranged, good order. Yards and grounds clean, no outbuildings. House well-kept. Yard clipped close; small, neat garden. General neatness good. Considerable attention apparently given to care of home.

Size

4 = Seven rooms, all rather small. Two-story house. Rooms convenient, although small. Propositus [= foster child], three younger children, mother and stepfather.

5 = Seven rooms, two-story house. Good-sized rooms. Plenty of room, conveniently arranged. Two adults, father and mother, propositus, and younger sister. Rather small front yard, good open porch. Large back yard as city yards go.

Parental Conditions

4 = Father a painter, in good health. Mother probably normal. Harmonious most of the time. Mother nags father some on account of irregular work. No separation. Father away at work during day.

5 = Father normal, has average success as a carpenter. Mother keeps home in fair condition. So far as known, there is harmony between the parents. Mother at home all of the time, father away at work most of day. (In practice, we never assigned a rating as high as 5 to this item if either parent tested with a mental age below 12.0.)

Parental Supervision

4 = Father apparently interested in welfare of boys. Fairly good control. Equally fair treatment as far as known. Father a colored preacher. Good habits and reputation.

5 = Parents interested in health, education and welfare of children. Kind and intelligent discipline. Complete fairness as far as known. Parents of good reputation and character, good example to children. Children kept at home evenings as a general rule.

The full frequency distributions were not given by Burks, but can be reconstructed approximately from the means and standard deviations. For example, to have a mean $\bar{x} = 4.7$ and a standard deviation $s = 0.4$ on a 5-point scale virtually requires that two-thirds of the families score 5; one-third score 4; and none score 1, 2, or 3. (To obtain this conclusion, let p_i = proportion of sample scoring i ($i = 1, \ldots, 5$). Then use the equations $\Sigma_{i=1}^{5} p_i = 1$, $\Sigma_{i=1}^{5} ip_i = \bar{x}$, $\Sigma_{i=1}^{5} i^2 p_i = s^2 + \bar{x}^2$, in conjunction with $0 \leq pi \leq 1$, to restrict the possible values of the p_i).

Actually Burks's team found that, for three items, the maximum value of 5 did not seem adequate. They extended the scales to a sixth point, described as follows: *Necessities.* 6 = Conspicuously superior to the level receiving 5 points. Seldom given to any home. Denotes unusually luxurious living conditions. *Parental Conditions.* 6 = Conspicuously superior to the level receiving 5 points. Both parents superior on the mental test, and *exceptionally* harmonious in their relationships. *Parental Supervision.* 6 = Care given the children and provision made for their welfare exceptional. On this extended scale, the item means changed to 4.9, 4.5, 4.7, 4.9, 4.9, and thus the Whittier Index mean increased to 23.9 on a 28-point scale. From this, it can be deduced that in 10–20 percent of the cases, a score of 5 was raised to a score of 6.

Material in this note was drawn from pp. 231–233, 269, of Burks. *See also* Kamin (1974, pp. 118–119).

2. Burks herself dealt with selectivity on pp. 222–223, saying that

> home environment cannot be expected to have as large a proportional effect upon the mental differences of the children we studied as though they were being reared in families unselected as to race or geographical location throughout the world.

She felt that the problem was not too severe, contending that:

> The distribution of homes of the children studied in this investigation was probably nearly as variable in essential features as homes of the

> general American white population (though somewhat skewed toward a superior level).

Her contention, of course, runs counter to the many indications of markedly superior environments noted in the text; the only evidence offered to support it is the fact that the variation of children's IQs was as large in the samples as it was in the standard population. This fact is difficult to explain. The difficulty is apparent if we take the position that environment is an important determinant of IQ, but it also arises if we, like Herrnstein, take a contrary position. A main line of argument in his book is that parental intelligence determines both the children's intelligence (via heredity) and their environment (via parental success, earnings, achievement, and so on). This implies that in the population, parents' socioeconomic status must be correlated with children's IQs, which, in turn, implies that, if we sample only families with high socioeconomic status, we should find reduced variation in children's IQs. But such a reduction does not appear in Burks's control families, although they were chosen to match the foster families, and thus had similarly superior environments: *see* Burks, pp. 263–277. For further analysis of the environmental and IQ distributions in Burks's samples, *see* Goldberger (1975).

3. For readers familiar with simple, but not with multiple, correlation: The multiple correlation of a variable y with a set of variables $x_1, \ldots, x_K$ can be interpreted as the simple correlation of y with a certain variable z. This variable z *is* constructed as a linear combination of the x's, namely $z = b_1x_1 + \ldots + b_Kx_K$, where the b's are chosen to maximize the correlation with y. These b's are regression coefficients; when all variables are measured in standard-deviation units, the b's are called standardized regression coefficients.

Burks tried some other explanatory variables but did not include them in her final multiple regression. She explains, on p. 287:

> To have gone through the operation of computing multiple correlations that utilized all nine of the variables would have been enormously time-consuming. To save labor, certain variables were eliminated, after first demonstrating, through multiples using three or four variables, that they contributed practically nothing to an estimate of the child's IQ not already contributed by variables retained for the final multiple. For example, in the foster multiple, income was retained, but Whittier and Culture indices were dropped out, because the multiple of IQ with all three together (.34) was only .01 higher than the correlation (.33) between IQ and income alone; again, mother's vocabulary was retained, but mother's mental age and mother's education were dropped out because the multiple of IQ with all three together (.254) was only .005 higher than the correlation (.249) between IQ and mother's vocabulary alone.

She was convinced that including all nine of the variables would not have raised the R by more than .02. But because she did not publish a full set of correlations, we cannot verify that.

4. This "Culture Index" is the sum of scores on five 5-point items referring to the parents' speech, education, interests, home library, and artistic taste. The respective mean scores in the foster group were 3.5, 3.7, 3.2, 3.1, 3.2 for a total of 16.9; in the control group, they were 3.4, 3.8, 3.1, 3.0, 3.0 for a total of 16.3. To suggest the meaning of these figures without going into the verbatim detail of note 1, I note the following: *Speech* is based on a vocabulary test; *Education* measures the average number of grades completed by the parents, with 1–3 grades scored 1, 4–6 grades scored 2, 7–9 grades scored 3, 10–12 grades scored 4, and more than 12 grades scored 5; *Interests* measures the quality of parent's hobbies and activities; *Home Library* measures the number of books in the home, with less than 10 books scored 1, . . . , more than 500 books scored 5. The content of the *Artistic Taste* component is perhaps best captured by some direct quotation:

Artistic Taste

2 = . . . trashy ornaments, such as kewpies and gaudy bric-a-brac scattered about.

3 = . . . the Victrola records and piano music are not of a high type. . . . Photographs of the family are usually abundant.

4 = . . . no trashy ornaments about, and family "photos" are absent or present in very moderate numbers.

5 = . . . Musical selections for piano or Victrola are from standard composers (though a little popular music or jazz may be included as well). . . .

For Wright's purposes, the Culture Index was preferable to the Whittier Index as a single measure of environment because it gave larger correlations. He felt unable to use a combination of several variables to represent environment because the correlations among them were not published by Burks. (Note indeed that for the foster group even the correlation of the Culture Index with midparental IQ is lacking.)

The correlations involving "midparent" P were constructed by Wright from the separate correlations involving father and mother, by a standard procedure.

Material in this note was drawn from pp. 234–235 and 269 of Burks and from correspondence with Wright.

5. The set of variables used in the control group was: father's IQ, father's vocabulary, mother's IQ, Whittier Index (*see* Burks, p. 287). This is not quite the same as her list for the foster group.

6. Having already invested some time in Burks, the reader might wish to proceed by checking out the reports of her study which have been given by Jensen (1972, pp. 128–130; 1973, pp. 196–197, 203–204, 240) and by Eysenck (1971, pp. 63–65). *See also* Goldberger (1975).

References

Bronfenbrenner, U. 1975. "Nature with Nurture: A Reinterpretation of the Evidence." *In Race and IQ*, ed. A. Montagu, pp. 114–144. London: Oxford University Press.

Burks, B. S. 1928. "The Relative Influence of Nature and Nurture upon Mental Development: A Comparative Study of Foster Parent–Foster Child Resemblance and True Parent–True Child Resemblance." In *Twenty-seventh Yearbook: Nature and Nurture, Part I, Their Influence on Intelligence*, pp. 219–316. Bloomington, Ind.: National Society for the Study of Education.

Eysenck, H. J. 1971. *Race, Intelligence and Education*. London: Temple-Smith.

Goldberger, A. S. 1975. "Professor Jensen, Meet Miss Burks." *Educational Psychologist*, 12.

Herrnstein, R. J. 1973. *I.Q. in the Meritocracy*. Boston: Little, Brown & Co.

Jensen, A. R. 1972. *Genetics and Education*. New York: Harper & Row.

———. 1973. *Educability and Group Differences*. New York: Harper & Row.

Kamin, L. J. 1974. *The Science and Politics of I.Q.* New York: Halstead Press.

Kelley, T. L. 1947. *Fundamentals of Statistics*. Cambridge: Harvard University Press.

Wright, S. 1931. "Statistical Methods in Biology." *Journal of the American Statistical Association* 26, Suppl. (March): 155–163.

———. 1934. "The Method of Path Coefficients." *Annals of Mathematical Statistics* 5 (September): 161–215.

———. 1954. "The Interpretation of Multivariate Systems." In *Statistics and Mathematics in Biology*, ed. O. Kempthorne et al., pp. 11–33. Ames: Iowa State College Press.

———. 1974. *Evolution and the Genetics of Populations*. vol. 3. Manuscript.

PART III

Social and Political Consequences

Introduction

THE ARTICLES IN THIS SECTION DEAL WITH THE IMPLICATIONS FOR society of the existence of characteristics measured by IQ tests and their possible genetic basis. The implications are basically of the following three kinds: First, the significance of IQ for the social-stratification patterns of a society (in particular, the possibility that a hereditary meritocracy will arise because social mobility is based on partly heritable characteristics); second, the use of IQ distribution statistics and explanatory hypotheses for this distribution as an ideological rationale for the existing distribution of wealth and power; and finally, implications concerning the possible distribution of educational resources among various groups in our society.

Noam Chomsky's "The Fallacy of Richard Herrnstein's IQ" at the beginning of this section is a criticism of Herrnstein's argument that American society is drifting toward a stable hereditary meritocracy based on IQ differences. Chomsky denies that the premises of Herrnstein's argument are true or, at least, known to be true. In particular, he denies Herrnstein's premise that individuals will labor only for transmittable wealth and power. Herrnstein, in his reply, argues that Chomsky is mistaken in imputing to him the assumptions that Chomsky thinks are required for the argument. Further, Herrnstein argues that any conceivable society, including one designed along lines Chomsky would approve, would itself have a tendency to stratify in the ways that Herrnstein suggests our society is doing. Chomsky, in a last word, insists that the premise that people labor only for material gain and transmittable reward is, indeed, required for Herrnstein's argument.

The article by Mary Jo Bane and Christopher Jencks is a short, popular attack on what they call five "myths" about IQ. The first myth is that IQ tests are the best measure of human intelligence. Their basic point is that in assessing IQ tests, one must ask exactly what kinds of intelligence are measured and what kinds fail to be measured. The second myth is that one's economic status in society is determined by one's IQ. Based on work by the authors in the book *Inequality*, they argue that IQ scores do not explain a significant part of the variation in either occupational status, income, or actual job performance of adults. The next myth they discuss is that IQ is overwhelmingly determined by genetic endowment. They argue that most estimates of the role of genes influencing IQ scores are overestimates and give several reasons why this is the case; that while genes probably account for something like 45 percent of variation in IQ scores, it does not follow that differences in actual learning capacity account for anything like this much variation. The next myth they discuss is that the main reason blacks and poor white children have low IQ scores is that they have "bad" genes. They suggest that the evidence available shows that genetic differences between classes play a rather small role in determining differences between children's IQs and that there is no very good evidence at all that any other difference between black and white scores is due to genetic factors. In any case, they argue that it does not make very much difference whether the IQ differences between blacks and whites are hereditary or environmental since IQ accounts for only a quarter of the income gap between blacks and whites. Finally, they attack the view that improving the quality of the schools will do much either in the way of abolishing differences in IQ and school achievements among children or in reducing differences in life chances.

The article by Clarence J. Karier examines the origins of the IQ-testing movement in early twentieth-century America. He argues that the state that emerged in the late nineteenth and early twentieth centuries required a means of rationalizing and standardizing manpower for the production and consumption of goods and services; and that the testing movement, financed largely by corporate foundations, was an essential part of that system. By looking at the writings of the leaders of the American testing movement, Karier attempts to uncover the nativist, racist, and elitist ideological viewpoints that themselves reflect the nature of IQ tests and the uses to which they were put.

Part II of Leon J. Kamin's article is also a discussion of the social and political backgrounds out of which intelligence testing arose in the United States. In particular, it is a case study of the use of science to provide a justification for a policy measure. In this case, it is the use of IQ testing and its results to justify and rationalize the passage by Congress in 1924 of a selective immigration bill. This bill had not only restricted the total number of immigrants but also established the practice of national-origin quotas. Immigrants from any European country would be allowed entry into America only to the extent that their fellow countrymen were already represented in the American population. Although this bill was passed in 1924, the census data used to establish the quotas were those established by the census of 1890. By using this cutoff point, the bill discriminated against immigrants from southern and eastern Europe.

The article by Carl Bereiter deals with the educational implications of individual differences in intelligence. What does heritability tell us about teachability? What are the prospects for reducing indivdual differences in intelligence? What are the educational implications of possible hereditary differences in intelligence associated with social class and race? Bereiter's conclusions are that the existence of individual differences in intelligence should encourage us to look for alternative methods of achieving educational objectives that do not rely so heavily on ability represented by IQ; that the apparently high heritability of IQ should influence our expectations as to what may be accomplished through allocation of existing environmental variations; that although reallocation may produce substantial gains in mean IQ, it should not be expected that the spread of individual differences will be greatly reduced; and finally, that with respect to social and racial differences, the educational policymaker need not concern himself with the question of whether these differences have a genetic basis.

The final article by the editors discusses the conceptual and empirical issues raised by IQ testing and the nature and political significance of the thesis that IQ has high heritability. It also deals with the morality of research and publication in the area of genetic racial differences.

The Fallacy of Richard Herrnstein's IQ

Noam Chomsky

HARVARD PSYCHOLOGIST RICHARD HERRNSTEIN'S BY NOW WELL-KNOWN *Atlantic* article, "IQ" (September 1971), has been the subject of considerable controversy. Unfortunately, this has tended to extend the currency of his ideas rather than to mitigate against them! Herrnstein purports to show that American society is drifting towards a stable hereditary meritocracy, with social stratification by inborn differences and a corresponding distribution of "rewards." The argument is based on the hypothesis that differences in mental abilities are inherited and that people close in mental ability are more likely to marry and reproduce,[1] so that there will be a tendency toward long-term stratification by mental ability, which Herrnstein takes to be measured by IQ. Second, Herrnstein argues that "success" requires mental ability and that social rewards "depend on success." This step in the argument embodies two assumptions: first, it is so in fact; and, second, it must be so for society to function effectively. The conclusion is that there is a tendency towards hereditary meritocracy, with "social standing" (which reflects earnings and prestige) concentrated in groups with higher IQs. The tendency will be accelerated as society becomes more egalitarian, that is, as artificial social barriers are eliminated, defects in prenatal (e.g., nutritional) environment are overcome, and so on, so that natural ability can play a more direct role in attainment of social reward. Therefore, as society becomes more egalitarian, social rewards will be concentrated in a hereditary meritocratic elite.

Herrnstein has been widely denounced as a racist for this argument, a conclusion that seems to me unwarranted. There is, however, an ideological element in his argument that is absolutely critical to it. Consider the second step, that is, the claim that IQ is a factor in attaining reward and that this must be so for society to function effectively. Herrnstein recognizes that his argument will collapse if, indeed, society can be organized in accordance with the "socialist dictum, 'From each according to his ability, to each according to his needs.' " His argument would not apply in a society in which "income (economic, social, and political) is unaffected by success."

Actually Herrnstein fails to point out that his argument requires the assumption not only that success must be rewarded, but that it must be rewarded in quite specific ways. If individuals are rewarded for success only by prestige, then no conclusions of any importance follow. It will only follow (granting his other assumptions) that children of people who are respected for their own achievements will be more likely to be respected for their own achievements, an innocuous result even if true. It may be that the child of two Olympic swimmers has a greater than average chance of achieving the same success (and the acclaim for it), but no dire social consequences follow from this hypothesis. The conclusion that Herrnstein and others find disturbing is that wealth and power will tend to concentrate in a hereditary meritocracy. But this follows only on the assumption that wealth and power (not merely respect) must be the rewards of successful achievement and that these (or their effects) are transmitted from parents to children. The issue is confused by Herrnstein's failure to isolate the specific factors crucial to his argument, and his use of the phrase "income (economic, social, and political)" to cover "rewards" of all types, including respect as well as wealth. It is confused further by the fact that he continually slips into identifying "social standing" with wealth. Thus he writes that if the social ladder is tapered steeply, the obvious way to rescue the people at the bottom is "to increase the aggregate wealth of society so that there is more room at the top"—which is untrue if social standing is a matter of acclaim and respect. (We overlook the fact that even on his tacit assumption redistribution of income would appear to be an equally obvious strategy.)

Consider then the narrower assumption that is crucial to his argument: Transmittable wealth and power accrue to mental ability and must do so for society to function effectively. If this assumption is false and society can be organized more or less in accordance with the socialist dictum, then nothing is left of Herrnstein's argument (except that it will apply to a competitive society in which his other factual assumptions hold). But the assumption is true, Herrnstein claims. The reason is that ability "expresses itself in labor only for gain" and people "compete for gain—economic and otherwise." People will work only if they are rewarded in terms of "social and political influence or relief from threat." All of this is merely asserted; no justification is given for these assertions. Note again that the argument supports the disturbing conclusions he draws only if we identify the "gain" for which people allegedly compete as transmittable wealth and power.

What reason is there to believe the crucial assumption that people will work only for gain in (transmittable) wealth and power, so that society cannot be organized in accordance with the socialist dictum? In a decent society, everyone would have the opportunity to find interesting work, and each person would be permitted the fullest possible scope for his talents. Would more be required—in particular, extrinsic reward in the form of wealth and power? Only if we assume that applying one's talents in interesting and socially useful work is not rewarding in itself, that there is no intrinsic satisfaction in creative and productive work, suited to one's abilities, or in helping others (say, one's family, friends, associates, or simply fellow members of society). Unless we suppose this, then even granting all of Herrnstein's other assumptions, it does not follow that there should be any concentration of wealth or power or influence in a hereditary elite.

For Herrnstein's argument to have any force at all, we must assume that people labor only for gain, and that the satisfaction in interesting or socially beneficial work, or in work well done, or in the respect shown to such activities is not a sufficient "gain" to induce anyone to work. The assumption, in short, is that without material reward people will vegetate. For this crucial assumption no semblance of an argument is offered. Rather Herrnstein merely asserts that if bakers and lumberjacks "got the top salaries and the top social approval"[2] in place of those now at the top of the social ladder, then "the scale of IQs would also invert," and the most talented would strive to become bakers and lumberjacks. This, of course, is no argument, but merely a reiteration of the claim that necessarily individuals work only for extrinsic reward. Furthermore, it is an extremely implausible claim. I doubt very much that Herrnstein would become a baker or lumberjack if he could earn more money in that way.

Similar points have been made in commentary on Herrnstein's article,[3] but in response, he merely reiterates his belief that there is no way "to end the blight of differential rewards." Continued assertion, however, is not to be confused with argument. Herrnstein's further assertion that "history shows . . ." in effect concedes defeat. Of course, history shows concentration of wealth and power in the hands of those able to accumulate it. One thought Herrnstein was trying to do more than merely put forth this truism. By reducing his argument finally to this assertion, Herrnstein implicitly concedes that he has no justification for the crucial assumption on which his argument rests, the unargued and unsupported claim that the talented must receive higher rewards.

If we look more carefully at what history and experience show, we find that if free exercise is permitted to the combination of ruthlessness, cunning, and whatever other qualities provide "success" in competitive societies, then those who have these qualities will rise to the top and will use their wealth and power to preserve and extend the privileges they attain. They will also construct ideologies to demonstrate that this result is only fair and just. We also find, contrary to capitalist ideology and behaviorist doctrine (of the nontautological variety), that many people often do not act solely or even primarily so as to achieve material gain or to maximize applause. As for the argument (if offered) that "history shows" the untenability of the "socialist dictum" that Herrnstein must reject for his argument to be valid, this may be assigned the same status as an eighteenth-century argument to the effect that capitalist democracy is impossible, as history shows.

One sometimes reads arguments to the effect that people are "economic maximizers," as we can see from the fact that given the opportunity, some will accumulate material reward and power.[4] By similar logic, we could prove that people are psychopathic criminals, since given social conditions under which those with violent criminal tendencies were free from all restraints, they might very well accumulate power and wealth while nonpsychopaths suffer in servitude. Evidently, from the lessons of history, we can reach only the most tentative conclusions about basic human tendencies.

Suppose that Herrnstein's unargued and crucial claim is incorrect. Suppose that there is, in fact, some intrinsic satisfaction in employing one's talents in challenging and creative work. Then, one might argue, this should compensate even for a diminution of extrinsic reward; and "reinforcement" should be given for the performance of unpleasant and boring tasks. It follows then that there should be a concentration of wealth (and the power that comes from wealth) among the less talented. I do not urge this conclusion, but merely observe that it is more plausible than Herrnstein's if his fundamental and unsupported assumption is false.

The belief that people must be driven or drawn to work by "gain" is a curious one. Of course, it is true if we use the vacuous Skinnerian scheme and speak of the "reinforcing quality" of interesting or useful work; and it may be true, though irrelevant to Herrnstein's thesis, if the gain sought is merely general respect and prestige. The assumption necessary for Herrnstein's argument, namely, that people must be driven or drawn to work by reward of wealth or power, obviously does not derive from science, nor does it appear to be supported by personal experience. I suspect that Herrnstein would exclude himself from the generalization, as already noted. Thus I am not convinced that he would at once apply for a job as a garbage collector if this were to pay more than his present position as a teacher and research psychologist. He would say, I am sure, that he does his work not because it maximizes wealth (or even prestige) but because it is interesting and challenging, that is, intrinsically rewarding; and there is no reason to doubt that this response would be correct. The statistical evidence, he points out, suggests that "if very high income is your goal, and you have a high IQ, do not waste your time with formal education beyond high school." Thus, if you are an economic maximizer, don't bother with a college education, given a high IQ. Few follow this advice, quite probably because they prefer interesting work to mere material reward. The assumption that people will work only for gain in wealth and power is not only unargued but quite probably false, except under extreme deprivation. But this degrading and brutal assumption, common to capitalist ideology and the behaviorist view of human beings, is fundamental to Herrnstein's argument.

There are other ideological elements in Herrnstein's argument, more peripheral, but still worth noting. He invariably describes the society he sees evolving as a "meritocracy," thus expressing the value judgment that the characteristics that yield reward are a sign of merit, that is, positive characteristics. He considers specifically

IQ, but, of course, recognizes that there might very well be other factors in the attainment of social success. One might speculate, rather plausibly, that wealth and power tend to accrue to those who are ruthless, cunning, avaricious, self-seeking, lacking in sympathy and compassion, subservient to authority, willing to abandon principle for material gain, and so on. Furthermore, these traits might very well be as heritable as IQ and might outweigh IQ as factors in gaining material reward. Such qualities might be just the valuable ones for a war of all against all. If so, then the society that results (applying Herrnstein's "syllogism") could hardly be characterized as a meritocracy. By using the word "meritocracy," Herrnstein begs some interesting questions and reveals implicit assumptions about our society that are hardly self-evident.

Teachers in ghetto schools commonly observe that students who are self-reliant, imaginative, energetic, and unwilling to submit to authority are often regarded as troublemakers and punished, on occasion even driven out of the school system. The implicit assumption that in a highly discriminatory society, or one with tremendous inequality of wealth and power, the meritorious will be rewarded is a curious one, indeed.

Consider further Herrnstein's assumption that, in fact, social rewards accrue to those who perform beneficial and needed services. He claims that the "gradient of occupations" is "a natural measure of value and scarcity," and that "the ties among IQ, occupation, and social standing make practical sense." This is his way of expressing the familiar theory that people are automatically rewarded in a just society (and more or less in our society) in accordance with their contribution to social welfare or "output." The theory is familiar, and so are its fallacies. Given great inequalities of wealth, we will expect to find that the "gradient of occupations" by pay is a natural measure of service to wealth and power—to those who can purchase and compel—and only by accident "a natural measure of value." The ties among IQ, occupation, and social standing that Herrnstein notes make "practical sense" for those with wealth and power, but not necessarily for society or its members in general.[5]

The point is quite obvious. Herrnstein's failure to notice it is particularly surprising, given the data on which he bases his observations about the relation between social reward and occupation. He bases these judgments on a ranking of occupations that shows, for example, that accountants, specialists in public relations, auditors, and sales managers tend to have higher IQs (hence, he would claim, receive higher pay, as they must if society is to function effectively) than musicians, riveters, bakers, lumberjacks, and teamsters. Accountants were ranked highest among seventy-four listed occupations, with public relations fourth, musicians thirty-fifth, riveters fiftieth, bakers sixty-fifth, truck drivers sixty-seventh, and lumberjacks seventieth. From such data, Herrnstein concludes that society is wisely "husbanding its intellectual resources"[6] and that the gradient of occupation is a natural measure of value and makes practical sense. Is it obvious that an accountant helping a corporation to cut its tax bill is doing work of greater social value than a musician, riveter, baker, truck driver, or lumberjack? Is a lawyer who earns a $100,000 fee to keep a dangerous drug on the market worth more to society than a farm worker or a nurse? Is a surgeon who performs operations for the rich doing work of greater social value than a practitioner in the slums who may work much harder for much less extrinsic reward? The gradient of occupations that Herrnstein uses to support his claims with regard to the correlation between IQ and social value surely reflects, in part at least, the demands of wealth and power; a further argument is needed to demonstrate Herrnstein's claim that those at the top of the list are performing the highest service to "society," which is wisely husbanding its resources by rewarding the accountants and public relations experts and engineers (e.g., designers of antipersonnel weapons) for their special skills. Herrnstein's failure to notice what his data immediately suggest is another indication of his uncritical (and apparently unconscious) acceptance of capitalist ideology in its crudest form.

Notice that if the ranking of occupations by IQ correlates with ranking by income, then the data that Herrnstein cite can be interpreted in part as an indication of an unfortunate bias towards occupations that serve the wealthy and powerful and away from work that might be more satisfying and socially useful. At least, this would certainly seem a plausible assumption, one that Herrnstein never discusses, given his unquestioning acceptance of the prevailing ideology.

There is, no doubt, some complex of characteristics conducive to material reward in a state capitalist society. This complex may include IQ and quite possibly other more important factors, perhaps those noted earlier. To the extent that these characteristics are heritable (and a factor in choosing mates), there will be a tendency towards stratification in terms of these qualities. This much is obvious enough.

Furthermore, people with higher IQs will tend to have more freedom in selection of occupation. Depending on their other traits and opportunities, they will tend to choose more interesting work or more remunerative work, these categories being by no means identical. Therefore, one can expect to find some correlation between IQ and material reward, and some correlation between IQ and an independent ranking of occupations by their intrinsic interest and intellectual challenge. Were we to rank occupations by social utility in some manner, we would probably find at most a weak correlation with remuneration or with intrinsic interest and quite possibly a negative correlation. Unequal distribution of wealth and power will naturally introduce a bias towards greater remuneration for services to the privileged, thereby causing the scale of remuneration to diverge from the scale of social utility in many instances.

From Herrnstein's data and arguments, we can draw no further conclusions about what would happen in a just society, unless we add the assumption that people labor only for material gain, for wealth and power, and that they do not seek interesting work suited to their abilities—that they would vegetate rather than do such work. Since Herrnstein offers no reason why we should believe any of this (and there is certainly some reason why we should not), none of his conclusions follow from his factual assumptions, even if these are correct. The crucial step in his syllogism, in effect, amounts to the claim that the ideology of capitalist society expresses universal traits of human nature, and that certain related assumptions of behaviorist psychology are correct. Conceivably, these unsupported assumptions are true. But once it is recognized how critical their role is in his argument and what empirical support they, in fact, have, any further interest in this argument would seem to evaporate.

I have assumed so far that prestige, respect, and so on might be factors in causing people to work (as Herrnstein implies). This seems to me by no means obvious, though even if it is true, Herrnstein's conclusions clearly do not follow. In a decent society, socially necessary and unpleasant work would be divided on some egalitarian basis, and beyond that, people would have, as an inalienable right, the widest possible opportunity to do work that interests them. They might be "reinforced" by self-respect if they do their work to the best of their ability, or if their work benefits those to whom they are related by bonds of friendship and sympathy and solidarity. Such notions are commonly an object of ridicule—as it was common, in an earlier period, to scoff at the absurd idea that a peasant has the same inalienable rights as a nobleman. There always have been, and no doubt always will be, people who cannot conceive of the possibility that things could be different from what they are. Perhaps they are right, but again one awaits a rational argument.

In a decent society of the sort just described—which, one might think, becomes increasingly realizable with technological progress—there should be no shortage of scientists, engineers, surgeons, artists, craftsmen, teachers, and so on, simply because such work is intrinsically rewarding. There is no reason to doubt that people in these occupations would work as hard as those fortunate few who can choose their own work generally do today. Of course, if Herrnstein's assumptions, borrowed from capitalist ideology and behaviorist belief, are correct, then people will remain idle rather than do such work unless there is deprivation and extrinsic reward. But no reason is suggested as to why we should accept this strange and demeaning doctrine.

Lurking in the background of the debate over Herrnstein's syllogism is the matter of race, though he himself barely alludes to it. His critics are disturbed, and rightly so, by the fact that his argument will surely be exploited by racists to justify discrimination, much as Herrnstein may personally deplore this fact. More generally, Herrnstein's argument will be adopted by the privileged to justify their privilege on the grounds that they are being rewarded for their ability and that such reward is necessary if society is to function properly. The situation is reminiscent of nineteenth-century racist anthropology. Marvin Harris notes:

> Racism also had its uses as a justification for class and caste hierarchies; it was a splendid explanation of both national and class privilege. It helped to maintain slavery and serfdom; it smoothed the way for the rape of Africa and the slaughter of the American Indian; it steeled the nerves of the Manchester captains of industry as they lowered wages, lengthened the working day, and hired more women and children.[7]

We can expect Herrnstein's arguments to be used in a similar way and for similar reasons. When we discover that his argument is without force, unless we adopt unargued and implausible premises that happen to incorporate the dominant ideology, we quite naturally turn to the question of the social function of his conclusions and ask why the argument is taken seriously, exactly as in the case of nineteenth-century racist anthropology.

Since the issue is often obscured by polemic, it is perhaps worth stating again that the question of the validity and scientific status of a particular point of view is, of course, logically independent from the question of its social function; each is a legitimate topic of inquiry, and the latter becomes of particular interest when the point of view in question is revealed to be seriously deficient on empirical or logical grounds.

The nineteenth-century racist anthropologists were no doubt quite often honest and sincere. They might have believed that they were simply dispassionate investigators, advancing science, following the facts where they led. Conceding this, we might, nevertheless, question their judgment, and not merely because the evidence was poor and the arguments fallacious. We might take note of the relative lack of concern over the ways in which these "scientific investigations" were likely to be used. It would be a poor excuse for the nineteenth-century racist anthropologist to plead, in Herrnstein's words, that "a neutral commentator . . . would have to say that the case is simply not settled" (with regard to racial inferiority) and that the "fundamental issue" is "whether inquiry shall (again) be shut off because someone thinks society is best left in ignorance." The nineteenth-century racist anthropologist, like any other person, is responsible for the effects of what he does, insofar as they can be clearly foreseen. If the likely consequences of his "scientific work" are those that Harris describes, he has the responsibility to take this likelihood into account. This would be true even if the work had real scientific merit—more so, in fact, in this case.

Similarly, imagine a psychologist in Hitler's Germany who thought he could show that Jews had a genetically determined tendency towards usury (like squirrels bred to collect too many nuts) or a drive towards antisocial conspiracy and domination, and soon. If he were criticized for even undertaking these studies, could he merely respond that "a neutral commentator . . . would have to say that the case is simply not settled" and that the "fundamental issue" is "whether inquiry shall (again) be shut off because someone thinks society is best left in ignorance"? I think not. Rather, I think that such a response would have been met with justifible contempt. At best, he could claim that he is faced with a conflict of values. On the one hand, there is the alleged scientific importance of determining whether, in fact, Jews have a genetically determined tendency towards usury and domination (as might conceivably be the case). On the other, there is the likelihood that even opening this question and regarding it as a subject for scientific inquiry would provide ammunition for Goebbels and Rosenberg and their henchmen. Were this hypothetical psychologist to disregard the likely social consequences of his research (or even his undertaking of research under existing social conditions, he would fully deserve the contempt of decent people. Of course, scientific curiosity should be encouraged (though fallacious argument and investigation of silly questions should not), but it is not an absolute value.

The extravagant praise lavished on Herrnstein's flimsy argument and the widespread failure to note its implicit bias and unargued assumptions[8] suggest that we are not dealing simply with a question of scientific curiosity. Since it is impossible to explain this acclaim on the basis of the substance or force of the argument, it is natural to ask whether the conclusions are so welcome to many commentators that they lose their critical faculties and fail to perceive that certain crucial and quite unsupported assumptions happen to be nothing other than a variant of the prevailing ideology. This failure is disturbing, more so, perhaps, than the conclusions Herrnstein attempts to draw from his flawed syllogism.

Turning to the question of race and intelligence, we grant too much to the contemporary investigator of this question when we see him as faced with a conflict of values: scientific curiosity versus social consequences. Given the virtual certainty that even the undertaking of the inquiry will reinforce some of the most despicable features of our society, the intensity of the presumed moral dilemma depends critically on the scientific significance of the issue that he is choosing to investigate. Even if the scientific significance were immense, we should certainly question the seriousness of the dilemma, given the likely social consequences. But if the scientific interest of any possible finding is slight, then the dilemma vanishes.

In fact, it seems that the question of the relation, if any, between race and intelligence has little scientific importance (as it has no social importance, except under the assumptions of a racist society). A possible correlation between mean IQ and skin color is of no greater scientific interest than a correlation between any two other arbitrarily selected traits, say, mean height and color of eyes. The empirical results, whatever they may be, appear to have little bearing on any issue of scientific significance. In the present state of scientific understanding, there would appear to be little scientific interest in the discovery that one partly heritable trait correlates (or not) with another partly heritable trait. Such questions might be interesting if the results had some bearing, say, on hypotheses about the physiological mechanisms involved, but this is not the case. Therefore, the investigation seems of quite limited scientific interest, and the zeal and intensity with which some pursue or welcome it cannot reasonably be attributed to a dispassionate desire to advance science. It would, of course, be foolish to claim in response that "society should not be left in ignorance." Society is happily "in ignorance" of insignificant matters of all sorts. And with the best of will it is difficult to avoid questioning the good faith of those who deplore the alleged "anti-intellectualism" of the critics of scientifically trivial and socially malicious investigations. On the contrary, the investigator of race and intelligence might do well to explain the intellectual significance of the topic he is studying and thus enlighten us as to the moral dilemma he perceives. If he perceives none, the conclusion is obvious, with no further discussion.

As to social importance, a correlation between race and mean IQ (were this shown to exist) entails no social consequences except in a racist society in which each individual is assigned to a racial category and dealt with, not as an individual in his own right, but as a representative of this category. Herrnstein mentions a possible correlation between height and IQ. Of what social importance is that? None, of course, since our society does not suffer under discrimination by height. We do not insist on assigning each adult to the category "below six feet in height" or "above six feet in height" when we ask what sort of education he should receive or where he should live or what he should do. Rather, he is what he is, quite independent of the mean IQ of people of his height category. In a nonracist society, the category of race would be of no greater significance. The mean IQ of individuals of a certain racial background is irrelevant to the situation of a particular individual, who is what he is. Recognizing this perfectly obvious fact, we are left with little, if any, plausible justification for an interest in the relation between mean IQ and race, apart from the "justification" provided by the existence of racial discrimination.

The question of heritability of IQ might conceivably have some social importance, say, with regard to educational practice. However, even this seems dubious, and one would like to see an argument. It is, incidentally, surprising to me that so many commentators should find it disturbing that IQ might be heritable, perhaps largely so.[9] Would it also be disturbing to discover that relative height, or musical talent, or rank in running the 100-yard dash is in part genetically determined? Why should one have preconceptions one way or another about these questions, and how do the answers to them, whatever they may be, relate either to serious scientific issues (in the present state of our knowledge) or to social practice in a decent society?

Notes

1. He does not specifically mention this assumption, but it is necessary to the argument.
2. Note again Herrnstein's failure to distinguish remuneration from social approval, though the argument collapses if the only reward is approval.
3. *Atlantic Monthly*, November 1971. *See* p. 110, first paragraph, for his rejoinder.
4. *See*, e.g., H. W. Blair, "The Green Revolution and 'Economic Man': Some Lessons for Community Development in South Asia." *Pacific Affairs*, Fall 1971, pp. 353–367.
5. To assume that society tends to reward those who perform a social service is to succumb to essentially the same fallacy (among others) that undermines the argument that a free market, in principle, leads to optimal satisfaction of wants—whereas when wealth is badly distributed, the system will tend to produce luxuries for the few who can pay rather than necessities for the many who cannot.
6. Misleadingly, Herrnstein states: "Society is, in effect, husbanding its intellectual resources by holding engineers in greater esteem and paying them more." But if he really wants to claim this on the basis of the ties between IQ and social standing that his data reveal, then he should conclude as well that society is husbanding its intellectual resources by holding accountants and PR men in greater esteem and paying them more. Quite apart from this, it is not so obvious as he apparently believes that society is wisely husbanding its intellectual resources by employing most of its scientists and engineers in military and space research and development.
7. M. Harris, *The Rise of Anthropological Theory* (New York: Thomas Y.

Crowell, 1968), pp. 100–101. By the 1860s, he writes, "anthropology and racial determinism had become almost synonymous."

8. *See* the correspondence in *Atlantic*, November 1971.

9. An advertisement in the *Harvard Crimson* (November 29, 1971), signed by many faculty members, refers to the "disturbing conclusion that 'intelligence' is largely genetic, so that over many, many years society might evolve into classes marked by distinctly different levels of ability." Since the conclusion does not follow from the premise, as already noted, it may be that what disturbs the signers is the "conclusion that 'intelligence' is largely genetic." Why this should seem disturbing remains obscure.

Whatever Happened to Vaudeville?

A Reply to Professor Chomsky

Richard J. Herrnstein

PROFESSOR CHOMSKY WINDS UP A DISCUSSION (1972) OF SOME OF MY views on social stratification (1971) by listing what he calls my assumptions: "that people labor only for material gain, for wealth and power, and that they do not seek interesting work suited to their abilities—that they would vegetate rather than do such work" (p. 292). Quite sensibly in the face of such strange assumptions, he rejects any conclusions that follow, especially since I offer "no reason why we should believe any of this (and there is certainly some reason why we should not)" (p. 292). I agree completely, except that I made no such assumptions, nor are they required by any of my conclusions. Nevertheless, to be misread by a person of Professor Chomsky's quickness of wit impels me to try to restate, perhaps to clarify, the part of my discussion that has eluded him. I believe that he was led astray because my conclusions run contrary to his political suppositions rather than by my prose style. In particular, my argument wrecks the egalitarianism to which Professor Chomsky, like many other American intellectuals, pays homage.

First, what is my argument? Concisely stated, it is that, (1) since people inherit their mental capacities (as indexed, for example, in intelligence tests) to some extent, and (2) since success in our society calls for those mental capacities, therefore, (3), it follows that success in our society reflects inherited differences between people. To readers who have not worked their way through the 15,000 or so words leading up to that conclusion, my syllogism may seem rash and unsubstantiated. To those readers, I can only commend my original article. Since Professor Chomsky did not question the facts behind the two premises, I will not bother reviewing them here. Nor did Professor Chomsky object to the logical form of my argument, to the inevitability of (3), given (1) and (2). Some readers may wonder, at this point, what other objection there can be to a syllogism, besides to the factual truth of its premises or the logical soundness of its steps. I will come to that in a moment. But first, I note that the second premise—that success in our society depends in some significant way on mental capacities—means specifically that mental competence is necessary, but not sufficient, for getting ahead. To put it yet another way, each level of success in our society, as defined by the members of society themselves, permits a range of intelligence, but the higher the level, the narrower the range. At the top of the scale of occupations, the intellectual requirement is set relatively high, but high intelligence does not guarantee success. This, too, is a factual assertion, which I therefore offer here without further substantiation.

I focus on premise 2 because Professor Chomsky's main argument centers thereon. As he puts it: "This step in the argument embodies two assumptions: first, it is so in fact; and second, it must be so, for society to function effectively." (p. 285) Professor Chomsky disputes only the inevitability of premise 2, not its current accuracy. But do I really assume that the second premise "must be so"? In fact, I find no such assumption in my article, nor can I see that I need it. For me, it is no disgrace if my argument holds merely for *existing* societies, not necessarily *all possible* ones. As regards societies about which we have some data, such as the American or the Japanese or the Western European or the Russian, Professor Chomsky apparently accepts premise 2, and so is inexorably carried on to my conclusion—that society will stratify itself increasingly by genetic factors as it divests itself of the unfair barriers of race, religion, family connections, inherited wealth, and so on. Or at least, he has not at this time chosen to dispute my actual conclusion. Instead, his dispute is with a possibility I offer for scrutiny, which is that hereditary social classes may stratify not only existing societies but also any conceivable society in which merit is a factor in social status. Professor Chomsky challenges, then, not my conclusion but my extrapolation, which is a distinction that I would like to register even as he overlooks it.

Now, having made the distinction, I contend that his criticism of my extrapolation has barely a shred of merit, which, it seems to me, strengthens my case, for it would be a subtle flaw indeed that would elude Professor Chomsky's sharp eye. Professor Chomsky notes that my argument "would not apply in a society in which 'income (economic, social, and political) is unaffected by success.' " (p. 286) That is correct, for in such a society, should it come to pass, the second premise may no longer hold. If people's accomplishments did not gain the rewards society has to dispense, then, indeed, success (as measured by social rewards) might not depend in any significant way on mental capacities, thereby violating premise 2. But now, Professor Chomsky does something odd. Instead of depicting a hypothetical society without differential rewards, he instead postulates a society in which people are rewarded for their accomplishments "only by prestige." If they are differentially rewarded for their accomplishments, even if only by prestige, my syllogism applies, as Professor Chomsky admits. The second premise is back, albeit only for success defined by prestige—not by money or dachas or limousines or winter vacations in the Caribbean or on the Black Sea. In his idea of the "decent society," says Professor Chomsky, "It will only follow (granting his other assumptions [with which, I note, he registers no dispute at any point]) that children of people who are respected for their achievements will be more likely to be respected for their own achievements, an innocuous result even if true" (p. 286).

Why does Professor Chomsky find it "innocuous" for prestige to run in families? He does not say, but I believe it is because he thinks that if the *sole* status distinction between people is based on the prestige or respect they earn from their fellows, they will not suffer for their failures as much as they do in our society, in which the penalty for failure is poverty, or at least, relative poverty as compared with our society's successful people. Let us, for argument's sake, suppose that Professor Chomsky's decent society could get its work done. It is, after all, possible (although rather unlikely, I wager) that a society using prestige as its only differential social reward could render prestige potent enough to sustain work no less well than do the rewards in our society, including money and power. But if prestige were that potent, then surely the lack of it would cause sadness and regret, just as the lack of money and power causes sadness and regret now. Perhaps it would be a kinder world than ours, if, in eliminating the suffering caused by poverty, it did not substitute equally or more painful psychic deprivations. But "innocuous" hardly seems like the right word for a society stratified by a mortal competition for prestige. Professor Chomsky did not postulate any such competition, but that is what he would find, nevertheless, if prestige could be made potent enough to replace the material rewards of existing societies. Yet, if prestige were not that potent, could Professor Chomsky's utopia get its work done?

Does anyone doubt that the differential rewards granted in society function like the potential difference in an electrical circuit—as a kind of labor pump? By attaching different outcomes for different jobs, or for jobs done well or poorly, society directs the flow of labor one way or the other—as, for example, out of vaudeville and into radio and motion pictures, which had captured its audience and the attendant multiple rewards. As a more timely example, consider the diminishing ranks of applicants for graduate schools and the lengthening queues for law and medical schools, precisely in tune with the shifting demands in society at large. Or remember that when the rewards for manufacturing spats disappeared, so did spats manufacturers. The inherent rewards of making spats, such as they were, could not have changed, but the extrinsic ones evaporated, and so did the industry. Now, this is not to suggest that society always distributes its rewards sensibly, humanely, or even attractively; merely that the distribution expresses something like a social consensus, which then gets converted into human effort. Sometimes, because of extraneous perturbations, or short-term influences, the consensus may be faulty, as Professor Chomsky notes. Thus, our society may be harming itself in the long run by paying public relations experts such high salaries (to the irritation of both Professor Chomsky and me), thereby attracting into the business the bright people who then sell us a bill of goods. No doubt, Professor Chomsky rightly notes that such high salaries often come from the wealthy few who have a stake in keeping the rest of us fooled. But all is not lost, for there is glory (if not also money) waiting for the fellow who sets the public straight—such as Ralph Nader, perhaps—showing that the system may have more resiliency than Professor Chomsky supposes. In any case, the merits of a given social consensus as compared with another is immaterial to the issue, except insofar as Professor Chomsky thinks I approve of the American consensus on the values of various occupations and therefore holds me accountable for what he considers its flaws. Whether I approve or not, whether the consensus is wise or not, the point is simply that labor flows toward the rewards, and that if a given reward successfully guides the flow of labor, then it is valuable enough to cause psychological pain by its absence. The relevant principle is one that Professor Chomsky, in his distaste for Professor Skinner's psychology, has apparently never grasped. If a reward can sustain effort by its acquisition, then it will punish by its deprivation. Or, to be concrete about it—if, in Professor Chomsky's hypothetical world, prestige and respect are strong enough to direct labor in accordance with the social consensus, then they are strong enough to bring unhappiness to those who fail to get it. It does not matter a bit whether the consensus comes from a free market, monopoly capitalists, or the central government.

In all likelihood, however, Professor Chomsky did not mean to allow prestige to be all that important in his decent society. Although a bit vague on this point, I believe that he intended the differential reward of prestige to be ineffective, which is to say, he intended that it have little or no effect on the distribution of labor. For the moment, let us grant that work would go on anyway—a supposition I firmly disbelieve and will later reconsider. With prestige unimportant, Professor Chomsky correctly infers that the lack of prestige would cause no pain and that distinctions would be innocuous. But, if so, he erred in concluding that prestige would then run in families. If prestige was an impotent reward (hence, causing no suffering by its lack), then the better-endowed people would not end up in the prestigious occupations. They would instead be randomly dispersed among the various occupational levels (except for another of Professor Chomsky's implausible postulations, to which I also return later), and their children would be randomly dispersed too. Premise 2 is either right or wrong as a factual matter. If it is right (assuming premise 1 is right also), then there will be a hereditary meritocracy *and* some people will suffer the pain of having lost out in the competition for society's rewards. If it is wrong, then we have no reason to infer a hereditary meritocracy *and* no reason to suppose anyone is hurting for the lack of society's impotent rewards.

Perhaps Professor Chomsky erred because he forgot that one reason we now find the better-endowed people in the highly rewarded occupations surely is that they enjoy a competitive advantage in the contest to fill them. The correlation between occupational level and intellect has something to do with the greater desirability of the better jobs, on the one hand, and the greater competence (on the average) of the intelligent, on the other. Professor Chomsky accepts the fact of a correlation now,* predicts that there would be a correlation in his hypothetical world, but disbelieves that the extrinsic social rewards—which he would ban—have anything to do with it. But why, if society did not create a gradient of potent lures toward the various occupations—by money, power, prestige, or whatever—would there be any sort of correlation between intellect and occupational level (defined by some secret consensus of social utility that was not being translated into differential rewards)?

Professor Chomsky apparently believes that the inherent pleasures of labor would yield a good strong correlation anyway. "In a decent society everyone would have the opportunity to find interesting work, and each person would be permitted the fullest possible scope for his talents. Would more be required, in particular, extrinsic reward in the form of wealth and power? Only if we assume that applying one's talents in interesting and socially useful work is not rewarding in itself, that there is no intrinsic satisfaction in creative and productive work, suited to one's abilities, or in helping others (say, one's family, friends, associates, or simply fellow members of society)" (pp. 287f.). In short, his hypothetical society would sort people roughly the way a beehive sorts bees—by a differentiation in the individuals rather than in their extrinsic rewards. Only steady-handed, nerveless intellectuals would yearn to be surgeons; only true-eared, sensitive artists would crave to sing in public; only masters of logical complexity would declare themselves chess players (and no one must keep score at matches, for the pleasures of victory are paid for by the pain of defeat). And at the other end of the range, the underendowed would cheerily and spontaneously designate themselves assistant clerks or plumbers' helpers. That is how Professor Chomsky must get his correlation, for there would be no differences in pay, no differences in power,

* Here and there, Professor Chomsky muses about the importance of ruthlessness and the like, as opposed to intellect, in the struggle for achievement in our society. He apparently does not know that the data on the matter show intellect, as indexed in IQ tests or just by schooling, to be a far better predictor than any measure of personality that might identify those who are "ruthless, cunning, avaricious, self-seeking, lacking in sympathy and compassion, subservient to authority, and willing to abandon principle for material gain" (p. 290), to quote Professor Chomsky's formula for success n America. No doubt, given two people of equal intellect, personality and character may spell the difference, but even here, the data do not bear out Professor Chomsky's gloomy vision. Instead, the ones who succeed tend, on the average, to be the buoyant, energetic, independent, healthy ones, although there are many interesting exceptions.

and such differences in respect as he grants would be innocuous, which is to say that they would not much affect the competition for jobs. There would only be differences in "intrinsic satisfaction in creative and productive work." Clearly, Professor Chomsky feels I have seriously underrated the power of that intrinsic satisfaction. And I grant that I hold it to be less important than he does, for, while he wants the world to run on it, I believe that human society can no more transform itself into a beehive than vice versa. I would say that the burden falls squarely on him to prove otherwise.

It is not necessary to prove that work is sustained by a mixture of intrinsic and extrinsic rewards, and that both contribute to its attractiveness. To that extent, I agree with Professor Chomsky when he says that the extrinsic rewards do not solely determine the distribution of labor. However, I know of no one who would disagree with him, not even Professor Skinner. Neither Professor Skinner nor I have any trouble understanding why house painters get more money for painting the outsides of houses than for painting the insides. They require extra extrinsic rewards to offset the intrinsic disadvantages of clinging to the twentieth or thirtieth story on windy days, as compared to the safety indoors. Throughout the scale of occupations, something similar operates. For a given level of social utility (not in any philosophical sense, but as measured by the prevailing consensus), the intrinsic and extrinsic benefits add up to something like a constant. Thus, Albert Schweitzer did not have to get paid a lot of money for his work in order to keep him going into old age, for the respect and the eternal reward he envisioned were apparently recompense enough. A society wisely praises such men "richly," if it is to have the fruits of great talent unstintingly dedicated. But such rich praise, as well as other sorts of riches, would have to be prohibited to inactivate my syllogism.

The main issue between Professor Chomsky and me finally boils down to this. Suppose all extrinsic rewards for labor, from gratitude to cash, were somehow held constant over all occupations. Now, let only the intrinsic satisfactions vary as they will. Professor Chomsky supposes that in such a world, in which my syllogism would truly be innocuous, everything would go just fine. There would, he assumes, be no more clumsy surgeons, suicidal airplane pilots, inarticulate teachers, rude salesmen, out-of-tune singers in his ideal world than there are in the one we are living in. He supposes, in other words, that the matching of talent to occupation need owe nothing to society's system of differential rewards. A remarkable supposition; to me, utterly unbelievable. Moreover, Professor Chomsky supposes that in his world there would be no shortage of labor; that somehow, the intrinsic rewards of coal mining, ditch digging, schooling, garbage collecting, poodle clipping, even house painting would keep the work at just the level society needs, neither too much nor too little. He supposes that the extrinsic rewards contribute nothing essential to the monitoring of labor. The reason Professor Chomsky must be supposing these outlandish things is that, as he well knows, as soon as he grants a role to the extrinsic rewards, my syllogism starts cranking away. It changes my argument not at all if intrinsic satisfaction could account for some of the distribution of labor, for the distribution of external rewards will compensate for such complications as the social consensus dictates. An onerous but important job will draw rich rewards; a pleasurable, insignificant one will draw little if any. As long as some of the distribution of labor depends at all significantly upon differential extrinsic rewards, and as long as the likelihood of success depends upon inherited mental differences (which, please recall, Professor Chomsky grants, or at least does not challenge), then social standing will depend upon inherited differences to some degree.

Professor Chomsky may find me lacking in imagination. Why, he may ask, can I not picture his revolutionary new man or woman, eager to serve the decent society for no differential rewards except the sense of a useful job well done? The answer is that I know what has happened before when the state has told its citizens to be good and productive henceforth for the sake of the state (usually in the name of the "people"), instead of for their own sakes, and then enforces its vision of "classlessness." History does not encourage further ventures of that sort; at least, it does not encourage me. Soon after the leaders discover that selflessness cannot be counted upon, they are most likely to impose a gradient of punishment, which may have about the same potential for producing labor as our society's gradient of reward, except that it is bound to be more, rather than less, cruel. It hardly looks like an improvement to substitute imprisonment or forced labor camps for poverty (especially when the poverty persists). Not that I think Professor Chomsky favors any such reign of terror, but he might have little to say at that point, for the revolution's visionaries are often among its first victims. Professor Chomsky surely knows that the persistent status differentials in all socialist states follow directly from individual differences in ability *and* the Skinnerian principles of reward and punishment that he so contemptuously, and repeatedly, keeps dismissing. He must therefore tell us how his decent society will steer its way between those venerable human limitations.

So much for the main issue between Professor Chomsky and me. Unlike some of my more vehement critics, Professor Chomsky does not accuse me of racism, for which I am grateful. However, he does hold me accountable for making an argument that "will surely be exploited by racists to justify discrimination" (p. 293). Unfortunately for the sake of further discussion, Professor Chomsky fails to say which of my arguments will have that unwholesome consequence, which makes his assertion difficult either to evaluate or refute. Professor Chomsky does, however, provide what he considers an analogous case to mine, with which I heartily disagree. Perhaps, if I refute his analogy, I will have dealt with his complaint.

First, the analogy:

> Imagine a psychologist in Hitler's Germany who thought he could show that Jews had a genetically determined tendency toward usury (like squirrels bred to collect too many nuts) or, a drive toward antisocial conspiracy and domination, and so on. If he were criticized for even undertaking these studies, could he merely respond that [Professor Chomsky is quoting me here] "a neutral commentator . . . would have to say that the case is simply not settled" and that the "fundamental issue" is "whether inquiry shall (again) be shut off because someone thinks society is best left in ignorance"? I think not. Rather, I think that such a response would have been met with justifiable contempt. At best, he could claim that he is faced with a conflict of values. On the one hand, there is the alleged scientific importance of determining whether in fact Jews have a genetically determined tendency toward usury and domination (an empirical question, no doubt). On the other, there is the likelihood that even opening this question and regarding it as a subject for scientific inquiry would provide ammunition for Goebbels and Rosenberg and their henchmen. [p. 294–295]

Presumably, then, because I did not deny the possibility of a racial difference in IQ, I am like the scientist in the analogy, studying innate Jewish usury in Hitler's Germany. One must make allowances for Professor Chomsky's tendency toward hyperbole.

America is not Hitler's Germany, and I was not proposing to study, nor was I asserting, a genetic flaw in any race comparable to inborn usuriousness. Let us, nevertheless, consider the analogy on its own merits. To begin with, I agree that, in Hitler's Germany, I might not study innate hoarding in Jews. But then, in Hitler's Germany, would I do any science at all (disregarding, for argument's sake, that Professor Chomsky and I would both be in some concentration camp)? I hope that I would have had the strength to cease being a scientist in such a society and the good sense to have been among those who fled from Germany in the 1930s.

I therefore share Professor Chomsky's contempt for his hypothetical scientist, but not for Professor Chomsky's reason. The scientist's specialty hardly matters, compared with his willingness to stay and work at all. Wolfgang Köhler, who voluntarily vacated the prime academic chair for a psychologist in Germany—the professorship at Berlin—worked on the physiological basis of perception. Was his gesture any less admirable because his research had no clear relevance to Nazi ideology? I think not. Goebbels and Rosenberg did not need Köhler's data; they needed Köhler's acquiescence in German society. And so it would be with Professor Chomsky's hypothetical scientist.

As a matter of fact, if Professor Chomsky's scientist is honest, Goebbels and Rosenberg would surely stop him from carrying out his research anyway. Professor Chomsky forgets that in honest research, one does not always know the answer beforehand. Goebbels and Rosenberg would worry that it would turn out that Jews do *not* have an innate tendency toward usury, which would be quite embarrassing for the party, if it got around. Instead, they would simply find some pseudoscientist who would invent more convenient findings. Professor Chomsky, who wants me to subject scientific findings to the test of political suitability (*see* his pp. 293–296), should expect no less from Goebbels.

Professor Chomsky's analogy proves to be quite revealing, although it has little bearing on my article and hardly proves that my argument will be "exploited by racists." Contrary, I am sure, to his intention, I draw the lesson that we should encourage *more* research on people, not less. And all of it of passable quality should be published, not picked over for symptoms of apostasy. I trust that, as always before, the truth will turn out to be more complex and subtle than called for in anyone's orthodoxy. Since society must cope with what people are really like, rather than with the fictions embodied in one political philosophy or the other, we would do well to learn as much as we can at every opportunity, limited, of course, by the rights of individuals to their privacy.

References

CHOMSKY, N. 1972. "Psychology and Ideology." *Cognition* 1:11–46. (Chomsky's "The Fallacy of Richard Herrnstein's IQ," reprinted in this volume, is a part of "Psychology and Ideology." Page numbers referred to by Herrnstein are keyed to the pagination of Chomsky's article as it appears in this volume. Eds.)

HERRNSTEIN, R. J. 1971. "IQ." *Atlantic Monthly,* September, pp. 43–64.

Comments on Herrnstein's Response

Noam Chomsky

In my paper "Psychology and Ideology" [*Cognition* 1, no. 1 (1972): 11–46] I considered Herrnstein's argument that a genetic component in IQ is producing a stable hereditary meritocracy and showed that the argument requires tacit premises which happen to incorporate basic elements of the prevailing ideology and unsubstantiated behaviorist doctrine. It is necessary not only to adopt his explicit premise that people labor only for gain but also the further premise that the gain sought is transmittable and cumulative, the obvious candidate being wealth. Once errors are cleared away, Herrnstein's reply adds supporting evidence to this conclusion and reveals still more clearly the assumptions about the human species that are implicit in his general conception.

Herrnstein objects: (A) That I was wrong to describe the assumption that people labor only for material gain in transmittable reward (say, wealth) as "[his] assumption"; and (B) that no such assumption is "required by any of [his] conclusions." Neither objection stands. As to (A), I did not attribute the assumption to Herrnstein; rather, my point was that the assumption was unacknowledged, and given his reply we may now add "rejected," but that once the hidden premise is made clear the argument loses all interest. Contrary to his objection (B), the assumption in question is indeed required for his conclusion that "there will be a hereditary meritocracy." The same assumption is required for his "extrapolation" that any viable society will be stratified in "hereditary social classes" based on inherited mental capacity.

Let us isolate for examination exactly what is at issue. There is no doubt that if some complex of characteristics C, partially heritable, is the factor that leads to "success," then success will partially reflect inherited differences. Furthermore, it is uncontroversial that there are vast differences in wealth, power, and other "rewards" in industrial societies (there is, furthermore, no detectable move towards equalization in the past several generations). When pressed, Herrnstein retreats to these truisms, but of course his article would never have been published unless it went beyond them, as it did in two respects. First, Herrnstein asserted that mental capacity as measured by IQ is a crucial component of the complex C. Furthermore, he claims to have identified the source of existing inequalities, and since he believes this factor to be essential for a viable society, he "extrapolates" to his conclusion that a hereditary meritocracy is virtually inescapable. Central to his argument is the thesis that ability "expresses itself in labor only for gain," that "labor flows towards the rewards," and that there is no way "to end the blight of differential rewards." Herrnstein does not qualify this assumption in his response. What he denies is that the gain in question must be transmittable and cumulative (say, wealth) for his argument to go through. Let us take up in turn the points at issue.

Consider Herrnstein's claim that IQ is a major factor in producing the "gain" which alone motivates labor. As I noted, Herrnstein presents no serious evidence that IQ is a major factor in determining social reward. But others have investigated the matter. Bowles and Gintis conclude from a careful analysis of the data that IQ is a minor factor and the heritable component in IQ an insignificant factor in determining income: "A perfect equalization of IQ's across social classes would reduce the intergenerational transmission of economic status by a negligible amount." Furthermore, when IQ, social class, and education are evaluated for their independent contribution to the determination of income, "IQ is by far the least important."[1] Jencks et al. give as their "best estimate" that there is "about 3 percent less income inequality in genetically homogenous subpopulations than in the entire American population."[2] Thus empirical investigations of such data as exist indicate that IQ is a minor component of the complex C which brings economic reward, and the heritable element in IQ, a negligible component. Herrnstein presents no evidence that prestige, acclaim, or any other "social reward" accrues to mental capacity in general, nor is there any to my knowledge (consider the annual list of the ten most admired men and women). Thus if mental capacity is measured by IQ and success by income (or, so far as we know, prestige, and so on), Herrnstein's factual premise (2) that success requires mental capacity has only limited validity, and his conclusion that success reflects inherited mental capacity has virtually no force.

These observations suffice to dismiss Herrnstein's argument. However, my concern was not its empirical inadequacies but rather its ideological assumptions and the further question why any interest has been aroused in work so plainly devoid of substance. Therefore, after noting that IQ was very likely only a minor element (and its inherited component, a negligible element) in determining measurable "success," I disregarded the issue and granted Herrnstein's factual claims, for the sake of argument, a fact which apparently led him to believe that I accepted these claims as correct.

Let us now turn to the second question. Still assuming, with Herrnstein, that ability expresses itself in labor only for gain and that this ability is partially heritable, let us ask what conclusion can be drawn if the gain that motivates effort is nontransmittable, say, prestige and acclaim. Herrnstein takes issue with my observation that the conclusions in this case are innocuous. I did not explain the point, assuming it to be obvious, but perhaps a comment is in order, given Herrnstein's failure to understand it.

Consider two persons with greater than average ability, who attain thereby an increment R of reward beyond the average. By hypothesis, their child is likely to have higher than average ability (though less so than the parents because of regression toward the mean, as Herrnstein correctly points out) and thus will be expected to attain, by virtue of his own ability, an increment R′ of reward beyond the average, where R′ is less than R. Suppose that reward is prestige and acclaim. Then the children of talented (hence, we assume, rewarded) parents will in general be less talented and, by assumption, less acclaimed. Conceivably, one might argue that "prestige" is itself transmittable and cumulative (like wealth). But this claim (which Herrnstein does not make) would be most dubious. One might just as easily argue that in a competitive society, the discrepancy between R′ and R might further diminish the child's total increment ("he didn't live up to expectations"). Noting further that the heritable component in IQ is a negligible component in C, and that there is no reason to suppose C itself to be a significant (or any) factor in choosing mates, we see that any tendency for perpetuation of "prestige" along family lines further diminishes. On these assumptions, there will be no significant tendency for reward to be concentrated in a "hereditary meritocracy."

Suppose, however, that reward is transmittable wealth (and the power that flows from wealth). Then the child's total increment is $R' + R_1 + R_2 + R_3$, where R_1 is the portion of R transmitted to the child, R_2 is the additional wealth generated by R_1 itself (say, by investment), and R_3 is the increment attained by the child beyond R' by virtue of the initial advantage afforded him by possession of R_1. If $R_1 + R_2 + R_3$ is greater than $R - R'$, one can deduce a tendency towards "hereditary meritocracy" on appropriate assumptions about selection of mates. As I noted in the original article, Herrnstein does continually fall into the assumption that wealth is the reward that motivates labor, as when he writes that if the social ladder is tapered steeply, the obvious way to rescue the people at the bottom is "to increase the aggregate wealth of society so that there is more room at the top"—which is untrue if "social standing" is a matter of prestige and acclaim (and is incorrect for different reasons if it is wealth, since redistribution is an equally obvious strategy).

I hope it is now clear that if we accept all of Herrnstein's assumptions but take the reward that alone motivates labor to be prestige and acclaim, then we cannot deduce that "there will be a hereditary meritocracy," nor will this result be increasingly likely given "the successful realization of contemporary political and social goals," nor can we "extrapolate" to the conclusion that in any viable society a "virtually hereditary meritocracy" will appear. To reach Herrnstein's major and "most troubling" conclusions, we must add the tacit premise that the motivating reward is transmittable gain, such as wealth (though conceivably one might invent some other new premise to resurrect the argument—I will disregard this possibility in ensuing discussion). Contrary to Herrnstein's assertion that the assumption that people labor only for transmittable (material) gain is not "required by any of [his] conclusions," we see that it is critical for his only nontrivial conclusions.

By assuming that prestige and acclaim suffice as the motivating reward for labor, we can distinguish two conclusions that Herrnstein continually confuses in presenting his "syllogism." In his response, Herrnstein argues initially that from his premises (1) that mental capacities are partially inherited, and (2) that success in our society calls for those mental capacities, it follows that (3) "success in our society reflects inherited differences between people." In the next paragraph, he states that by accepting (1) and (2) one is "inexorably carried on" to the conclusion (4) "that society will stratify itself increasingly by genetic factors as it divests itself of the barriers commonly held to be unfair." And shortly after, he claims that from (1) and (2) it follows directly that (5) "there will be a hereditary meritocracy." In fact, (3) does follow from (1) and (2);[3] taking prestige as "success" in the "syllogism," it follows from (1) and (2) that prestige partially reflects inherited differences. But (5) does not follow from (1) and (2) and cannot be deduced unless we add the assumption that people labor for transmittable gain, which measures "success." As to (4), it does not follow from (1) and (2) if we understand by the phrase "will stratify itself increasingly by genetic factors" something like "will stratify itself in a hereditary meritocracy," as presumably intended. Now the conclusions that Herrnstein and his readers regarded as important were (5) and (4), under the suggested interpretation, namely, the two conclusions that do not follow from his premises. What was "most troubling" was his claim that "the growth of a virtually hereditary meritocracy will arise out of the successful realization of contemporary political and social goals" and that "social classes [will] not only continue but become ever more solidly built on inborn differences," namely IQ. But, as I noted, no significant social consequences follow from the fact that the children of two Olympic swimmers are likely to receive acclaim for their swimming ability (though less so, on the average, than their parents). Herrnstein's claims about existing societies (as well as his "extrapolation") require that the reward that alone motivates labor must be transmittable gain (and other assumptions about the extent of transmittability and choice of mates). Thus the major point in his response is clearly false.

At this stage of the discussion, Herrnstein shifts his ground, introducing another assumption which is as revealing as it is unwarranted. He argues that if prestige were the reward that motivated labor, then there would be "a society stratified by a mortal competition for prestige." How does he reach this conclusion?

Evidently, prestige differs from wealth not only in that it is not transmittable and cumulative in the same way (or at all, as already noted), but also in that it is not in short supply. By granting more of this reward to one individual, we do not correspondingly deprive another. Still assuming that individuals labor only for gain, then if the reward is prestige and acclaim, high performance could be assured generally by granting each individual prestige to the extent that he achieves in accordance with his abilities, whatever his task. But now consider some individual who will work only if his reward in prestige is not only greater than what he would attain by not working or working less well but also greater than the prestige given to others for their accomplishments. Such an individual might find himself in a "mortal competition for prestige" and might suffer "sadness and regret" (as Herrnstein predicts) or even "painful psychic deprivations" if others are successful—say, if someone else writes an outstanding novel or makes a scientific discovery or does a fine job of carpentry and is acclaimed for his achievement. Rather than take pleasure in this fact, such a person will be pained by it. Herrnstein assumes, without argument, that this psychic malady is characteristic of the human race.

By hypothesis, individuals labor only for "differential reward" in prestige. But Herrnstein confuses two senses of this term. Under sense 1, the phrase "differential prestige" refers to the increment an individual receives beyond others; under sense 2, it refers to the increment he obtains beyond what he would receive were he not to work or were he not to work in a way commensurate with his talents. Herrnstein speaks of the consequences of "attaching different outcomes for different jobs [sense 1], or for jobs done well or poorly [where a single individual is involved, sense 2]." But under sense 2, there will be no "mortal competition for prestige," no pain if others are successful, and no tendency for prestige to accrue to talent. The "deprivation" that makes "differential prestige" an operative reward will be the lack of the prestige that would be obtained by working or working well. Herrnstein reaches the opposite conclusion in each of these respects because he slips unconsciously into sense 1 of "differential prestige," thus assuming that only reward beyond one's fellows can motivate labor and that individuals are not pained (indeed, may be rewarded) by noting the "sadness and regret" or "painful psychic deprivations" that they cause for others by gaining "differential prestige" themselves. This assumption is even more extraordinary than the explicit assumptions in his original article; e.g., if bakers and lumberjacks were to get the top salaries and the top social approval, then the most talented would strive to become bakers and lumberjacks rather than, say engineers, surgeons, and research scientists.

To summarize, if prestige and acclaim suffice as motivating reward, then contrary to Herrnstein's belief, we cannot conclude that there will be a tendency towards social stratification in a hereditary meritocracy. Furthermore, Herrnstein's belief that there must then be a "mortal competition for prestige" follows only if we assume that differential prestige in sense 2 does not suffice and that an individual will labor only for the pleasure of being acclaimed beyond his fellows (thus causing them sadness and regret or perhaps painful psychic deprivations).

Continuing, let us assume that there is a complex of characteristics C,[4] partially heritable and a factor in choosing mates, that brings transmittable and cumulative social reward (say, wealth and power) sufficient to induce a tendency towards a hereditary meritocracy. One who finds this troubling might simply conclude that we should change the pattern of reward. We now turn to Herrnstein's "extrapolation" to the conclusion that this decision is impossible because no viable society can function otherwise. A society in which everyone has the opportunity to find interesting work suited to his talents—what Herrnstein predictably calls a "beehive society" —cannot survive. Herrnstein raises two objections to the beehive society. First, no one will be willing to do such work as "coal mining, ditch digging, schooling, garbage collecting, etc." He considers it an "outlandish" claim that people might voluntarily choose to be, say, teachers or students, unless given "differential reward" in his sense (sense 1). Suppose that he is correct and that in the beehive society no one would willingly choose to do what Herrnstein apparently regards as the dirty work of society—no problem in the society he depicts, of course, because the untalented are driven to such work by need. By assumption, the talented would find interesting work in the beehive society, but the dirty work would not get done. In my original discussion, I mentioned several possible ways to deal with this problem, which Herrnstein ignores in his reply. The obvious suggestion is that in the beehive society onerous tasks should be shared equally. But suppose that extrinsic motivation is necessary to guarantee this and that prestige and respect do not suffice. Then a differential reward of material gain must be provided to ensure that those who do not have the talent or inclination for work that is intrinsically satisfying (or that brings acclaim) will undertake the dirty work. Herrnstein's syllogism now "starts cranking away" to give a result precisely opposite to the one he foresees, namely, material reward will accrue to those lacking in "mental capacity," and, on his further assumptions about heritability and selection of mates, there will be a long-term tendency towards social stratification in a hereditary "antimeritocracy." I emphasize again that I do not believe any of this, since I see no reason to doubt that in a decent (i.e., "beehive") society onerous work will be shared out of a sense of social responsibility or perhaps because of the respect afforded to achievement or socially useful work. But on Herrnstein's assumptions, a hereditary antimeritocracy is quite a plausible outcome. Contrary to his beliefs, then, his syllogism cranks away to give virtually any long-term result, depending on one's implicit ideological commitments. In short, the syllogism itself is once again seen to be empty of significant consequences.

Herrnstein then turns to his second objection: He argues that "the burden [of proof] falls squarely on [me] to prove" that the "beehive society" is possible. This is a transparent evasion. My point was that Herrnstein's claims, even admitting his fallacious factual assertions, rest entirely on the assumption that there can be no such decent society, that his argument collapses unless we add tacit and entirely unsubstantiated premises borrowed from behaviorist doctrine and capitalist ideology. Accordingly, I made no effort to prove that these tacit assumptions are false but rather pointed out that without them there is nothing left of his nontrivial conclusions. Since he is the one who is claiming that society is moving (and must move) in a certain direction, clearly the burden of proof is on him to show that the assumptions required for his argument are plausible. It will not do to evade the issue by asking his critic to prove that the tacit assumptions are wrong. In another context, I would, indeed, argue that a decent ("beehive") society is possible, increasingly so as technological progress eliminates the need for onerous and stultifying work and provides means for democratic decision-making and for freely undertaken productive labor under conditions of voluntary association. But this is another matter entirely.

Herrnstein's final recourse, once again, is that history proves. . . . But as I noted in the original article, this argument in itself has all the cogency of an eighteenth-century argument that capitalist democracy is impossible, as history proves. For the argument to have any force, a reason must be adduced to explain the alleged fact that in existing societies ability "expresses itself in labor only for gain," further, for gain in transmittable and cumulative reward—a necessary assumption for Herrnstein's argument, as we have seen. One might argue, say, that it is a fact about human nature that humans will work only for such gains; that they will vegetate rather than seek an interesting outlet for their abilities; that they will never sacrifice for ideals or to help others; perfect nonsense, I am sure. The point is that for the argument from history to have any force as a projection for the future, one must be able to explain the historical facts in terms of some factors that are expected to endure. Since Herrnstein presents nothing even plausible in this regard, this argument goes the way of the others he has offered.

It remains a political decision, not a conclusion of science, or "reason," or history, to accept forms of social organization that permit accumulation of material reward, decision-making power, control of productive resources. Those who are "troubled" by the consequences of permitting free play to the complex of characteristics C that brings reward in a competitive society will therefore argue that social conditions be changed to prevent accumulation of wealth and power by those who have these characteristics. Herrnstein offers no hint of a counterargument. Rather, his uncritical and apparently unconscious acceptance of prevailing ideology in its most vulgar form leads him to conclude that things could not be different.

It is striking that Herrnstein can conceive of only one alternative to existing society. The only alternative that occurs to him is a society in which the leaders of the state order citizens to do this or that "for the sake of the state." This is what he takes to be the characteristic of "socialist states." On his weird assumptions about human nature, it may be that this is the only alternative to capitalism. I need hardly emphasize that it is not the socialist alternative; nor do I (or other libertarian socialists) believe that a socialist society exists today, though the germs of socialism, democracy, and freedom exist in some measure in various societies.

Herrnstein's response contains numerous other errors.[5] For example, he believes that I "keep dismissing" the "Skinnerian principles of reward and punishment." On the contrary, as I have repeatedly pointed out, I exclude from my criticism the actual scientific work (e.g., on partial reinforcement) which has no bearing on any of the issues raised in the work I was discussing. As to the trivialities and tautologies that are invoked to support such argument as there is, I merely urge that they be labeled for what they are; and more generally, that Skinner's general approach to explanatory theory, models, "inner states," and so on be recognized for what *it* is, namely, a rejection of the general approach of the sciences and even of engineering practice. Herrnstein, like some of his colleagues, interprets this reaction as a "contemptuous dismissal" of some scientific material. And, like others of his persuasion, he does not attempt to show some error of fact or reasoning, or to meet the simple challenge: Produce some nontrivial hypothesis on the "causation of behavior" that has empirical support and has any demonstrable social consequences or any bearing on the positions that are presented.

Perhaps this is enough, without belaboring the point further, to show that Herrnstein's argument is indeed an intellectual shambles. But I was interested in showing that it is equally dismal from a moral point of view. Let us now turn to Herrnstein's treatment of this matter.

I observed that Herrnstein's argument will surely be used by racists to justify discrimination and, more generally, by the priviledged to justify their privilege. Herrnstein responds that he does not see why his argument "will have that unwholesome consequence" (of being exploited by the racist) and pleads that he "was not proposing to study . . . a genetic flaw in any race." This seems rather disingenuous. He begins his seven-paragraph discussion of IQ and race by noting that "it is the relation between heritability and racial differences that raises the hackles" and then asks whether "the well-established, roughly fifteen-point black-white difference in IQ . . . arises in the environment or the genes," concluding finally that "a neutral commentator . . . would have to say that the case is simply not settled" and that the "fundamental issue" is "whether inquiry shall [again] be shut off because someone thinks society is best left in ignorance."

I take it that Herrnstein does understand why his argument will be used to justify privilege in general—at least, he does not question this. With regard to the matter of race, it is surely plain that a racist would be quite pleased with the statements just quoted. In a racist society, no one interprets Herrnstein as saying that "a neutral commentator would have to say that the case of White inferiority to Blacks is not settled" (as indeed it is not). And surely there is a far more "fundamental issue" than the one he identifies, namely, that the scientist, like anyone else, is responsible for the foreseeable consequences of his acts. Suppose that even opening certain questions for inquiry is likely to have malicious social consequences (as in the example in question). Then the scientist has the clear moral responsibility to show that the importance of his inquiry outweighs these malicious consequences. It would take a degree of moral imbecility to fail to perceive even that there is a conflict of values.

In the case in question, the conflict is easily resolved, if only because the possible correlation between partially heritable traits is of little scientific interest (in the present state of our understanding), and were someone interested in pursuing this matter as a scientist, he would certainly select traits more amenable to study than race and IQ. Furthermore there are no social consequences except under the assumptions of a racist society. What calls for explanation, then, is the zeal and intensity with which some investigators pursue scientifically trivial and socially malicious investigations. Herrnstein contests none of these observations, so I will elaborate them no further.

To illustrate my point I offered two analogies, one real, one hypothetical. The real example is the racist anthropology of the nineteenth century, and the second, the case of a hypothetical psychologist in Hitler Germany who tried to show that Jews had a genetically determined tendency towards usury, unperturbed by the fact that even opening this question for inquiry would provide ammunition for Goebbels and Rosenberg. Herrnstein attempts to "refute" the second analogy, with reasoning even more defective than that already illustrated.

Herrnstein observes that "America is not Hitler's Germany," that the hypothetical scientist should have left Germany, and that Goebbels and Rosenberg would have stopped the research. If I had given the hypothetical psychologist the name "Hans," Herrnstein could, with equal logic, have argued that he should have changed his name to "Peter." Obviously, the only relevant question is whether the analogy brings out the point at issue, namely, the conflict of values that should arise for a scientist who opens a question for inquiry with the knowledge that by doing so he may reinforce some of the most despicable features of his society. The point of the analogy, which Herrnstein evades, is that it would have been a poor excuse for the hypothetical psychologist (or the far-from-hypothetical racist anthropologist) to argue, with Herrnstein, that the question is not settled and the "fundamental issue" is "whether inquiry shall (again) be shut off." The reason is the one I have already stated: the inescapable conflict of values, which Herrnstein never faces.

In a final misrepresentation of the issue, Herrnstein states that contrary to what he takes to be my intention, he concludes that we should encourage "*more* research on people, not less." But what I asserted was that "Of course, scientific curiosity should be encouraged (though fallacious argument and investigation of silly questions should not), but it is not an absolute value"; and where a conflict of values arises, it can not simply be swept under the rug. Herrnstein objects to what he perceives as a demand that he "subject scientific findings to the test of political suitability." As I noted in "Psychology and Ideology," this is the typical maneuver of those who are attempting to evade the moral issue and to place the onus of "antiintellectualism" on those who ask the reasonable question: How do you justify your research, given its likely social consequences? Herrnstein fails utterly to understand the simple, elementary point that the scientist is responsible for the foreseeable consequences of his acts and therefore must justify his pursuit of research with predictable malicious consequences (and in this case, no scientific merit). I had assumed that Herrnstein was simply thoughtless in overlooking this trivial point and misidentifying the "fundamental issue" so grossly. That he seems so incapable of perceiving the point even when it is spelled out in detail seems to me most remarkable.

Finally, a word on the question why there is such interest in work so clearly lacking in substance. Here we can only speculate, and the natural question to raise is: What is the social import of the conclusions presented? In "Psychology and Ideology," I mentioned the social import of nineteenth-century racist anthropology and pointed out that "the tendencies in our society that lead toward submission to authoritarian rule may prepare individuals for a doctrine that can be interpreted as justifying it." In the case of Herrnstein's presentation, apart from the fact that his observations with regard to the study of race and IQ will be welcomed by the racist (whatever the consequences of the investigation—which are sure to be uncertain, confused, and to leave open the possibility that . . .), the hidden premise that individuals will labor only for material gain and "differential reward" beyond their fellows appears to give scientific credibility to prevailing ideology. And furthermore, the privileged will naturally delight in the conclusion that a hereditary meritocracy results as society wisely husbands its intellectual resources, that it may even be inevitable if society is to function properly. In short, the underprivileged and dispossessed must accept, even welcome, their fate, since things could not really be otherwise. Even the most benign social policy will only accentuate the problem, and they themselves gain by the fact that the social order is prejudiced against them. Herrnstein's argument thus is a contribution to what has elsewhere been called "the new assault on equality."[6]

This is a particularly important matter in a period when one of the standard and more effective techniques of social control may diminish in effectiveness or disappear, namely, the belief that as the economy grows every individual share will grow so that it is irrational, for culturally determined economic man, to urge revolution, reform, or redistribution, with all of their attendant uncertainties and dislocation. Walter Heller, Chairman of the Council of Economic Advisers under the liberal American administrations of the early 1960s, expressed the point this way: "When the cost of fulfilling a people's aspirations can be met out of a growing horn of plenty—instead of robbing Peter to pay Paul—ideological roadblocks melt away, and consensus replaces conflict."[7] But the social costs of the "growing horn of plenty" are now becoming apparent, as is the fact that laws of physics and chemistry impose certain limits on growth that may not be too far distant for the industrial societies, where the problem is most severe. Noting many uncertainties, still the problem cannot be simply dismissed, and it troubles economic and political elites as well as many scientists. But if the technique of social control based on the assumption of limitless growth loses its efficacy, some new device must be constructed to ensure that privilege is not threatened. What could serve better than the theory that the poor and weak must accept and welcome their status, which results from the wise decision to reward the talented in a "hereditary meritocracy"?

I note again, as in the original article, that questions of fact and questions of ideological commitment are, in principle, separable. If there were any merit to arguments such as those presented by Skinner and Herrnstein, one could not reject them on the ground that their consequences are unpleasant. Since the arguments are entirely lacking in merit, however, we turn rather to the questions I have just raised, and I think that the speculation just offered is a reasonable one.

Notes

1. S. Bowles and H. Gintis, "I.Q. in the U.S. Class Structure," *Social Policy* (November–December, 1972). They also present an important and, I believe, persuasive discussion of the social and ideological factors that lead Herrnstein and others to stress false claims in this matter.

2. C. Jencks et al., *Inequality: A Reassessment of the Effect of Family and Schooling in America* (New York: Basic Books, 1972), p. 221. Jencks et al. also carry out a detailed study of heritability of IQ (Appendix A), suggesting that Herrnstein accepts an estimate of heritability that is far higher than what the data warrant.

3. To be precise, what (1) and (2) imply is only (3′): Success in our society requires a characteristic which is partially inherited. Further premises are needed for the conclusion that Herrnstein apparently intended in his vaguely worded (3), namely, that successful people are genetically distinguishable in the population at large. Assuming (1) and (2), we can conclude nothing about genetic distinctiveness of successful people. Suppose, say, that the heritable component in "mental capacity" is negatively correlated with some characteristic P that far outweighs it as a factor in success. Then it might result that successful people are deficient in the heritable component of "mental capacity," while (1) and (2) remain true.

   I am indebted to Ned Block for clarification on this point.

4. From the observation that IQ is only a minor factor in determining income, it follows that one must seek other characteristics that play the role of C in what remains of Herrnstein's syllogism, if income is taken as index of "success." Jencks et al., *Inequality*, point out that there is little empirical evidence on this matter. I noted a number of factors that seem plausible candidates, e.g., ruthlessness, avarice, lack of concern for others, and so on, and pointed out that the term "meritocracy" seems to beg some interesting questions. It is a measure of Herrnstein's uncritical acceptance of prevailing ideology that he takes this suggestion to be refuted by the fact that "the data" show that those who succeed are "buoyant, energetic, independent, healthy ones." Evidently, there would be no reason for surprise if, in a competitive society with an ideological commitment to labor only for material gain, those who are ruthless, avaricious, and the like, and thus well constituted for a war of all against all, are regarded by conventional standards as "healthy," or if they prove to be "bouyant, energetic, independent."

5. Along with at least one improvement over the original. He now says that high rewards to those who serve "the wealthy few" may harm society, whereas his original claim was that society was "husbanding its intellectual resources" with the existing "gradient of occupations" which "is a natural measure of value and scarcity" and makes "practical sense." His remarks about "structural inadequacies," as "consensus of social utility," and so on suggest that he does not see the true force of the point, however.

6. For discussion, *see* the May–June 1972 issue of *Social Policy* and the references cited in notes 1 and 2.

7. W. W. Heller, *New Dimensions of Political Economy* (Cambridge, Mass.: Harvard University Press, 1966), p. 12, cited by R. B. Du Boff and E. S. Herman, "The New Economics: Handmaiden of Inspired Truth," *Review of Radical Political Economics* 4 (August 1972), an important discussion of economics and ideology.

Five Myths About Your IQ

Mary Jo Bane and Christopher Jencks

STANDARD IQ TESTS PURPORT TO MEASURE "INTELLIGENCE," WHICH IS widely viewed as the key to adult success. As a result, children with low IQ scores* are the subject of anxious solicitude from their parents, while groups that test badly, notably blacks, are constantly on the defensive. This is doubly true when, as usually happens, those who do poorly on IQ tests also do poorly on school achievement tests that measure things like reading comprehension and arithmetic skills.

Parents and teachers' anxieties have been further intensified as a result of claims that IQ scores are largely determined by heredity. If an individual's genes determine his IQ, and if IQ then determines his chances of adult success, it is a short step to the conclusion that there is nothing he can do to improve his prospects. Moreover, if life chances are determined at birth, many recent efforts at social reform have obviously been doomed from the start.

The controversy over IQ and achievement tests has become so bitter that it is almost impossible to discuss the subject rationally. Neither social scientists nor laymen seem to have much interest in

* An intelligence quotient is computed by ascertaining a person's mental age on the basis of a standardized intelligence test, and multiplying the result by 100. That result is then divided by the person's chronological age, to yield the IQ. Thus, the average IQ of the population is (and arithmetically must be) 100. About one person in six has an IQ under 85, and about one in six has an IQ over 115. About one in forty is under 70, and about one in forty is over 130.

the actual facts, which are extremely complex. The best currently available evidence suggests that:

- IQ tests measure only one rather limited variety of intelligence, namely, the kind that schools (and psychologists) value. Scores on the tests show remarkably little relationship to performance in most adult roles. People with high scores do a little better in most jobs than people with low scores, and they earn somewhat more money, but the differences are surprisingly small.
- The poor are seldom poor because they have low IQ scores, low reading scores, low arithmetic scores, or bed genes. They are poor because they either cannot work, cannot find adequately paying jobs, or cannot keep such jobs. This has very little to do with their test scores.
- Claims that "IQ scores are 80 percent hereditary" appear to be greatly exaggerated. Test results depend almost as much on variations in children's environments as on variations in their genes.
- While differences in the environments that children grow up in explain much of the variation in their test scores, differences in their school experiences appear to play a relatively minor role. But even socioeconomic background has a quite modest impact on test scores. Many factors that influence the scores seem to be unrelated to either school quality or parental status. At present, nobody has a clear idea what these factors are, how they work, or what we can do about them.
- If school quality has a modest effect on adult test scores, and if test scores then have a modest effect on economic success, school reforms aimed at teaching basic cognitive skills are likely to have minuscule effects on students' future earning power.

Each of these conclusions contradicts a commonly accepted myth about IQ.

Myth 1: *IQ tests are the best measure of human intelligence.*

When asked whether IQ tests really measure intelligence, psychologists are fond of saying that this is a meaningless question. They simply define intelligence as "whatever IQ tests measure." This is rather like Humpty Dumpty, for whom words meant whatever he wanted them to mean, and it was just a question of who was to be master. The trouble is that psychologists are *not* the masters of language, and they cannot assign arbitrary meanings to words without causing all kinds of confusion. In the real world, people cannot use a term like intelligence without assuming that it means many different things at once—all very important. Those who claim that "intelligence is what intelligence tests measure" ought logically to assume for example, that "intelligence is of no more consequence in human affairs than whatever intelligence tests measure." But people do not think this way. Having said that intelligence is what IQ tests measure, psychologists always end up assuming that what IQ tests measure *must* be important, because intelligence is important. This road leads through the looking glass.

What then, does the term "intelligence" really mean? For most people, it includes all the mental abilities required to solve whatever theoretical or practical problems they happen to think important. At one moment, intelligence is the ability to unravel French syntax. At another, it is the intuition required to understand what ails a neurotic friend. At still another, it is the capacity to anticipate future demand for hog bristles. We know from experience that these skills are only loosely related to one another. People who are "intelligent" in one context often are remarkably "stupid" in another. Thus, in weighing the value of IQ tests, one must ask exactly what *kinds* of intelligence they really measure and what kinds they do not measure.

The evidence we have reviewed suggests that IQ tests are quite good at measuring the kinds of intelligence needed to do school work. Students who do well on IQ tests are quite likely to get good grades in school. They are also likely to stay in school longer than average. But the evidence also suggests that IQ tests are *not* very good at measuring the skills required to succeed in most kinds of adult work.

Myth 2: *The poor are poor because they have low IQs. Those with high IQs end up in well-paid jobs.*

The fact is that people who do well on IQ and achievement tests do not perform much better than average in most jobs. Nor do they earn much more than the average. There have been more than a hundred studies of the relationship between IQ and people's performance on different jobs, using a wide variety of techniques for rating performance. In general, differences in IQ account for less than 10 percent of the variation in actual job performance. In many situations, there is no relationship at all between a man's IQ and how competent he is at his job. IQ also plays a modest role in determining income, explaining only about 12 percent of the variation. Thus, 88 percent of the variation in income is unrelated to IQ.

Nor do IQ differences account for much of the economic gap between blacks and whites. Phillips Cutright of the University of Indiana has conducted an extensive investigation of blacks who were examined by the Selective Service System in 1952. These men all took the Armed Forces Qualification Test, which measures much the same thing as an IQ test. In 1962, the average black in this sample earned 43 percent less than the average white. Blacks with AFQT scores as high as the average white's earned 32 percent less than the average white. Equalizing black and white test scores, therefore, reduced the income gap by about a quarter. Three-quarters of the gap had nothing to do with test scores. This same pattern holds for whites born into working-class and middle-class families. Whites with middle-class parents earn more than whites with working-class parents, but only 25–35 percent of the gap is traceable to test-score differences between the two groups.

None of this means that a child with a high IQ has no economic advantage over a child with a low IQ, nor that a child with high reading and math scores has no economic advantage over a child with low scores. It just means that the economic effect is likely to be much smaller than anxious parents or educational reformers expect. Among white males, those who score in the top fifth on standardized tests appear to earn about a third more than the national average. Those who score in the bottom fifth earn about two-thirds of the national average. These differences are by no means trivial. But they do not look very impressive when we recall that the best-paid fifth of all workers earns six or seven times as much as the worst-paid fifth. Most of that gap has nothing to do with test scores, and cannot be eliminated by equalizing test scores.

How can this be? We know that test scores play a significant role in determining school grades, in determining how long students stay in school, and in determining what kinds of credentials they eventually earn. We also know that credentials play a significant role in determining what occupations men enter. Occupations, in turn, have a significant effect on earnings. But at each stage in this process there are many exceptions, and the cumulative result is that exceptions are almost commonplace. A significant number of students with relatively low test scores earn college degrees, for example. In addition, a significant number of individuals without college degrees enter well-paid occupations, especially in business. Finally, people in relatively low-status occupations (such as plumbers and electricians) often earn more than professionals (think of teachers and clergymen). Overall, then, there are a lot of people with rather low test scores who nonetheless make above-average incomes, and a lot of people with high IQs but below-average incomes.

The limited importance of test scores is also clear if we look at the really poor—those who have to get by on less than half the national average. Nearly half of all poor families have no earner at all, either because they are too old, because they are headed by a woman with young children, or because the father is sick, alcoholic, mentally ill, or otherwise incapacitated. These problems are a bit more common among people with low IQs, but that is not the primary explanation for any of them.

This does not mean that financial success depends primarily on socioeconomic background, as many liberals and radicals seem to believe. Socioeconomic background has about the same influence as IQ on how much schooling a person gets, on the kind of occupation he enters, and on how much money he makes. Thus, we can say that *neither* socioeconomic background *nor* IQ explains much of the variation in adult occupational status or income. Most of the economic inequality among adults is due to other factors.

Unfortunately, we do not know enough to identify with much precision the other factors leading to economic success. All we can do is suggest their complexity. First, there is a wide variety of skills that have little or no connection with IQ but have a strong relationship to success in some specialized field. The ability to hit a ball thrown at high speed is extremely valuable, if you happen to be a professional baseball player. The ability to walk along a narrow steel beam 600 feet above the ground without losing your nerve is also very valuable, if you happen to be a construction worker. In addition, many personality traits have substantial cash value in certain contexts. A man who is good at figuring out what his boss wants, or good at getting his subordinates to understand and do what he wants, is at a great premium in almost any large hierarchical organization. While these talents are doubtless related to IQ, the connection is obviously very loose. Similarly, a person who inspires confidence is likely to do well regardless of whether he is a doctor, a clergyman, a small businessman, or a Mafioso; and inspiring confidence depends as much on manner as on mental abilities.

Finally, there is the matter of luck. America is full of gamblers, some of whom strike it rich while others lose hard-earned assets. One man's farm has oil on it, while another man's cattle get hoof-and-mouth disease. One man backs a "mad inventor" and ends up owning a big piece of Polaroid, while another backs a mad inventor and ends up owning a piece of worthless paper. We cannot say much about the relative importance of these factors, but when it comes to making a dollar, IQ is clearly only a small part of a big, complicated picture.

Myth 3: *Your IQ is overwhelmingly determined by your genetic endowment.*

Over the past decade, an enormous number of school reform programs have attempted to raise the scores of those who do poorly on standardized tests. These programs have involved preschool education, curriculum development, teacher training, compensatory education, administrative reorganization, and many other innovations. None appears to have produced the promised results on a permanent basis. This has led many people to the conclusion that variations in IQ scores must reflect innate genetic differences between individuals and groups. This is a logical non sequitur. But more important, the theory that IQ scores are determined at the moment of conception is not supported by the evidence. Genes clearly have a significant influence on IQ and school achievement scores, but so does environment. The reason reform programs have failed to improve test scores is not that the environment is irrelevant but that the reforms have not altered the most important features of the environment.

Much of the continuing furor over IQ scores derives from Arthur Jensen's controversial claim that genes "explain" something like 80 percent of the variation in children's performance on IQ tests. We have reviewed the same evidence as Jensen, and while it certainly shows that genes have *some* effect on IQ, we believe that his 80 percent estimate is much too high. The details of the argument are extremely complicated, but the basic reasons that Jensen overestimated the role of heredity and underestimated the role of environment are fairly easy to understand.

First, Jensen estimated the influence of genes on IQ scores largely by using data from studies of twins. Some of these studies dealt with identical twins who had been-separated early in life and brought up by different parents. Identical twins have exactly the same genes. When they are brought up in different environments, all differences between them can be attributed to the effects of their environments.

Other studies compared identical twins who had been reared together with fraternal twins who had been reared together. Fraternal twins have only about half their genes in common; identical twins have all their genes in common. Thus, if identical twins were no more alike on IQ tests than fraternal twins, we would have to conclude that genetic resemblance did not affect the children's test scores. But identical twins are, in fact, considerably more alike than fraternal twins, so it seems reasonable to suppose that genes have a significant effect on test scores. (Identical twins may also be treated somewhat more alike than fraternal twins, but the effect of this appears to be small.)

It is perfectly legitimate to use twin studies to estimate the relative influence of heredity and environment on test scores. But we can also estimate the effects of environment by measuring the degree of resemblance between adopted children reared in the same home. When we do this, environment appears to have somewhat more effect than it does in twin studies, while genes appear to have somewhat less effect. No one has ever offered a good explanation for this discrepancy, but that does not justify ignoring it. The most reasonable assumption is that the true effect of heredity is somewhat less than that suggested by twin studies but somewhat more than that suggested by studies of unrelated children in the same home.

A second difficulty with Jensen's estimate is that it is based on twin studies in England as well as in the United States. When we separate the American and English studies, we find that genetic factors appear to be more important in England than in America. This suggests that children's environments are more varied in the United States than they are in England. Other evidence, as well as common-sense observation of the two cultures, supports this interpretation. Consequently, when Jensen pools English and American data to arrive at his estimate of the effects of genes on IQ scores, he overestimates the relative importance of genes in America and underestimates their importance in England.

A third problem: Jensen assumes that the effects of genes and those of environment are completely independent of one another. In fact, since parents with favorable genes tend to have above-average cognitive skills, they tend to provide their children with unusually rich home environments. Our calculations suggest that this double advantage accounts for about a fifth of the variation in IQ scores.

After correcting all these biases, our best estimate is that genes explain 45 rather than 80 percent of the variation in IQ scores in contemporary America. This 45 percent estimate could easily be off by 10 percent either way, and it might conceivably be off by as much as 20 percent either way. The estimate would change if the range of environments were to increase or decrease for any reason. Genes are relatively more important in small homogeneous communities, where children's environments are relatively similar, than in America as a whole. By the same token, genes are relatively less important among groups whose environments are unusually diverse. If, for example, there were a sharp increase in the number of children suffering from acute malnutrition, or if large numbers of children were excluded from schools, environmental inequality would increase, and the relative importance of genes in determining IQ scores would decrease.

While genes probably account for something like 45 percent of the variation in IQ scores, it does not follow that genetic differences in actual learning capacity account for anything like this much variation. Genes influence test scores in two quite different ways. First, they influence what an individual learns from a given environment. Placed in front of the same TV program, one child may remember more of what he sees than another. Confronted with subtraction, one child may "catch on" faster than another. These differences derive partly from genetically based differences in learning capacity. In addition, however, genes can influence the environments to which people are exposed. Imagine a nation that refuses to send children with red hair to school. Under these circumstances, having genes that cause red hair will lower your reading scores. This does not tell us that children with red hair cannot learn to read. It tells us only that, in this particular situation, there is a socially imposed relationship between genes and opportunities to learn. In America, the genes that affect skin color have an indirect influence on an individual's opportunities and incentives to learn many skills. So, too, hereditary appearance and athletic ability influence a youngster's chance of getting into many colleges, and thus affect his or her later test scores.

Beyond all that, a person's genes may influence his actual learning capacity, which may then affect his opportunities and incentives to learn. If an individual has low test scores for genetic reasons, he may be assigned to a "slow" class where he learns less. Or he may be excluded from college. Such practices tend to widen the initial test-score gap between the genetically advantaged and the genetically disadvantaged. The resulting inequality is thus due *both* to genes *and* to environment. Yet conventional methods of estimating heritability impute the entire difference to genes.

When we say that genes "explain" 45 percent of the variation in test scores, we are talking about their overall effect, including their effect both on the capacity to learn and on opportunities and incentives to learn. No one has yet devised a method for separating these two effects. But if opportunities and incentives to learn were absolutely equal, genetically determined differences in learning capacity would account for considerably less than 45 percent of the variation in IQ scores.

Myth 4: *The main reason black children and poor white children have low IQ scores is that they have "bad" genes.*

Children from poor families tend to get lower scores on both IQ and school achievement tests than children from middle-class families. This difference is apparent when children enter school, and it does not seem to change much as children get older. Many liberals argue that the reason poor children do badly on these tests is that the tests are biased. Most of the tests contain items that are culturally loaded, in the sense that they presume familiarity with certain objects or assume the correctness of certain attitudes. The bias in these items always appears to favor children from middle-class backgrounds. Yet when psychologists have examined children's answers to these "loaded" items, they have not found that poor children did particularly badly on them. Nor have they found that eliminating such items from tests reduced the disparity in overall performance between poor and middle-class children. Middle-class children outscore poor children by as much on "culture-free" tests as on "culturally loaded" tests. This suggests that what poor children lack is not specific information but more basic skills that are relevant to many different kinds of tests.

These findings seem to support the theory that test-score differences between rich and poor derive from genetic differences between rich and poor children. Like the "cultural bias" explanation, this "genetic" explanation has considerable logical appeal. Everyone who has studied the matter agrees that genes have *some* influence on test scores, that test scores have *some* influence on education attainment, and that education has *some* influence on adult success. It follows that there must be *some* genetic difference, however small, between economically successful and unsuccessful adults. If this is true, there must also be some genetic difference between children with successful and unsuccessful parents.

The evidence suggests, however, that genetic differences between successful and unsuccessful families play a very minor role in determining children's IQs. Studies of adopted children indicate that genes may account for as much as half the observed correlation between parental status and children's test scores. Indirect evidence, derived from the relationship between test scores and parental success, suggests that the relationship is even weaker. Overall, our best guess is that genetic differences between social classes explain no more than 6 percent of the variation in IQ scores, while cultural differences between social classes explain another 6–9 percent.

This conclusion means that the average middle-class child may have a small genetic advantage over the average working-class child. But it also means that there are more working-class children than middle-class children with high genetic potential. This is because there are more working-class children to begin with. While their average score is a little lower than that of middle-class children, the difference is very small, and nearly half of all working-class children are above the middle-class average.

Furthermore, while differences between rich and poor whites are probably partly genetic, this tells us nothing about the origins of differences between whites and blacks. Blacks have lower IQ and achievement scores than whites, even when their parents have similar economic positions. But blacks also grow up in very different social and cultural environments from whites, even when their parents have the same occupations and incomes. We have no way of measuring the effects of these cultural differences. Our personal feeling is that black-white cultural differences could easily explain the observed IQ difference, which is only about 15 points. Differences of this magnitude are often found between white subcultures. Both black and white scores on military tests rose about 10 points between World War I and World War II, for example. Whites in eastern Tennessee improved by almost this much between 1930 and

1940, apparently as a result of the introduction of schools, roads, radios, and so on.

The key point is that *it doesn't much matter whether IQ differences between blacks and whites are hereditary or environmental.* IQ accounts for only a quarter of the income gap between blacks and whites. Therefore, even if genes accounted for *all* the IQ gap between blacks and whites, which is hardly likely, they would account for only a quarter of the economic gap. In all probability their role is far smaller. The widespread obsession with possible genetic differences between races is thus a diversion from the real problem. We ought to worry about eliminating the discrimination that still accounts for most of the observed economic difference between blacks and whites. If this could be done, the average black would be earning almost as much as the average white, and the pointless debate about possible genetic differences would no longer seem important to most sensible people.

Myth 5: *Improving the quality of the schools will go a long way toward wiping out differences in IQ and school achievement and, therefore, in children's life chances.*

Whether we like it or not, the quality of a child's school has even less effect than his social class on his test scores. The best evidence on this still comes from the 1965 Equality of Educational Opportunity Survey, whose first and most famous product was the Coleman Report. This survey did not give individual IQ tests to children, but it did give "verbal ability" tests that are very similar. It also gave reading and math tests. The results of this survey have aroused all sorts of controversy, but they have been confirmed by several other large surveys, notably the national study of high schools conducted by Project Talent throughout the 1960s. These surveys show that the differences among students in the same school are far greater than the difference between the average student in one school and the average student in another.

The surveys also show that test score differences between the alumni of different schools are largely due to differences between the entering students. Those from high-status families who enter school with high test scores tend to end up with high scores no matter what school they attend. Conversely, students from low-status families with low initial scores tend to end up with low scores even in what most people define as "good" schools. It follows that even if all schools had exactly the same effects on students' scores, the variation in students' IQ and achievement scores would decline very little. Qualitative differences among elementary schools seem to account for less than 6 percent of the variation in IQ test scores. Qualitative differences among high schools account for less than 2 percent.

In theory, of course, we could give students with low initial scores *better* schooling than students with high initial scores. This would allow us to reduce initial differences by more than 6 percent. The difficulty is that nobody knows how to do this. There is no consistent relationship between the amount of money we spend on a school and the rate at which children's test scores improve after they enter. Indeed, while school expenditures nearly doubled during the 1960s, a recent Project Talent survey shows that eleventh graders' school achievement scores hardly changed at all during that decade. Nor is there a consistent relationship between any specific school resource and the rate at which students' test scores rise. Neither small classes, well-paid teachers, experienced teachers, teachers with advanced degrees, new textbooks, nor adequate facilities have a consistent effect on students' scores.

Compensatory educational programs aimed at boosting the test scores of disadvantaged students have also produced discouraging results. Some studies report big gains, but others show that students who were not in the program gained more than those who were. Taken as a group, these studies suggest that students' scores do not improve any faster in compensatory programs than elsewhere. Thus, while there are good theoretical reasons for assuming that we can improve the test scores of those who enter school at a disadvantage, there is also strong evidence that educators simply do not know how to do this at the present time. (Neither, we should add, do educational critics, including ourselves.)

Racial and economic segregation may have slightly more effect than school expenditures on IQ test scores. The evidence, however, is by no means conclusive. Blacks who attend what we might call "naturally" desegregated schools, that is, schools in racially mixed neighborhoods, generally have higher test scores than blacks who attend segregated schools. These differences are apparent when students enter first grade, but they increase over time. This suggests that attending a racially mixed school boosts test scores somewhat faster than attending an all-black school. But the cumulative difference over six years of elementary school is small enough so that the effect in any one year is likely to be almost undetectable.

When we turn from "naturally" desegregated schools to schools that have been desegregated by busing, the evidence is more ambiguous. Some busing studies report that blacks showed appreciable gains. Very few report losses. Most show no statistically reliable difference. Since most of the studies involve small samples and short periods of time, this is not surprising. Taken together, the studies suggest that, *on the average*, busing probably increases black students' test scores, but that there are plenty of exceptions. Our best guess, based on evidence from both studies of busing and studies of naturally desegregated schools, is that desegregated schools would eventually reduce the test-score gap between blacks and whites by about 20 percent. This is, however, only an educated guess. The evidence on this question remains inconclusive, despite some recent extravagant claims to the contrary.

Nor is there any obvious reason to suppose that either decentralization or community control will improve students' test scores. Among whites, relatively small districts score at about the same level as large ones, once other factors are taken into account. Neither decentralization nor community control has been tried on a large enough scale in black communities to prove very much. But predominantly black suburban school districts, like Ravenswood and Compton in California, do not appear to have produced particularly impressive results. Neither have they done particularly badly.

None of this means that we should spend less on schools, stop trying to desegregate them, or reject decentralization or community control. Quite the contrary. If additional expenditures make schools better places for children, then they are justified, regardless of their effects on reading comprehension or IQ scores. If school desegregation reduces racial antagonism over the long run, then we should desegregate even though the students' test scores remain unaffected. And if community control gives parents the feeling that the schools belong to "us" rather than "them," this, too, is worthwhile for its own sake. Given the slim connection between test scores and adult success, it would be myopic to judge any sort of school reform primarily in terms of its effect on either IQ or school achievement.

Because a student's mastery of the skills taught in school provides a very poor measure of how well he will do once he graduates, reforms aimed at teaching these skills more effectively are not likely to take us very far toward economic prosperity or equality among adults. We have noted that only 12 percent of the variation in men's earnings is explained by their test scores. We have also seen that reducing inequality in test scores is very difficult. Under these circumstances, it makes little sense for economic egalitarians to concentrate on equalizing IQ and achievement scores. Instead, they should concentrate on eliminating the other sources of income inequality, which cause 88 percent of the problem.

To be sure, this is more easily said than done. Those who do well in the present economic system inevitably resist reforms that would reduce their privileges. Those who do poorly in the present system are for the most part too demoralized to protest in any effective way. So long as this persists, there will be little chance of reducing economic inequality.

The complacency of the rich and the demoralization of the poor are reinforced by theories that attribute economic success to genetic superiority and economic failure to genetic deficiency. These theories are nonsense. In 1968, the income difference between the best- and worst-paid fifth of all workers was about $14,000. Of this, perhaps $500 or $1,000 was attributable to genetically determined differences in IQ scores. The idea that genetic inequality explains economic inequality is thus a myth. Like the divine right of kings, such myths help legitimize the status quo. But they should not be taken seriously by those who really want to understand the modern world, much less those who want to change it.

Testing for Order and Control in the Corporate Liberal State

Clarence J. Karier

IN 1897, THE TOTAL CAPITALIZATION OF ALL CORPORATIONS, INDIVIDUALLY valued at a million dollars or more, came to only $170 million. Just three years later, the same figure for total capitalization stood at $5 billion, and in 1904, at over $20 billion.[1] With this rapid consolidation of corporate capital, there emerged mass-production industries with their own needs for standardized producers and consumers.[2] During this same time, the last great wave of immigrants flooded into the burgeoning urban centers, manning the rapidly expanding mass production industries. The complex problems which resulted from urbanization, industrialization, and immigration were eventually faced by large corporate interests following the lead of Germany in utilizing the state to ameliorate social, economic and political problems. Unlike Germany, however, American social-service institutions at the city, state, and federal level were not as responsive to efficient management. Neither the populist nor the immigrant in the urban ghetto was easily managed.[3] The thrust of progressive reformers, working at the city, state, and eventually national level, was to reorganize a pluralistic America for efficient, orderly production and consumption of goods and services.

The corporate liberal state which thus emerged during the Progressive Era[4] has withstood the varied tests and challenges of two world wars, the Great Depression, and the Cold War. Each decade of the century brought new and unexpected problems which, in turn, helped shape the direction and growth of that state. While the Depression served to stimulate the growth of corporate-state power along the lines of welfare capitalism, the effect of the Cold War has been to stimulate the growth of that same power along the lines of a military capitalism. In spite of the massive growth and significant changes in direction that have occurred, there still remain certain characteristics of that state which appear fairly constant. One such characteristic is the cooperative working relationship between big labor unions, corporate wealth, and government which had its origin in the National Civic Federation, founded in 1900.

To most members of the National Civic Federation representing labor, government, and corporate wealth, the nineteenth-century notion of laissez-faire capitalism was a self-destructive concept which had to be replaced by a more comprehensive view of enlightened self-interest. Such a view clearly recognized that, if large corporate interests were to survive and effectively prosper in the twentieth century, a cooperative alliance with labor, government, and corporate wealth was necessary for the efficient development of both production and consumption of goods and services. Although cutthroat competition had proven to be destructive for large corporate interests, and, therefore, those interests had taken effective political action by 1918 to protect their own vital concerns, most small entrepreneurs continued to express the rhetoric and live the actuality of laissez-faire capitalism to their own demise well into the twentieth century. In general, the larger corporations took the more enlightened view that ultimately their own best interests were served through a regulatory system which eliminated conflict and stabilized the economic-social system. George W. Perkins of International Harvester put it best when, in arguing for industrial compensation for laborers, he said that cooperation in business "is taking and should take the place of ruthless competition . . . and that if this new order of things is better for capital and better for the consumer, then in order to succeed permanently it must demonstrate that it is better for the laborer."[5] While the old liberalism justified individualism and cutthroat competition, the new corporate liberalism which emerged in the thinking of such men as Herbert Croly, Edward Ross, John R. Commons, and others would protect the basic structure of wealth and power in the "new order" by increasing the standard of living for a larger middle class. Herbert Croly expressed these sentiments well when he suggested that progressive democracy was "designed to serve as a counter poise to the threat of working class revolution."[6]

The corporate liberal state, which emerged by World War I, included an array of bureaucratic regulatory agencies, which co-operatively worked with business and labor to achieve that optimal balance of interests for all concerned. The logical thrust of corporate industry, as well as the progressive liberals who tended to dominate the new social sciences, was toward the development of a new scientific management in order to socially engineer for control and order.[7]

Whether it was John R. Commons, Edward A. Ross, and Richard T. Ely at the University of Wisconsin; or Samuel Gompers and Andrew Carnegie in the National Civic Federation; or perhaps, Jane Addams and Walter Rauschenbusch in the settlements of Chicago and New York, all thought and worked toward a larger, more orderly corporate state, utilizing knowledgeable experts to ameliorate the many varied problems of that state. On the one hand, old public institutions had to be reorganized to increase the effectiveness of administrative and bureaucratic functions at each level of government.[8] On the other hand, in the private sector, relatively new organizations were created which effectively channeled corporate wealth toward the support of liberal progressive reform. Philanthropic foundations became a major stimulus for political as well as educational reform. For example, many of the major municipal research bureaus whose urban studies provided the evidence that led to progressive reform of city governments (which indirectly but effectively disenfranchized many immigrant groups) were originally funded by large foundations and later taken over by the city councils.[9] The practice of foundations initiating various kinds of activity and then allowing the public sector to assume control became a common practice of the major foundations dealing with policy formation in America. The profound influence of foundations issues from their ability to flexibly employ large blocks of wealth for research, initiate new activities, and facilitate existing programs.[10] As Fred M. Hechinger aptly put it:

> The ideal foundation-sponsored enterprise is one that blazes a new trail, thrives for a while on sponsored dollars, gathers momentum, and is quickly taken over as a permanent program by the local school board, the state education authority, or a university's own budget.[11]

The emergence of foundations to a key role in the policy formation of the corporate liberal state reflected a development of virtually a fourth branch of government, which effectively represented the interest of corporate wealth in America. That interest, however, was usually broadly interpreted to include financial support for projects which varied in range all the way from hospital development to research on urban slums. The breadth of support reflected the liberal influence in the dissemination of foundation wealth in attempting to maintain a flexible growing society. One does not need to conjure up a conspiracy theory of history to recognize that the foundations did not consciously, over any extended period of time, support that which threatened to destroy the basic framework of the corporate liberal state. Looking back over the support policy of the foundations in the twentieth century, one can conclude that, for the most part, the philosophy behind the policy makers for the foundations appears to have been that of a liberal pragmatist who appreciated the need for survival.

To be sure, America has had a long history of philanthropy; however, the development and creation of large corporate foundations was very much a twentieth-century phenomena. Foundations of varied sorts grew rapidly in numbers from 21 in 1900, to a total of 4,685 in 1959.[12] To the chagrin of many congressmen and taxpayers, the tax-exempt foundations in the United States also grew from 12,295 at the close of 1952 to 42,124 by the end of 1960.[13] From the very beginning of the century, the new philanthropic endeavors of corporate wealth were directed at influencing the course of educational policy. John D. Rockefeller's General Education Board, which received its national charter in 1903, greatly influenced and shaped educational policy for Black America in the South, while the Carnegie Institute of Washington (1904) and the Carnegie Foundation for the Advancement of Teaching (1906) came to play a major role in shaping educational policy in both the South and the North. By 1912, however, John D. Rockefeller and Andrew Carnegie faced an increasingly hostile Congress at the national level and sympathetic legislative support at the state level. Thus, the incorporation of philanthropic foundations proceeded along lines of least resistance, i.e., at the state rather than the national level. By 1913, a concerned Sixty-second Congress directed the Industrial Relations Commission to investigate the role of foundations. Charles W. Eliot testified as to the noble purposes and activities of all foundations. As he said:

> I have never known a charitable or educational corporation to do anything which threatened the welfare or the liberties of the American people. I have had no observation of any such corporation—of any such attempts. And I have, on the other hand, seen a great deal of the activity and intelligent promotion of the public welfare by such corporations. There is in them, so far as my experience teaches me, not only no menace, but a very great hope for the Republic.[14]

Nevertheless, after a year of testimony, the majority of the Commission concluded that, "The domination of men in whose hands the final control of a large part of American industry rests is not limited to their employees, but is being rapidly extended to control the education and social service of the nation."[15] They went on to point out that the policies of the foundations would inevitably be those of the corporations which sponsored them. While some members of the Commission called for the confiscation of the foundations, liberals such as John R. Commons and Florence Harriman cautioned against such precipitous action. Even though the findings of the commission cut very close to the heart of the problem of power in the corporate liberal state, the commission's findings were ignored as the attention of the Congress and the nation shifted to the war that was developing in Europe.

America's entry into the war brought to a head certain trends which were evolving within the Progressive Era. Radical populists and socialists were jailed in the name of national unity and a native American Left was demoralized, the larger corporations found more profitable ways to work with government, and a large cadre of social science experts tried out their new-found techniques for the management of the new corporate state. John Dewey, for example, saw in the war the great possibility for "intelligent administration"[16] based eventually on a solid social science. The corporate liberal state emerged from the war stronger than ever. Progressive liberal reform did not come to an end with the war but rather became institutionalized. Henceforth, most social change would be institutionally controlled and the interest of government, corporate wealth, and labor more securely managed. The state which thus emerged included a mass system of public schools which served the manpower needs of that state. One of the more important ways that system served the needs of the state was through the process of rationalizing and standardizing manpower for both production and consumption of goods and services.

For many "professional educators," the school as a trainer of producers and consumers necessarily led to a view of the school as a business model or factory.[17] Ellwood P. Cubberley spoke for many professional educators in the twentieth century when he said:

> Every manufacturing establishment that turns out a standard product or a series of products of any kind maintains a force of efficiency experts to study methods of procedure and to measure and test the output of its works. Such men ultimately bring the manufacturing establishment large returns, by introducing improvements in processes and procedure, and in training the workmen to produce larger and better output. Our schools are, in a sense, factories in which the raw products (children) are to be shaped and fashioned into products to meet the various demands of life. The specifications for manufacturing come from the demands of twentieth-century civilization, and it is the business of the school to build its pupils according to the specifications laid down. This demands good tools, specialized machinery, continuous measurement of production to see if it is according to specifications, the elimination of waste in manufacture, and a large variety in the output.[18]

The testing movement, financed by corporate foundations, helped meet the need for "continuous measurement" and "accountability." It also served as a vital part of the hand which helped fashion the peculiar meritocracy within that state.

Although the testing movement is often viewed as getting under way with the mass testing of 1.7 million men for classification in the armed forces in World War I, the roots of the American testing movement lie deeply imbedded in the American progressive temper which combined its belief in progress, its racial attitudes, and its faith in the scientific expert working through the state authority to ameliorate and control the evolutionary progress of the race. While America has had a long history of eugenics advocacy, some of the key leaders of the testing movement were the strongest advocates for eugenics control. In the twentieth century, the two movements often came together in the same people under the name of "scientific" testing and, for one cause or the other, received foundation support.

One such leader of the Eugenics Movement in America was Charles Benedict Davenport who, having seriously studied Galton and Pearson, sought to persuade the new Carnegie Institution of Washington to support a biological experiment station with himself as director. In 1904, he became director of such a station at

Cold Spring Harbor on Long Island. As his interest in experiments in animal breeding began to wane, he used his influence as secretary of "the Committee on Eugenics of the American Breeders Association" to interest others in the study of human heredity. Supported by the donations of Mrs. E. H. Harriman, Davenport founded the Eugenics Record Office in 1910, and by 1918, the Carnegie Institution of Washington assumed control. The work of the Record Office was facilitated by the work of committees on: "Inheritance of Mental Traits" which included Robert M. Yerkes and Edward L. Thorndike; "Committee on Heredity of Deaf-mutism" with Alexander Graham Bell; "Committee on Sterilization" with H. H. Laughlin; and "Committee on the Heredity of the Feeble-minded" which included, among others, H. H. Goddard.

These committees took the lead in identifying those who carried defective germ plasm and disseminating the propaganda which became necessary to pass sterilization laws. For example, it was Laughlin's "Committee to Study and Report on the Best Practical Means of Cutting off the Defective Germ-Plasm in the American Population," which reported that "society must look upon germ-plasm as belonging to society and not solely to the individual who carries it."[19] Laughlin found that approximately 10 percent of the American population carried bad seed and called for sterilization as a solution. More precisely, he defined these people as "feeble-minded, insane, criminalistic (including the delinquent and wayward), epileptic, inebriate, diseased, blind, deaf, deformed and dependent (including orphans, ne'er-do-wells, the homeless, tramps, and paupers)."[20] Social character, from murder to prostitution, was associated with intelligence and the nature of one's germplasm. The first sterilization law was passed in Indiana in 1907, followed in quick succession by fifteen other states. In Wisconsin, such Progressives as Edward A. Ross and Charles R. Van Hise, president of the University of Wisconsin, took strong public stands supporting the passage of sterilization laws. America pioneered in the sterilization of mental and social defectives twenty years ahead of other nations.

Between 1907 and 1928, twenty-one states practiced eugenical sterilization involving over 8,500 people. California, under the influence of the Human Betterment Foundation which counted Lewis M. Terman and David Star Jordan as its leading members, accounted for 6,200 sterilizations. California's sterilization law was based on race purity as well as criminology. Those who were "morally and sexually depraved" could be sterilized. Throughout the sterilization movement in America ran a Zeitgeist, reflecting the temper of pious reformers calling for clean living, temperance, fresh air schools, as well as sterilization (*see* figs. 2, 3, and 4, p. 372). The use of sterilization for punishment reached the point where laws were introduced which called for sterilization for chicken stealing, car theft, as well as prostitution.[21]

H. H. Goddard, fresh from G. Stanley Hall's seminars at Clark University, translated the Binet-Simon Scale (1908), using the test to identify feeble-minded at the training school at Vineland, New Jersey. Various scales and tests which were freely used and patterned after the original scale were later proven to lack reliability to the extent that, according to some testers, upward of half the population was feeble-minded. From the Binet scale, Goddard went on to publish *The Kallikak Family*, which showed the family history of Martin Kallikak as having sired both a good and bad side to his family tree. The bad side, which began with his involvement with a feeble-minded girl, contributed such "social pests" as "paupers, criminals, prostitutes and drunkards." Goddard's next book, *Feeble-mindedness: Its Causes and Consequences*, gave further "scientific" justification to the notion of the relationship between feeble-mindedness and moral character.

Interestingly enough, the liberal tradition in America from Jefferson on usually assumed a positive relationship between "talent and virtue." It was then not surprising to find people assuming that anyone with less talent will have less virtue. This relationship was assumed in the passage of most sterilization laws. Society would rid itself of not only the genetic defective, but more importantly, the socially undesirable. Laughlin, Goddard, Terman, and Thorndike all made similar assumptions. Terman argued that the feeble-minded were incapable of moral judgments and, therefore, could only be viewed as potential criminals. He said:

> . . . all feeble-minded are at least potential criminals. That every feeble-minded woman is a potential prostitute would hardly be disputed by anyone. Moral judgment, like business judgment, social judgment or any other kind of higher thought process, is a function of intelligence.[22]

The same thinking which guided Terman to find a lower morality among those of lesser intelligence had its mirror image in the work of Edward L. Thorndike, who found a higher morality among those with greater intelligence. Thorndike was convinced that, "To him that hath a superior intellect is given also on the average a superior character."[23] The sterilization solution to moral behavior problems and the improvement of intelligence continued to be advocated by Thorndike, as well as by his pupils. By 1940, in his last major work, he concluded that:

> By selective breeding supported by a suitable environment we can have a world in which all men will equal the top ten per cent of present men. One sure service of the *able* and *good* is to beget and rear offspring. One sure service [about the only one] which the inferior and vicious can perform is to prevent their genes from survival.[24]

The association of inferior with vicious and intelligence with goodness continued to appear in the psychology textbooks. Henry E. Garrett, a former student and fellow colleague with whom Thorndike was associated, who won a "reputation for eminence,"[25] continued to project the story of Martin Kallikak in terms of goods and bads in his textbook on *General Psychology* as late as 1955. Just in case someone might miss the point, the children of the feeble-minded tavern girl were pictured as having horns, while the "highest types of human beings," were portrayed as solid Puritan types *(see* fig. 5, p. 373).[26]

This view of the Kallikaks was no accident. As chairman of Columbia's department of psychology for sixteen years, and past president of the American Psychological Association as well as a member of the National Research Council, Garrett was in sympathy with Thorndike's views on the place of the "inferior and vicious" in American life. By 1966, as a Professor Emeritus from Columbia, in the midst of the civil rights movement, he produced a series of pamphlets which drew out what he believed to be the implications of sixty years of testing in America. Sponsored by the Patrick Henry Press, over 500,000 copies of his pamphlets were distributed free of charge to American teachers. In *How Classroom Desegregation Will Work; Children Black and White*; and especially in *Breeding Down*, Garrett justified American racism on "scientific" grounds. Going back to Davenport and the Eugenics Record Office as well as Terman's work and others, Garrett argued:

> You can no more mix the two races and maintain the standards of White civilization than you can add 80 (the average I.Q. of Ne-

> groes) and 100 (average I.Q. of Whites), divide by two and get 100. What you would get would be a race of 90s, and it is that 10 per cent differential that spells the difference between a spire and a mud hut; 10 per cent—or less—is the margin of civilization's "profit"; it is the difference between a cultured society and savagery.
>
> Therefore, it follows, if miscegnation would be bad for White people, it would be bad for Negroes as well. For, if leadership is destroyed, all is destroyed.[27]

He went on to point out that the black man is at least 200,000 years behind the whites and that intermarriage, as well as desegregation, would destroy what genetic lead the white man had achieved through "hard-won struggle" and "fortitudinous evolution." The state, he argued, "can and should prohibit miscegenation, just as they ban marriage of the feeble-minded, the insane and various undesirables. Just as they outlaw incest."[28]

The style and content of Garrett's arguments were but echoes of similar arguments developed earlier by Davenport, Laughlin, Terman, Brigham, Yerkes, and Thorndike. For example, C. C. Brigham spoke of the superior Nordic draftees of World War I and seriously worried about inferior germplasm of the Alpine, Mediterranean, and Negro races in *A Study of American Intelligence*.[29] What disturbed Brigham as well as the United States Congress was, of course, the fact that 70 percent of the total immigration in the early 1920s was coming from Alpine and Mediterranean racial stock. H. H. Laughlin of the Carnegie Foundation of Washington provided the scientific evidence to the Congress in his report, "An Analysis of America's Melting Pot." Using information from the Army tests, and from his Eugenics Record Office dealing with the insane and feeble-minded, Laughlin built a case that the new immigrant from southern Europe was of inferior racial stock by virtue of the numbers that appeared as wards of the state.[30]

Supported by the Commonwealth Fund, Lewis M. Terman reported similar evidence from his study. Addressing the National Education Association at Oakland, California on July 2, 1923, he expressed concern about the fecundity of the superior races. As he put it:

> The racial stocks most prolific of gifted children are those from northern and western Europe, and the Jewish. The least prolific are the Mediterranean races, the Mexicans and the Negroes. The fecundity of the family stocks from which our gifted children come appears to be definitely on the wane.... It has been figured that if

> the present differential birth rate continues, 1,000 Harvard graduates will at the end of 200 years have but 50 descendants, while in the same period 1,000 South Italians will have multiplied to 100,000.[31]

It was this kind of "scientific" data derived from the testing and eugenics movement which entered into the dialogue which led to the restrictive immigration quota of 1924, that clearly discriminated against southern Europeans.

After World War I, America had moved toward a more restrictive immigration policy. While small manufacturers, represented by the National Association of Manufacturers and Chamber of Commerce, tended to favor a sliding-door policy which would open according to the labor needs of the small manufacturers, most larger manufacturers and labor unions, represented by the National Civic Federation, favored restricting immigration. Perhaps the motivation was best stated by Edward A. Filene, a pioneer in employee management, of Boston, when he said:

> Employers do not need an increased labor supply, since increased use of labor-saving machinery and elimination of waste in production and distribution will for many years reduce costs more rapidly than wages increase, and so prevent undue domination of labor.[32]

The Carnegie money that Laughlin used in his campaign for greater restrictions was ultimately money well spent in the interest of the larger manufacturer. Nevertheless, the rhetoric of the times and of the testers was, perhaps, best put when President Coolidge proclaimed, "America must be kept American."

The nativism, racism, elitism, and social class bias which were so much a part of the testing and Eugenics Movement in America were, in a broader sense, part of that *Zeitgeist* which was America. This was the land of the Ku Klux Klan, the Red scare, the Sacco-Vanzetti and Scopes trials as well as the land of real opportunity for millions of immigrants. It was this kind of contradictory base in which the corporate liberal state took firm root, building a kind of meritocracy that even Plato could not have envisioned. Just as Plato ascribed certain virtues to certain occupational classes, so, too, Lewis Terman assigned numbers which stood for virtue to certain occupational classes. It was clear to Terman that America was the land of opportunity, where the best excelled, and the inferior found themselves on the lower rungs of the occupational order. Designing the Stanford-Binet Intelligence Test, Terman developed questions which were based on presumed progressive difficulty in performing tasks which he believed were necessary for achievement in ascending the hierarchical occupational structure. He then proceeded to find that, according to the results of his tests, the intelligence of different occupational classes fit his ascending hierarchy. It was little wonder that IQ reflected social class bias. It was, in fact, based on the social class order. Terman believed that, for the most part, people were at that level because of heredity and not social environment. He said:

> Preliminary investigations indicate that an I.Q. below 70 rarely permits anything better than unskilled labor; that the range from 70 to 80 is preeminently that of semi-skilled labor, from 80 to 100 that of the skilled or ordinary clerical labor, from 100 to 110 or 115 that of the semi-professional pursuits; and that above all these are the grades of intelligence which permit one to enter the professions or the larger fields of business. Intelligence tests can tell us whether a child's native brightness corresponds more nearly to the median of (1) the professional classes, (2) those in the semi-professional pursuits, (3) ordinary skilled workers, (4) semi-skilled workers, or (5) unskilled laborers. This information will be of great value in planning the education of a particular child and also in planning the differentiated curriculum here recommended.[33]

Plato had three classes and Terman had five; both maintained the "myth of the metals," and both advocated a differentiated curriculum to meet the needs of the individuals involved. Terman so completely accepted the assumption of the social class meritocracy and the tests which were based on that meritocracy that he never seemed to even wonder why, in his own study of the gifted: "The professional and semi-professional classes together account for more than 80 percent. The unskilled labor classes furnish but a paltry 1 percent or 2 percent."[34]

Social class was not the only problem with the tests. Whether one reads Terman's Stanford-Binet or his Group Test of Mental Ability or the Stanford Achievement Tests, the army tests, or the National Intelligence Tests,[35] certain characteristics emerge. They all reflect the euphemisms, the homilies, and the morals which were, indeed, the stock and trade of *Poor Richard's Almanac*, Noah Webster's blue-back speller, and McGuffey's readers. The child who grew up in a home and attended a school where these things were in common usage stood in distinct advantage to the newly arrived immigrant child. At a time when over half the children in American schools were either immigrants or children of immigrants, this movement represented discrimination in a massive way.

By 1922, Walter Lippmann, in a series of six articles for *The New Republic,* questioned whether intelligence is fixed by heredity and whether the tests actually measure intelligence.[36] While Lippmann challenged the validity of the test, he did not attack the presumption of meritocracy itself. Although Lippmann seemed to get the best of the argument, Terman fell back to the high ground of the condescending professional expert who saw little need to debate proven "scientific" principles.[37]

Conscious of the social implications of their work, Goddard, Terman, and Thorndike viewed themselves as great benefactors of society. The concern for social order and rule by the intelligent elite was ever present in their writings. Goddard put it bluntly when he argued that, "The disturbing fear is that the masses—the seventy or even eighty-six million—will take matters into their own hands."[38] The "four million" of "superior intelligence" must direct the masses. Throughout the literature of this period, the fear of the masses appears as a constant theme. Under such circumstances, one could hardly turn to the masses for an enlightened solution. The assumed role of the "professional" scientific expert was to lead the masses out of the irrational morass of ignorance. The definition of democracy had changed. It no longer meant rule by the people, but rather rule by the intelligent. As Thorndike put it, "The argument for democracy is not that it gives power to men without distinction, but that it gives greater freedom for ability and character to attain power."[39]

Luckily, mankind's wealth-power-ability and character were positively correlated. This, indeed, was not only Plato's ideal, but the testers' view of the meritocracy which they in fact were fashioning. Late in life, Thorndike reflected on these concerns and said:

> It is the great good fortune of mankind that there is a substantial positive correlation between intelligence and morality, including good will toward one's fellows. Consequently our superiors in ability are on the average our benefactors, and it is often safer to trust our interests to them than to ourselves. No group of men can be expected to act one hundred per cent in the interests of mankind, but this group of the ablest men will come nearest to the ideal.[40]

To be sure, there have been and, still are, inequities between men of intelligence and of wealth, Thorndike argued, but through the "beneficence of such men as Carnegie and Rockefeller," this discrepancy had been somewhat overcome.[41]

Although Thorndike was directly involved in the Army classification testing during World War I and the creation of the National Intelligence Test after the war, all of which skyrocketed the testing movement in American schools, perhaps his most profound influence on American schools came through his work in organizing the classroom curriculum. His name appears on approximately 50 books and 450 monographs and articles, including the much-used *Thorndike Dictionary*. He wrote many of the textbooks, tests, achievement scales, and teacher's manuals. In short, he told the schoolteachers what to teach and how to teach it, and in turn, how to evaluate it. Much of his work was, indeed, made possible through the beneficence of Carnegie. The Carnegie Foundation from 1922 to 1938 had made grants supporting his work totaling approximately $325,000.[42] It was men like Thorndike, Terman, and Goddard, supported by corporate wealth,[43] who successfully persuaded teachers, administrators, and lay school boards to classify and standardize the school's curriculum with a differentiated track system based on ability and values of the corporate liberal society. The structure of that society was based, then, on an assumed meritocracy, a meritocracy of white, middle-class, management-oriented professionals.

The test discriminated against members of the lower class—southern Europeans and blacks—indirectly by what they seemed to leave out, but more directly by what they included; for example: On a Stanford-Binet (1960 revision), a six-year-old child is asked the question, "Which is prettier?"[44] and must select the Nordic Anglo-Saxon type to be correct. If, however, the child is perhaps a Mexican American or of southern European descent, has looked at himself in a mirror, and has a reasonably healthy respect for himself, he will pick the wrong answer. Worse yet, is the child who recognizes what a "repressive society" calls the "right" answer and has been socialized enough to sacrifice himself for a higher score. (See fig. 6). The same is true in the case of the black six-year-old. (See fig. 7). Neither blacks nor southern Europeans were beautiful according to the authors of the Stanford-Binet, but then, there was no beauty in these people when Goddard, Laughlin, Terman,[45] Thorndike, and Garrett called for the sterilization of the "socially inadequate," the discriminatory closing of immigration, the track-

L–M IV–6, 1 Card C

Figure 6. *From Stanford-Binet Intelligence Scale, 1960.*

L–M IV–6, 1 Card A

Figure 7. *From Stanford-Binet Intelligence Scale, 1960.*

ing organization of the American school, or for that matter, defined their place in the meritocracy.

The test, then, discriminated in content against particular groups in the very questions that were used, as well as the questions that were not used, with respect to particular minority experiences.

While some educational psychologists sought to eliminate bias from the content of the test, as well as introduce a broader cultural basis for the test, others sought the impossible: a culturally free IQ test. Still other educational psychologists, hard-pressed to define intelligence, fell back to the assertion that it was simply that which the tests measured. Although many gave up their concern about intelligence, others argued that the various intelligence tests were achievement tests which could also be good predictors of success within both the corporate society and the bureaucratic school system which served that society. At this point, the testers had come full circle, ending up where Terman started.

Terman's tests were based on an occupational hierarchy which was, in fact, the social class system of the corporate liberal state which was then emerging. The many varied tests, all the way from IQ to personality and scholastic achievement, periodically brought up-to-date, would serve a vital part in rationalizing the social class system. The tests also created the illusion of objectivity which, on the one side, served the needs of the "professional" educators to be "scientific," and on the other side, served the need of the system for a myth which could convince the lower classes that their station in life was part of the natural order of things. For many, the myth had apparently worked. In 1965, the Russell Sage Foundation issued a report entitled *Experiences and Attitudes of American Adults Concerning Standardized Intelligence Tests*.[46] Some of the major findings of that report indicated that the effects of the tests on social classes were "strong and consistent" and that, while "the upper class respondent is more likely to favor the use of tests than the lower class respondent," the "lower class respondent is more likely to see intelligence tests measuring inborn intelligence."[47]

The lower-class American adult was, indeed, a product of fifty years of testing. He had been channeled through an intricate bureaucratic educational system which, in the name of meeting his individual needs, classified and tracked him into an occupation appropriate to his socioeconomic class status. The tragic character of this phenomenon was not only that the lower class learned to believe in the system, but worse, through internalizing that set of beliefs, made it work. It worked because the lowered self-image which the school and society reinforced[48] on the lower-class child did result in lower achievement. A normal child objectified as subnormal and treated by the teacher and the school as subnormal will almost surely behave as a subnormal child. Likewise, the lower-class child, who is taught in many ways to doubt his own intelligence, can be expected to exhibit a lower achievement level than those children who are repeatedly reminded that they are made of superior clay,[49] and therefore, are of superior worth.

Intelligence and achievement tests used throughout American society are a vital part of the infrastructure which serves to stabilize and order the values of that society.[50] Arthur R. Jensen put it well when he said: "Had the first IQ tests been devised in a hunting culture, 'general intelligence' might well have turned out to involve visual acuity and running speed, rather than vocabulary and symbol manipulation."[51] Jensen, as Terman and others, argued that

> what we now "mean" by intelligence is something like the probability of acceptable performance (given the opportunity) in occupations varying in social status.
>
> So we see that the prestige hierarchy of occupations is a reliable objective reality in our society. To this should be added the fact that there is undoubtedly some relationship between the levels of the hierarchy and the occupations' intrinsic desirability, or gratification to the individual engaged in them. Even if all occupations paid alike and received equal respect and acclaim, some occupations would still be viewed as more desirable than others, which would make for competition, selection, and again, a kind of prestige hierarchy.[52]

The hierarchy, Jensen argued, was inevitable because, "Most persons would agree that painting pictures is more satisfying than painting barns, and conducting a symphony orchestra is more exciting than directing traffic."[53] While the hierarchy was culturally determined, it is clear that certain values which Jensen preferred appeared to him more intrinsic than others. Nevertheless, he admitted that, "We have to face it: the assortment of persons into occupational roles simply is not 'fair' in any absolute sense. The best we can hope for is that true merit, given equality of opportunity, act as the basis for the natural assorting process."[54] Herein lies the crucial weakness of the argument. Given the current racist, economic, and socially elitist society where wealth, power, and educational privilege are so unevenly distributed, what does it mean, then, to assume "equality of opportunity" and "hope" that "true merit" will somehow result from a "natural assorting process"?

Jensen, like Thorndike and Terman before him, assumed an ideal liberal community where "equality of opportunity" balanced with lively competition produced a social system where "true" merit was rewarded. Although the "Jefferson-Conant" ideal of the good community is in itself questionable, the problem was compounded when the ideal society was confused with the real society. In spite of Terman, Thorndike, and Jensen's idealized assumptions about equal opportunity and the natural assorting process, all based their objective data on the real world of economic and social privilege. With highly questionable sociological data, they proceeded to even more questionable biological conclusions. The leap from sociology to genetics was an act of faith.

Most testers refused to admit the possibility that they were, perhaps, servants of privilege, power, and status, and preferred, instead, to believe and "hope" that what they were measuring was, in fact, true merit. This was also an act of faith, a faith based on the belief that somehow the "prestige hierarchy of occupations" and the people in it who provided the objective standard upon which the tests were based, were there not because of privilege, wealth, power status, and violence, but because of superior talent and virtue. This was a fundamental axiom in the liberal's faith in the meritocracy which emerged in twentieth-century American education.

Throughout this century, within this liberal faith, there emerged a series of doctrinal disputes engaging the attention of millions of people. The nature-nurture argument was one such continuous dispute from Galton to Jensen. The course of this dispute reflected little more than increasing refinement of statistical techniques and accumulation of data on both sides of the issue. Given the extent of unprovable propositions on both sides of the issue, one finds the choice between heredity or environment more a matter of faith than hard evidence. In many respects, the nature-nurture argument is a misleading issue. One can accept a strong hereditarian position and still advocate political, economic, and social equality just as one might accept a strong environmentalist position and still argue for political, economic, and social inequality. There is, in fact, no inherent logic either in the mind of man or in the universe which predetermines that differences in intellectual ability necessarily should mean differences in social power. Why, for example, should one more favorably be rewarded because, through no effort of his own, he happened to inherit a superior intelligence, or because he happened to be born into a superior social environment. Repeatedly, from Terman to Herrnstein, psychologists have attempted to link ability to the meritocracy without questioning the values inherent in the meritocratic principle itself.[55]

Most psychologists did not take their position to its logical conclusion, for to do so would be to question not only the ideal assumption upon which the meritocracy rests (such as equal opportunity and the inherent value of competition), but to further question the hierarchy of values which undergird the work of the professional knowledgeable expert in the liberal society. The professional expert, with his esoteric knowledge, is a vital element in both the creation and maintenance of the corporate meritocracy. His economic and political self-interest, as well as his very survival, is at stake. Thus, it is understandable that so few in the professional middle class are disturbed by the presumed meritocracy, or are seriously inclined to question it. Those who have done well by the system can hardly be expected to be its best critics. To be sure, some are willing to suggest that we ought to look at the social system from the bottom up, and out of humanitarian motives, or perhaps survival motives, allow more opportunity for those who have been cut out of the system, but few are willing to doubt critically the validity, if not the equity of the system itself.

It is also understandable that those who have been severely cut out of the system should provide the most vehement source of criticism. The unusual thing, however, is that in the past half century there have been so few real challenges. The meritocracy issue was bound to surface as the United States Supreme Court attacked segregation of children in the public schools in the 1954 Brown decision. The same effect of segregation on the basis of race could be achieved in most communities through segregation on the basis of "tested" ability. If, in fact, the tests were based on socioeconomic class, then the net result of segregation was possible.

In *Bolling v. Sharpe* (1954), the court ordered the desegregation of the District of Columbia schools, and in 1956, the Board of Education adopted a tracking system for the Washington public schools based on so-called *"ability grouping."* In *Hobson v. Hansen* (District Court case, 1967) Skelly Wright, a United States circuit court judge, wrote the opinion which challenged not only the use of ability tracking in the Washington public schools to circumvent desegregation, but went further by questioning the basis of tracking in the first place. While the school board insisted that the tracking system was based on meeting the needs of individuals through curricular adjustment according to their ability, they also denied racial bias, but admitted that enrollment in the tracks was related to socioeconomic status of the students.

The Washington, D.C., school system was operating a four-track system—which included an honors track for the gifted, a college preparatory track, a general education track, and a special track for the "slow learners." The system was highly rationalized, and objectified, and supported by empirical data derived from an extensive testing program. It was a fairly typical track system which, since the work of Terman and others, had come to be a standard curricular design of the American school by midcentury. Terman's prediction, made in 1924, had come true; at that time, he said, "I predict that within a decade or two something like the three-track plan of Oakland, Berkeley and Detroit, supplemented by opportunity classes for defectives and for the very gifted, will become standard."[56] The track system had become standard, and interestingly enough, had been propelled on Terman's argument that this was a step toward "educational democracy." By recognizing individual differences and adjusting the curriculum according to individual needs, he argued that every child will have the ". . . opportunity to make the most of whatever ability he possesses."[57] He went on to say that he had little patience with those who condemned his tracking plan as undemocratic and labeled them as "educational sentimentalists." Forty-three years later, Skelly Wright, in reviewing the Washington, D.C., tracking system which was similar to the system espoused by Terman, concluded:

> Even in concept the track system is undemocratic and discriminatory. Its creator admits it is designed to prepare some children for white-collar, and other children for blue-collar jobs. Considering the tests used to determine which children should receive the blue-collar special, and which the white, the danger of children completing their education wearing the wrong collar is far too great for this democracy to tolerate.[58]

Carefully considering the basis for ability grouping in both theory and practice, the court found that, in theory, the tests which presumably measured ability and the tracks which supposedly served individual needs, in practice, actually measured past socioeconomic advantage as much as presumed ability, while the curricular track served to lock the child into the socioeconomic class from which he came. Few children ever crossed tracks. Once the child was objectified or tracked, a self-fulfilling prophecy seemed to take place.

> The real tragedy of misjudgements about the disadvantaged student's abilities is, as described earlier, the likelihood that the student will act out that judgement and confirm it by achieving only at the expected level. Indeed it may be even worse than that, for there is strong evidence that performance in fact *declines*.[59]

The court went on to warn that:

> . . . a system that presumes to tell a student what his ability is and what he can successfully learn incurs an obligation to take account of the psychological damage that can come from such an encounter between the student and the school; and to be certain that it is in a position to decide whether the student's deficiencies are true, or only apparent.[60]

As the court raised issues about "true" or "apparent" deficiencies, equal opportunity, and the credibility of the tests, and further recognized the fact that mental-ability grouping corresponded to social class, they proceeded to declare the tracking system as employed in the Washington schools a violation of the plaintiff's constitutional rights.[61] Even though the court came close, it did not question the social basis of the meritocracy in the liberal state, nor did they contest the right of that state, to use a Jeffersonian phrase, ". . . rake from the rubbish annually."

Although the Skelly Wright decision had generally little effect on the nation's schools, it did threaten the security of some testers, especially those who had come to believe in their own myths. Those, however, who realized that the tests were based on a meritocracy which, in turn, was based on a socioeconomic hierarchy, remained secure. Clearly, it would take nothing short of a fundamental socioeconomic revolution to shake the hold of testing in the corporate liberal state. To chip away at the tests was almost akin to chipping away at an iceberg. While the blacks in the ghetto schools might do away with standardized tests in outrage against the bias and the white middle-class values they reflected, they still could not do away with the socioeconomic hierarchy which controlled entrance into virtually every occupation within the corporate liberal state. And that, after all, was what the tests represented. For the most part, the tests were not arbitrary but rather validly represented the requirements of a corporate America. One of the central values of that state reflected in its schools as well as its social and political life was "efficiency." Virtually all educational debate seemed to end with that criteria. The need to classify and standardize an entire nation so as to efficiently maximize productivity had become a reality by 1970.[62] There were problems, however, in both the way in which people were tested and classified and the way in which this classification system was used to restrict employment opportunities within the corporate liberal state. As various institutions within that state increasingly used tests and educational requirements more as a vehicle for keeping people out of specific occupations rather than a test of actual job performance, the credentialing system became strained and inflated to the point where employers could discriminate against minority groups with relative ease.[63] By 1971, the United States Supreme Court struck down the personnel policies of Duke Power Company's Dan River power station which required a high school diploma or acceptable performance on two standardized tests. The company had difficulty showing any clear relationship between these educational requirements and actual job performance. Even though this decision sent a cold chill through those private companies who serviced the 55 million dollar educational testing business, the court did not attack the testing procedure but merely the test's relationships to actual job performance. The court declared that the employer who uses personnel tests must show a "manifest relationship" of such tests to the "employment in question." Tests, then, were used to discriminate against children in their schools and against adults in their occupations. While some might see in these events minor problems in Condorcet's dream of progress, other saw it as an Orwellian nightmare of "double think." Perhaps they were the same.

By 1965, the Carnegie Foundation under the leadership of Francis Keppel, John Gardner, and James B. Conant initiated the development of a national testing program called, "A National Assessment of Educational Progress." Reminiscent of Ellwood Cubberley's concern for "continuous measurement of production," Conant's national assessment filled the current efficiency need for "accountability." Somewhat akin to the school's surveys of the earlier efficiency movement, which concentrated on local educational production, national assessment would give indications of regional productivity. Whle the earlier movement was characterized by the stopwatch and the efficiency expert, the new movement was characterized by the computer and the systems analyzer. Employing the same statistical sampling techniques used to measure industrial productivity, some systems analyzers suggested that eventually a "growth" national product index could be developed to measure the progress of American schools. National assessment was advocated by Conant, in part at least, because of the difficulty he had in obtaining hard data when he did the earlier reports on the *American High School, Junior High School* and *Teacher Education.* This concern, coupled with his interest in the development of a commission which would serve as a vehicle to shape national educational policy, led him to propose, while financially supported by the Carnegie and Danforth Foundations, the development of the "Compact of the States."[64] The "Compact" provided the "quasi-public" vehicles through which national testing proceeded without too much public interference. Reflecting back over his many and varied educational accomplishments, Conant in *My Several Lives* counted the "Compact" as his "greatest" achievement.

On the surface, it might appear that American education is locally controlled by elected boards and that, at the national level, a power vacuum exists due to the reluctance of the American people to create a national system of education. This, however, is misleading. There does exist within the corporate liberal state quasi-public bureaucracy of boards, compacts, councils, and commissions which serve to shape policy by giving and withholding both public and private funds at key points in the system. The American Council on Education is one such agency through which hundreds of philanthropic foundations, private businesses, and public colleges and universities work in establishing nationwide educational policy.[65] In many ways, the Council has acted as a meeting ground for what appears disparate interests, but also as a conduit for channeling funds into selected areas of higher education, and thereby effectively shaping practice as well as policy.

While World War I provided the national stimulus for the testing movement, foundations such as the Carnegie Foundation for the Advancement of Teaching, the Commonwealth Fund, the Graduate Records Office of the Carnegie Foundation, and others provided the funds which sustained and propelled the movement; it was World War II which demonstrated the usefulness of systems analysis, the efficient need for systematic overall manpower planning,[66] and the crucial need for a centralized testing service.

In 1946–1947, the American Council on Education received a report from the Carnegie Foundation for the Advancement of

Teaching, recommending the consolidation of a number of testing units which had developed over the years such as the College Entrance Examination Board and the Graduate Records Office.[67] Under a self-perpetuating board of trustees which included people drawn from the American Council on Education, the foundations, public and private colleges, as well as business and government, the Educational Testing Service was born with a grant from the Carnegie Foundation. The rapid growth of ETS is reflected in its operating budget. Operational expenses for ETS increased from approximately $2 million in 1947 to $29.7 million by 1969.[68] By 1969, ETS, a private nonprofit organization, provided the tests taken by millions of Americans which determined their eligibility, all the way from the service academies to the Peace Corps, colleges and universities, as well as the professions.[69] The doorway, to virtually every profession in the corporate liberal state came under the influence and control of the new organization. It is interesting, but not unusual, that such power would be exercised by private rather than public boards. This is one of the key characteristics of this state.

That fourth branch of government which was born in the Progressive Era, representing liberal corporate interests, i.e., foundations, has flexibly and effectively served to maintain the interests of corporate wealth through the support and maintenance of the liberal state. Through the varied shifting support policy of the foundations in the twentieth century, one need not conjure up a conspiracy theory in order to perceive some rather consistent practices which emerge. The key is enlightened self-interest. The foundation trustees and executives were intelligent men who recognized their own interests in what they viewed as a progressively developing society. It was in the interest of what they perceived as racial progress that the Carnegie Foundation of Washington supported the Laughlin studies used to close immigration, and in the interest of educational progress that they supported the testing movement after the war and the standardization of American education along the values laid down by Thorndike. Whether it was Terman calling for special education for the gifted or Conant founding the Educational Testing Service, all served to classify, standardize, and rationalize human beings to serve the productive interests of a society essentially controlled by wealth, privilege, and status.

The tests, whether measuring intelligence or achievement, as well as the meritocracy itself, served to so mask power as to effectively immobilize any real revolutionary opposition. If a man truly believes that he has a marginal standard of living because he is inferior, he is less likely to take violent measures against that social system than if he believes his condition a product of social privilege. In the nineteenth century, Daniel Webster said that "Public education is a wise and liberal system of police, by which property, and life, and the peace of society are secured." In the twentieth century, a similar condition prevails. In this sense, the foundations' deep involvement in educational policy, whether it was the Ford Foundation in educational television, or the Carnegie Foundation in testing, or the Rockefeller Foundation in black education, all had an interest in an effective, efficiently managed system. The foundations' management of educational policy in the twentieth century has been clearly at the cutting edge of every educational reform from the "Carnegie Unit" to the "open classroom."

Even the rhetoric which engages the professional educators seems fairly well managed. Throughout the last four decades, the pendulum of educational rhetoric has swung from the child-centered discussion of the Thirties to the society-determined needs of the Fifties, then again, to the child-centered needs of the Seventies. In the Thirties, the Carnegie Foundation supported the Progressive Education Association with over $4 million, while in the Fifties, James Conant's study of American schools was supported by Carnegie, as was the project which culminated in Charles Silberman's *Crisis in the Classroom* in the 1970s. It is interesting to note that during periods of labor surplus, our educational rhetoric tends to be child-centered, while in periods of shortage, the rhetoric shifts to society-oriented needs. This may be the propelling factor. It is interesting, however, that when the rhetoric became so heated that people could be heard suggesting that we do away with the system or radically change it, Carnegie Foundation supported James Conant, who, in effect, said the system was basically sound but then co-opted the rhetoric of the attackers to recommend limited change. It was, after all, the survival of the system which Conant had in mind when he spoke of social dynamite in the ghettoes. By the 1970s, when most manpower projections clearly indicated surplus of labor for the next decade, the educational reform rhetoric shifted from training scientists and engineers to open classrooms. Again, critics could be heard suggesting that the system be radically altered if not abolished; and once again, the Carnegie Foundation supported a study by Silberman which, in effect, said that the system was basically sound but needed some reforming. Once again, the rhetoric of the attackers was co-opted for limited change. While the Carnegie Foundation obviously does not control the pendulum, they have played a major role in managing the rhetoric at critical points when the system is in acute danger.[70] It is this function as governor of the educational machinery which prevents destructive unmanaged revolution that foundations have performed so well. One, then, is left to ponder the question of whether Charles W. Eliot was right when testifying before Congress (1913) that he had "never known a charitable or educational corporation to do anything which threatened the welfare or the liberties of the American people." Or was the majority of that committee which heard his testimony perhaps more correct when they concluded that the policies of the foundations would inevitably be those of the corporations which sponsored them, and that, "The domination of men in whose hands the final control of a large part of American industry rests is not limited to their employees, but is being extended to control the education and social services of the nation."

Notes

For much of the research in this essay, I am indebted to the research assistance of Russell Marks at the University of Illinois. For calling my attention to the material used in this article which reflects the racial bias of the currently used Stanford-Binet Tests, I am also indebted to La Monte Wyche.

1. C. Forcey, *The Crossroads of Liberalism* (London: Oxford University Press, 1961), p. xiv.
2. R. H. Wiebe, *The Search for Order* (New York: Hill & Wang, 1967).
3. See N. Pollack, *The Populist Response to Industrial America* (New York: W. W. Norton & Co., 1962). *Also see* William Appleman Williams, *The Roots of the Modern American Empire* (New York: Random House, 1969).
4. See J. Weinstein, *The Corporate Ideal in the Liberal State* (Boston: Beacon Press, 1968). *Also see* G. Kolko, *The Triumph of Conservatism: A Reinterpretation of American History 1900–1916* (New York: Free Press of Glencoe, 1963).
5. As quoted in Weinstein, *ibid.*, p. 45.
6. *Ibid.*, p. xi.
7. See L. Baritz, *The Servants of Power* (New Fork: John Wiley & Sons, 1960).
8. Increased efficiency in city, state, and federal government often meant a corresponding decrease in political influence on the part of the poor and

the disinherited. *See* S. P. Hays, "The Politics of Reform in Municipal Government in the Progressive Era," in *American Urban History,* ed. A. B. Callon, Jr. (New York: Oxford University Press, 1969).

9. *See* N. N. Gill, *Municipal Research Bureaus* (Washington, D.C.: American Council on Public Affairs, 1944), pp. 15–17.

10. *See* D. W. Eakins, "The Development of Corporate Liberal Policy Research in the United States, 1885–1965," unpublished Ph.D. dissertation, University of Wisconsin, 1966. *Also see* Weinstein, *Corporate Ideal in the Liberal State.*

11. W. Weaver, *U.S. Philanthropic Foundations* (New York: Harper & Row, 1967), p. 245.

12. J. Lankford, *Congress and the Foundations in the Twentieth Century* (River Falls: Wisconsin State University, 1964), pp. 6, 92.

13. *Ibid.*, p. 92.

14. As quoted in *ibid.*, p. 29.

15. As quoted in *ibid.*, p. 31.

16. *See* J. Dewey, "A New Social Science," *New Republic,* April 16, 1918.

17. For a discussion of the efficiency movement in business and industry, *see* S. Haber, *Efficiency and Uplift: Scientific Management in the Progressive Era 1890–1920* (Chicago: University of Chicago Press, 1964). For a similar discussion in education, *see* R. Callahan, *Education and the Cult of Efficiency* (Chicago: University of Chicago Press, 1962).

18. E. P. Cubberley, *Public School Administration* (Boston: Houghton Mifflin Co., 1916), p. 338.

19. As quoted in M. H. Haller, *Eugenics* (New Brunswick, N.J.: Rutgers University Press, 1963), p. 133.

20. *Ibid.*, p. 133. For one proposed segregation and sterilization program, *see* figure 1, pp. 370–371 of this book.

21. State authority is still being used to sterilize people. On March 4, 1971, a bill was introduced into the Illinois Legislature which required sterilization of a mother who had two or more children while on welfare rolls before that mother could draw further support. The argument, however, is no longer based on racial purity or punishment, but rather more on the economic burden to society. While the constitutionality of pauper sterilization might be questionable, the right of the state to sterilize for eugenics purposes was settled in *Buck v. Bell* when Justice Holmes argued: "The principle that sustains compulsory vaccination is broad enough to cover the cutting of the Fallopian tubes . . . three generations of imbeciles are enough," as quoted in Haller, *Eugenics,* p. 129.

22. L. M. Terman, *The Measurement of Intelligence* (Boston: Houghton Mifflin Co., 1916), p. 11.

23. E. L. Thorndike, "Intelligence and Its Uses," *Harper's,* January 1920, p. 233.

24. E. L. Thorndike, *Human Nature and the Social Order* (New York: Macmillan Co., 1940), p. 957. Italics added.

25. G. Joncich, *The Sane Positivist* (Middleton, Conn.: Wesleyan University Press, 1968), p. 443.
26. See Russell Marks' paper delivered at the AERA, Spring 1971, "Testing for Social Control."
27. H. E. Garrett, *Breeding Down* (Richmond: Patrick Henry Press, undated), p. 10.
28. *Ibid.*, p. 17.
29. C. C. Brigham, *A Study of American Intelligence* (Princeton: Princeton University Press, 1923), pp. 159, 207, 210. It should be noted that in the decade of the thirties, however, Brigham went through considerable effort to refute his former work.
30. *See* U.S. Congress, House Committee on Immigration and Naturalization, *Europe as an Emigrant-Exporting Continent and the United States as an Immigrant-Receiving Nation,* by H. H. Laughlin, 68th Congress, 1st Session, March 8, 1924, p. 1311.
31. L. M. Terman, "The Conservation of Talent," *School and Society* 19, no. 483 (March 29, 1924): p. 363.
32. As quoted by R. DeC. Ward, "Our New Immigration Policy," *Foreign Affairs,* September 1924, p. 104.
33. L. M. Terman, *Intelligence Tests and School Reorganization* (New York: World Book Co., 1923), pp. 27–28.
34. Terman, "The Conservation of Talent," p. 363.
35. The National Intelligence Test, interestingly enough, was standardized on army officers and used in the school after World War I.
36. *See* W. Lippmann, "A Future for the Tests," *New Republic,* November 29, 1922. *Also see* "The Mental Age of Americans," October 25, 1922; "The Mystery of the 'A' Men," November 1, 1922; "The Reliability of Intelligence Tests," November 8, 1922; "The Abuse of Tests," November 15, 1922; "Tests of Hereditary Intelligence," November 22, 1922.
37. *See* L. M. Terman, "The Great Conspiracy," *New Republic,* December 27, 1922, p. 117.
38. H. H. Goddard, *Human Efficiency and Levels of Intelligence* (Princeton: Princeton University Press, 1920), p. 97. It is interesting to note that the total population of the United States in 1920 was 105.7 million.
39. Thorndike, "Intelligence and Its Uses," p. 235.
40. E. L. Thorndike, "How May We Improve the Selection, Training and Life-Work of Leaders?" *Addresses Delivered Before the Fifth Conference on Educational Policies* (New York: Teachers College, Columbia University Press, 1939), p. 32.
41. *Ibid.*, p. 31.
42. U.S. House of Representatives, *Special Committee to Investigate Tax-Exempt Foundations—Summary of Activities,* June 9, 1954, p. 20.

43. It should be noted here that up to 1954, Carnegie Foundation alone had invested $6,424,000 in testing. *Ibid.*, p. 78.

44. It should be noted here that this is the latest revision of the Stanford-Binet Intelligence Test.

45. Of this group, only Terman wavered from the original position. When he wrote his autobiography in 1932 he had stated his belief, "That the major differences between children of high and low IQ and the major differences in the intelligence test-scores of certain races, as Negroes and whites will never be fully accounted for on the environmental hypothesis." By 1951, he penciled in around that statement, "I am less sure of this now," and in 1955, again another note said, "I'm still less sure." See E. R. Hilgard, "Lewis Madison Terman," *American Journal of Psychology*, 1957.

46. O. G. Brim, J. Neulinger, and D. C. Glass. *Experiences and Attitudes of American Adults Concerning Standardized Intelligence Tests*, Technical Report no. 1 on the Social Consequences of Testing (New York: Russell Sage Foundation, 1965).

47. *Ibid.*, p. 89.

48. *Ibid.*, p. 89. One of the findings was that a member of the lower class "estimates his intelligence as inferior to others."

49. For an analysis of the way the idea of self-fulfilling prophecy works, *see* R. Rosenthal and L. Jacobson, "Self-Fulfilling Prophecies in the Classroom: Teachers' Expectations as Unintended Determinants of Pupils' Intellectual Competence," in M. Deutsch et al., *Social Class, Race and Psychological Development* (New York: Holt, Rinehart & Winston, 1967).

50. For an analysis of the role of psychologists in business, *see* Baritz, *Servants of Power*.

51. A. R. Jensen, "How Much Can We Boost IQ and Scholastic Achievement?" *Harvard Educational Review* 39, no. 1 (Winter 1969):14.

52. *Ibid.*, pp. 14–15.

53. *Ibid.*, p. 15.

54. *Ibid.*

55. For these particular ideas on the implications of the nature-nurture controversy, I am indebted to Russell Marks. For Herrnstein's association of meritocracy and IQ *see* R. J. Herrnstein, "IQ," *Atlantic Monthly*, September 1971. *See also* A. Alland, Jr., *Human Diversity* (New York: Columbia University Press, 1971), pp. 177–207.

56. Terman, "The Conservation of Talent," p. 364.

57. *Ibid.*

58. Hobson v. Hansen, Civil Action, No. 82–66. *Federal Supplement*, V. 269, p. 515.

59. *Ibid.*, p. 491.

60. *Ibid.*, p. 492.

61. The Skelly Wright decision was viewed by many, in 1967, as the probable direction the Warren Court would take when it eventually faced the thorny problem of de facto segregation. Since the development of the new court, this prediction is questionable.

62. *See* H. C. Hand, "The Camel's Nose"; and R. W. Tyler, "Assessing the Progress of Education," *Phi Delta Kappan*, September 1965, pp. 8–18.

63. For the questionable relationship between education and actual job performance, *see* I. E. Berg, *Education and Jobs: The Great Training Robbery* (New York: Praeger Publishers, 1970).

64. *See* J. B. Conant, *Shaping Educational Policy* (New York: McGraw-Hill Book Co., 1964). The financial report of the "Compact," as of May 1966, showed Carnegie and Danforth contributed $300,000 of a $318,000 budget. *See* F. Raubinger and H. C. Hand, "It Is Later Than You Think," unpublished manuscript, 1967, p. 29.

65. For the membership in that council, *see A Brief Statement of the History and Activities of the American Council on Education* (Washington, D.C., 1918–1959). There is some similarity between the kind of make-up and function of the American Council and the earlier National Civic Federation which also served disparate interests, and at the same time, worked for the common good of the liberal state.

66. It should be noted that it was Conant's concern for manpower planning which led him to recommend the GI Bill. One of the arguments used to support the GI Bill legislation was that it would effectively check severe unemployment in the immediate postwar period.

67. For the connection, I am indebted to Professor Fred Raubinger.

68. *See ETS Developments*, no. 4 (May 1970):2.

69. The kinds of tests used: the Preliminary Scholastic Aptitude Test, the Advanced Placement Examinations, the College Placement Tests, the College-Level Examination Program, and the Comparative Guidance and Placement Program—the last for two-year colleges.

ETS also administers other admissions tests: the Graduate Record Examinations for admission to the graduate study; more specialized tests for admission to Architectural Schools and to the Colleges of Podiatry; and now a test for admission to grades 7 through 11 of independent secondary schools. As one staff member remarked, "We have a test for admission to everything except heaven!"

Achievement Tests. In addition to the achievement tests of the College Board, there are the numerous Cooperative Tests (which now extended even to a Preschool Inventory); the Undergraduate Program (replacing the Graduate Record Examinations as formerly used by colleges for testing undergraduates); and the Graduate School Foreign Language Testing Program (to satisfy the language requirement for advanced degrees).

In a special category is the Test of English as a Foreign Language to measure the English proficiency of foreign students, government trainees, and others who apply for education in an English-speaking country.

Professional Examinations. Among the many professional examinations offered by ETS are the National Teacher Examinations, the National Council of Architectural Registration Boards Examinations, the Chartered Life Underwriter Examinations, the American Board of Obstetrics and Gynecology Examination, and Examinations in Speech Pathology and Audiology. Perhaps the stiffest of all examinations offered by ETS is called the Written Examination for Appointment as Foreign Service Officer (for Department of State and Information Agency Assignments), from *ETS Developments* 17, no. 4 (May 1970):2.

70. This seems to occur more as a condition of whom the foundations support rather than any overt attempt to control the rhetoric directly.

Figure 1. *From* Proceedings of the First National Conference on Race Betterment, *January 8–12, 1914, Battle Creek, Michigan. Published by the Race Betterment Foundation, p. 479.*

RATE OF EFFICIENCY OF THE PROPOSED SEGREGATION AND STERILIZATION PROGRAM

EFFECTIVENESS OF THE PROGRAM DEPENDS UPON:

1. *The length of time required to send to State institutions—regardless of length of commitments—all of the breeding stock of the varieties sought to eliminate.*
2. *The sterilization upon release from State custody of all individuals of such varieties possessing reproductive potentialities.*

The following table shows the reasonable expectation of the approximate working out of this program—if consistently followed—for eliminating the defective and antisocial varieties from the American population, under the following specific conditions:

1. *That the varieties (i.e., the breeding stock of defectives, including both affected and unaffected individuals of all ages) sought to exterminate now (1915) constitutes 10 percent of the total population.*
2. *That those portions of the defective varieties permitted to reproduce, increase at a rate equal to that of the general population, plus 5 percent each half decade.*
3. *That the group of state institutions for the socially inadequate, whose inmates by actual count constituted in 1890, .590 percent; in 1900, .807 percent; and in 1910, .914 percent of the total population, provide for and receive inmates at a rate equal to that of the increase of the general population, plus 5 percent each half decade.*
4. *That the inmates of institutions continue to be drawn from the two sexes and from each age group in even proportions to the total numbers of individuals of the corresponding sex and age group in the total population of the varieties sought to eliminate.*
5. *That the average "institution generation" (i.e., average period of commitment) be five years.*
6. *That the federal government cooperate with the states to the extent of prohibiting the landing of foreigners of potential parenthood of innate traits lower than the lowest of the better 90 percent of the blood already established here, i.e., belonging to the varieties of the lower 10 percent sought to cut off.*

Table 1.

1 *Half Decade Ending*	*1915*	*1920*	*1925*	*1930*	*1935*	*1940*	*1945*	*1950*	*1955*
2 Per cent. of total population constituting the varieties sought to eliminate	10.00	9.70	9.35	8.94	8.46	7.91	7.28	6.57	5.77
3 Per cent. of total population in institutions	.800	.840	.882	.926	.972	1.021	1.072	1.126	1.185
4 Probable total population of the U.S.		110,000,000		131,000,000		155,000,000		181,000,000	
5 Probable total numbers in the varieties sought to eliminate if program is carried out		10,670,000		11,711,400		12,260,500		11,891,700	
6 Total number of inmates in institutions called for by proposed program		924,000		1,213,060		1,582,550		2,038,060	
7 Number of sterilization operations required per year per 100,000 population	80.0	84.0	88.2	92.6	97.2	102.1	107.2	112.6	118.5
8 Probable total number of sterilizing operations required per year for the entire population		92,400		121,306		158,255		203,806	

The Race Betterment Movement Aims

To Create a New and Superior Race thru EUTHENICS, or Personal and Public Hygiene and EUGENICS, or Race Hygeine.

A thoroughgoing application of PUBLIC AND PERSONAL HYGIENE will save our nation annually:

1,000,000 premature deaths.

2,000,000 lives rendered perpetually useless by sickness.

200,000 infant lives (two-thirds of the baby crop)

The science of EUGENICS intelligently and universally applied would in a few centuries practically

WIPE OUT

Idiocy Insanity Imbecility Epilepsy

and a score of other hereditary disorders, and create a race of HUMAN THOROUGHBREDS such as the world has never seen.

Figure 2. *From* Official Proceeding of the Second National Conference on Race Betterment, *August 4–8, 1915, Battle Creek, Michigan. Published by the Race Betterment Foundation, p. 147.*

Methods of Race Betterment

Simple and Natural Habits of Life.

Out-of-Door Life Day and Night, Fresh-Air Schools, Playgrounds, Out-of-Door Gymnasiums, etc.

Total Abstinence from the Use of Alcohol and Other Drugs.

Eugenic Marriages.

Medical Certificates before Marriage.

Health Inspection of Schools.

Periodical Medical Examinations.

Vigorous Campaign of Education in Health and Eugenics.

Eugenic Registry.

Sterilization or Isolation of Defectives.

Figure 3. *From* Official Proceeding of the Second National Conference on Race Betterment, *August 4–8, 1915, Battle Creek, Michigan. Published by the Race Betterment Foundation, p. 160.*

Evidences of Race Degeneracy

Increase of Degenerative Diseases - - - - { Cancer; Insanity; Diseases of Heart and Blood Vessels; Diseases of Kidneys; Most Chronic Diseases; Diabetes }

Increase of Defectives { Idiots; Imbeciles; Morons; Criminals; Inebriates; Paupers }

Diminishing Individual Longevity

Diminished Birth Rate

Disappearance, Complete or Partial, of Various Bodily Organs - - - - { According to Wiedersheim there are more than two hundred such changes in the structures of the body }

Figure 4. *From* Official Proceeding of the Second National Conference on Race Betterment, *August 4–8, 1915, Battle Creek, Michigan. Published by the Race Betterment Foundation, p. 150.*

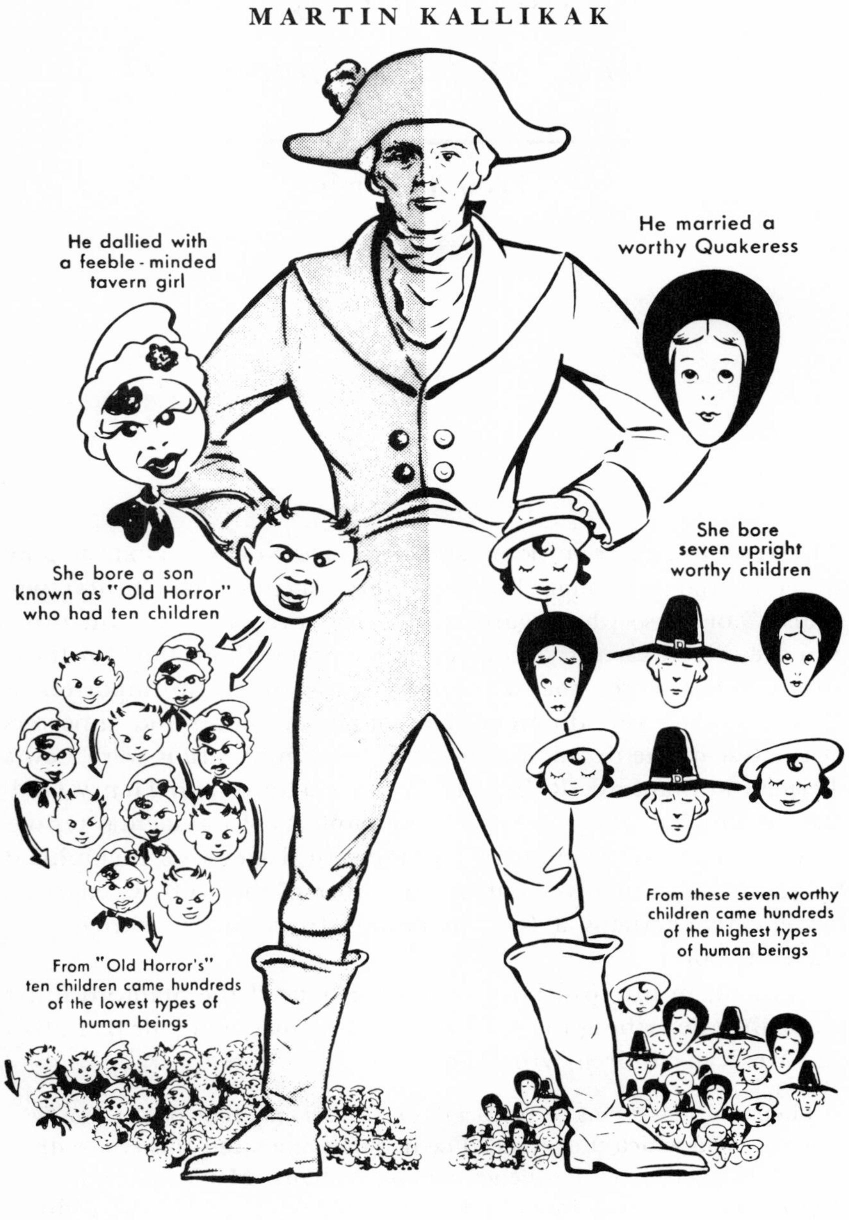

MARTIN KALLIKAK

He dallied with a feeble-minded tavern girl

He married a worthy Quakeress

She bore a son known as "Old Horror" who had ten children

She bore seven upright worthy children

From these seven worthy children came hundreds of the highest types of human beings

From "Old Horror's" ten children came hundreds of the lowest types of human beings

Figure 5. *The influence of heredity is demonstrated by the "good" and the "bad" Kallikaks. From Henry E. Garrett,* General Psychology *(New York: American Book Co., 1955), p. 65.*

Heredity, Intelligence, Politics, and Psychology: II

Leon J. Kamin

THE FIRST USABLE TEST OF GENERAL INTELLIGENCE WAS PUBLISHED BY Binet in 1905. Though Binet protested against the "brutal pessimism" of those who regarded the test score as a fixed quantity, and prescribed corrective courses in "mental orthopedics" for those with low test scores, the orientation of the American importers of Binet's test was very different. The major translators and importers of the test in the decade following Binet's publication were Lewis Terman at Stanford, Robert Yerkes at Harvard, and Henry Goddard at Vineland, New Jersey. These pioneers of the mental testing movement shared a number of sociopolitical views, as exemplified by their joint involvement in the turn-of-the-century eugenics movements. Perhaps a few quotations from their writings will make the point.

Terman, in his 1916 book which introduced the Stanford-Binet Test, after describing the poor test performance of a pair of Indian and Mexican children, wrote the following:

> Their dullness seems to be racial, or at least inherent in the family stocks from which they come. The fact that one meets this type with such extraordinary frequency among Indians, Mexicans, and negroes suggests quite forcibly that the whole question of racial differences in mental traits will have to be taken up anew . . . there will be discovered enormously significant racial differences . . . which cannot be wiped out by any scheme of mental culture.
>
> Children of this group should be segregated in special classes. . . .

> They cannot master abstractions, but they can often be made efficient workers.... There is no possibility at present of convincing society that they should not be allowed to reproduce... they constitute a grave problem because of their unusually prolific breeding. [Terman, 1916]

Professor Terman should not be thought of as a racist. His stern eugenical judgment was applied even-handedly to poor people of all colors. Writing in 1917 under the heading "The Menace of Feeble-Mindedness," he declared,

> only recently have we begun to recognize how serious a menace it is to the social, economic and moral welfare of the state... it is responsible... for the majority of cases of chronic and semi-chronic pauperism... organized charities... often contribute to the survival of individuals who would otherwise not be able to live and reproduce.... If we would preserve our state for a class of people worthy to possess it, we must prevent, as far as possible, the propagation of mental degenerates... the increasing spawn of degeneracy. [Terman, 1917]

The squandering of charitable moneys on the degenerate poor similarly caught the attention of Henry Goddard, who lectured to a Princeton audience in 1919 on the new science of "mental levels." That new science, he pointed out, had invalidated the arguments of gentlemen socialists who "in their ultra altruistic and humane attitude" were embarrassed that their own shoes cost $12.00, while those of a laborer cost only $3.00.

> Now the fact is, *that workman* may have a ten year intelligence while you have a twenty. To demand for him such a home as you enjoy is as absurd as it would be to insist that every laborer should receive a graduate fellowship. How can there be such a thing as social equality with this wide range of mental capacity?...
>
> ... The man of intelligence has spent his money wisely, has saved until he has enough to provide for his needs in case of sickness, while the man of low intelligence, no matter how much money he would have earned, would have spent much of it foolishly... during the past year, the coal miners in certain parts of the country have earned more money than the operators and yet today when the mines shut down for a time, those people are the first to suffer. They did not save anything, although their whole life has taught them that mining is an irregular thing and that... they should save.... [Goddard, 1920]

To be diagnosed as feeble-minded was not a light matter in a period when discriminations between the criminal, the poor, the insane, and the dull were not clearly drawn. The public institutions to provide for such degenerates were, in many states, administered by a single functionary, the Commissioner of Charities and Corrections. Further, prodded by the eugenicists, many states passed laws providing for the compulsory sterilization of the inmates of such taxpayer-supported institutions before their release. The preamble of the first such law, passed by Indiana, in 1907, was typical in its assertion: "Whereas, heredity plays a most important part in the transmission of crime, idiocy, and imbecility." To this list of genetically determined traits, the New Jersey legislature added, in 1911, "feeble-mindedness, epilepsy . . . and other defects," and Iowa, in 1913, contributed "lunatics, drunkards, drug fiends . . . moral and sexual perverts, and diseased and degenerate persons."

The lot of those officially diagnosed as feeble-minded was not enviable, and it is of interest to read Yerkes' caution:

> . . . never should such a diagnosis be made on the IQ alone. . . . We must inquire further into the subject's economic history. What is his occupation; his pay . . . we must learn what we can about his immediate family. What is the economic status or occupation of the parents? . . . When . . . this information has been collected . . . the psychologist may be of great value in getting the subject into the most suitable place in society. . . . [Yerkes and Foster, 1923]

The genetic interpretation of socioeconomic class differences in test scores, fostered by Terman, Goddard, and Yerkes, could clearly serve to legitimate the existing social order. Perhaps the first major practical effect of the testing movement, however, lay in its contribution to the passage and rationalization of the overtly racist immigration law of 1924. This disgraceful chapter in the history of American psychology is not without contemporary relevance.

Prior to the First World War, though certain classes of undesirables were excluded, there was no numerical limitation on immigration to the United States, nor were geographic distinctions drawn among European countries. But as early as 1912, the United States Public Health Service invited Henry Goddard to Ellis Island to apply the new mental tests to arriving European immigrants. Goddard (1913) reported that, based upon his examination of the "great mass of average immigrants," 83 percent of Jews, 80 percent of Hungarians, 79 percent of Italians, and 87 percent of Russians were "feeble-minded." He was later able to report (1917) that the use of mental tests "for the detection of feeble-minded aliens" had vastly increased the number of aliens deported.

The significance of these scientific findings was not lost upon the members of the Eugenics Research Association, who, in 1917, appointed Yerkes as chairman of their Committee on Inheritance of Mental Traits. The biologist, Harry Laughlin, secretary of the Association and editor of its journal, *Eugenic News*, wrote under the heading "The New Immigration Law": "When the knowledge of the existence of this science [mental testing] becomes generally known in Congress, that body will then be expected to apply the direct and logical test . . ." (Laughlin, 1917).

Within months, American entry into the war had brought the science of mental testing to a new level of public recognition. Intelligence tests were applied to some 2,000,000 draftees—under the direction of Colonel Robert M. Yerkes, with the assistance of many of the leading experimental psychologists. The influence of Yerkes may, perhaps, be detected in the massive influx of leading experimentalists into the Eugenics Research Association in 1920. But in any event, the results of the Army's testing program were published, under Yerkes's editorship, by the National Academy of Sciences (1921). The data provided the first large-scale evidence that blacks scored lower than whites. But the chapter of most immediate significance in 1921 was that on the foreign-born. The test performance of immigrant draftees was analyzed by country of origin, as reproduced in figure 1. The data presented in figure 1 were succinctly summarized: "The Latin and Slavic countries stand

Figure 1. *The mean intelligence tests scores of immigrant draftees from various countries fell in the following order (read down columns):*

England	Canada	Turkey
Holland	Belgium	Greece
Denmark	White Draft	All Other Foreign Countries
Scotland	Norway	Russia
Germany	Austria	Italy
Sweden	Ireland	Poland

From *Memoirs* of the National Academy of Sciences (1921).

low." The Poles, it was reported, did not score significantly higher than the blacks.

These scientific data speedily became "generally known in Congress," with the considerable assistance of the scientists of the Eugenics Research Association, and of Yerkes, now employed by the National Research Council. The secretary of the ERA was appointed "Expert Eugenics Affairs Agent" of the House Committee on Immigration and Naturalization of the United States Congress; in 1923, the psychological and biological scientists of the ERA elected as their organization's chairman the Honorable Albert Johnson. That gentleman, by a fortunate coincidence, was the congressman who chaired the House Committee on Immigration and Naturalization. Meanwhile, under Yerkes's leadership, the NRC's Division of Anthropology and Psychology established a Committee on Scientific Problems of Human Migration. That committee, in an effort to take the national debate over immigration "out of politics," and to place it on "a scientific basis," began to support relevant research. The first research supported was that of Carl Brigham, then Assistant Professor of Psychology at Princeton. The Princeton University Press published, in 1923, Brigham's *A Study of American Intelligence* with a foreword by Yerkes praising the book's contribution to the scientific study of immigration.

The unique contribution of Brigham's book was an intensive reanalysis of the Army data on immigrants. Brigham demonstrated that—pooling across all countries of origin—immigrants who had been in the country sixteen to twenty years before being tested were as bright as native-born Americans; that immigrants who had been in America only zero to five years when tested, were virtually feeble-minded. "We must assume," Brigham wrote, "that we are measuring *native or inborn intelligence*." The psychologists who devised the tests had, after all, constructed special tests for the illiterate. The explanation for the correlation of test score with years of American residence proved to be simple. Twenty years before, immigrants had flowed into the country from England, Scandinavia, Germany; five or ten years before the war, the massive "New Immigration" from southeastern Europe had begun—Italians, Poles, Russians, Jews. The decline of immigrant intelligence, Brigham noted, paralleled precisely the decrease in the amount of "Nordic blood," and the increase in the amount of "Alpine" and "Mediterranean" blood in the immigrant stream—a nice example of the power of correlational analysis as applied to intelligence test data. The Jew, Brigham declared, "is an Alpine Slav." The concluding paragraphs of Brigham's book pointed out that

> we are incorporating the negro into our racial stock, while all of Europe is comparatively free from this taint.... The steps that should be taken ... must of course be dictated by science and not by political expediency.... And the revision of the immigration and naturalization laws will only afford a slight relief.... The really important steps are those looking toward the prevention of the continued propagation of defective strains in the present population.

With this contribution behind him, Brigham moved on to the secretaryship of the College Entrance Examination Board, where he devised and developed the Scholastic Aptitude Test; and at length, to the secretaryship of the American Psychological Association.

The political usage of Brigham's book and of the Army data was immediate and intense. The book and the data figured prominently and repeatedly in Congressional committee hearings and debates on the new immigration law. I cite only a very few examples.

Dr. Arthur Sweeney, to the House Committee, January 24, 1923:

> The fact that the immigrants are illiterate or unable to understand the English language is not an obstacle.... "Beta" ... is entirely objective. . . . We . . . strenuously object to immigration from Italy ... Russia ... Poland ... Greece ... Turkey. The Slavic and Latin countries show a marked contrast in intelligence with the western and northern European group ... we shall degenerate to the level of the Slav and Latin races....

Mr. Francis Kinnicutt, to the Senate Committee, February 20, 1923:

> The immigration from [Poland and Russia] consists largely of the Hebrew elements ... some of their labor unions are among the most radical in the whole country. . . . The recent Army tests show ... these classes rank far below the average intelligence.... See "A Study of American Intelligence" by Carl C. Brigham.... Col. Robert M. Yerkes . . . vouches for this book, and he speaks in the highest terms of Prof. Carl C. Brigham, now assistant professor of psychology in Princeton University.

Mr. Madison Grant, to the Senate Committee, January 10, 1924:

> The country at large has been greatly impressed by . . . the Army intelligence tests . . . carefully analyzed by . . . Yerkes . . . Brigham. The experts . . . believe . . . the tests give as accurate a measure of intelligence as is possible. . . . The questions . . . were selected with a view to measuring innate ability. . . . Had mental tests been in operation . . . over 6,000,000 aliens now living in this country . . . would never have been admitted. . . .

The Congress passed, in 1924, a law not only restricting the total *number* of immigrants, but also assigning *national origin quotas.* That is, immigrants from any European country would be allowed entry into America only to the proportionate extent that their countrymen were already represented in the American population—*as determined by the Census of 1890.* The Congressional proponents of the law frankly asserted that the 1890 (rather than the 1920) census was used in order to curtail biologically inferior immigration from southeastern Europe. That is the law which led ultimately to the deaths of tens of thousands of victims of the Nazi terror, denied entry to the United States because the "German quota" was filled, though other quotas were undersubscribed.

The biological partitioning of the European Continent did not appease some ardent intelligence testers, who continued to perform relevant research. Nathaniel Hirsch's work, under McDougall at Harvard, was also supported by the National Research Council. To demonstrate the genetic basis of low IQs in immigrant stock, Hirsch tested the native-born *children* of immigrants. He reported in *Genetic Psychology Monographs:*

> That part of the law which has to do with the non-quota immigrants should be modified. . . . All mental testing upon children of Spanish-Mexican descent has shown that the average intelligence of this group is even lower than the average intelligence of the Portuguese and Negro children . . . in this study. Yet Mexicans are flowing into the country. . . .
>
> From Canada . . . we are getting . . . the less intelligent of working-class people . . . the increase in the number of French Canadians is alarming. Whole New England villages and towns are filled with them. The average intelligence of the French Canadian group in our data approaches the level of the average Negro intelligence.
>
> I have seen gatherings of the foreign-born in which narrow and

> sloping foreheads were the rule.... In every face there was something wrong—lips thick, mouth coarse . . . chin poorly formed . . . sugar-loaf heads... goose-bill noses... a set of skew-molds discarded by the Creator.... Immigration officials... report vast troubles in extracting the truth from certain brunette nationalities. [Hirsch, 1926]

That was the voice of *Genetic Psychology Monographs* in 1926. What shall we say of the voices of today's mental testers? The moral of this history seems to me sufficiently clear—and contemporary developments in mental testing too well known to us all—for explicit comment to be necessary. From this much, however, I cannot forbear. *The* domestic issue confronting the country in the 1920s was the problem of immigration. *The* domestic issue confronting us today—at least, until the recent Watergate amusements broke upon us—is what our politicians euphemistically refer to as "the welfare mess." The intertwining of profound social, economic, and racial conflicts within each of these great issues is obvious enough. We know now that the psychologists who offered "expert" and "scientific" testimony relevant to explosive social issues in the 1920s did so on the basis of pitifully inadequate data, data which I believe all of us would now reject as irrelevant to the question of the possible inheritance of intelligence. We have to ask, how much more surely grounded are today's psychological equivalents to the Expert Eugenics Affairs Agents of the 1920s than were their predecessors?

References

Brigham, C. C. 1923. *A Study of American Intelligence.* Princeton: Princeton University Press.

Goddard, H. H. 1913. "The Binet Tests in Relation to Immigration." *Journal of Psycho-asthenics* 18:105–107.

———. 1917. "Mental Tests and the Immigrant." *Journal of Delinquency* 2:243–277.

———. 1920. *Human Efficiency and Levels of Intelligence.* Princeton: Princeton University Press.

Hirsch, N. D. M. 1926. "A Study of Natio-racial Mental Differences." *Genetic Psychology Monographs* 1, whole nos. 3 and 4.

Laughlin, H. H. 1917. "The New Immigration Law." *Eugenic News* 2:22.

National Academy of Sciences, *Psychology Examining in the United States Army,* ed. R. M. Yerkes, vol. 15, *Memoirs.* Washington, D.C.

Terman, L. M. 1916. *The Measurement of Intelligence.* Boston: Houghton-Mifflin Co.

———. 1917. "Feeble-minded Children in the Public Schools of California." *School and Society* 5:161–165.

Yerkes, R. M., and Foster, J. C. 1923. *A Point Scale for Measuring Mental Ability.* Baltimore: Warwick & York.

Genetics and Educability

Educational Implications of the Jensen Debate

Carl Bereiter

In this paper I shall consider several educational issues growing out of A. R. Jensen's paper, "How Much Can We Boost IQ and Scholastic Achievement?" (Jensen, 1969). The first deals with the question of how education should adjust to the incontestable fact that approximately half the children in our schools are and always will be below average in IQ. Following this, I take up some of the more moot points of the "Jensen controversy"—what does heritability tell us about teachability? What are the prospects for reducing the spread of individual differences in intelligence? And what are the educational implications of possible hereditary differences in intelligence associated with social class and race?—ending with some implications that these issues have for educational research.

Intelligence as an Excuse for Poor Teaching

When children fail to learn, one may find fault either with the teaching or with the children. If the children who fail are in a minority, even if a fairly large minority, one has convenient grounds for a case that it is the children who are defective.

Every student of mental measurement knows the story of how the intelligence test was born, how French education authorities asked Alfred Binet to devise means of identifying children who were too dull to profit from regular schooling (c.f. Vernon, 1960).

Thus from its outset intelligence testing has been rooted in the effort to locate the causes of school failure in the child rather than in the way he was taught. It is interesting to speculate that if a different type of instruction had been used in French schools in 1904, so that a different type of child failed, we might today have a different concept of intelligence. It is even more interesting to speculate that intelligence testing may have served to perpetuate the kind of instruction that happened to be in use in France at the turn of the century.

The rationale for this latter speculation is as follows. If children of adequate IQ are found not to learn, then it is supposed that something must be wrong with the teaching, and the teaching is accordingly changed. If, however, the failing children have low IQs, then there is no cause for altering the method of teaching, because such children are expected to fail. The result is not necessarily a perpetuation of the exact forms of traditional instruction. Style and content may change. What remains constant is the complex of mental abilities that the child must have in order to succeed in school—the complex that is variously called "scholastic aptitude," *g*, or IQ.

To challenge the IQ, as some of Jensen's critics have done, on grounds that it represents a limited or biased conception of intelligence, is easy but pointless. Jensen has taken the much more daring and constructive approach of challenging the instructional methods that make IQ significant in the first place. He has raised the question, is it really necessary to teach in such a way that only children of average or above-average IQ can learn?

There is a head-on and a dodge-to-the-left way of meeting this question. The dodge-to-the-left way is to say that schools should not expect the same things of all children, that children should be free to pursue their own goals in their own ways, rather than having to pursue the particular achievement goals set forth by the schools. Although the proposal may have merits, it dodges the question of whether the traditional goals could be achieved by children who are deficient in scholastic aptitude as it is conventionally measured, if they were taught by different methods. The original assumption that "some kids just can't get it" remains unchallenged by this digression to other goals.

Jensen has tried to meet the question head-on, but there has not been much evidence for him to bring to the encounter. He has expressed the faith that basic scholastic skills could be learned in a variety of ways that make use of different mental abilities (pp. 116–117).* Schools, however, have adhered to methods and criteria which allow only the child possessed of abstract, verbal cognitive abilities to succeed. He points out that functional mastery of skills such as arithmetic computation is not regarded as a sufficient criterion of success, that "understanding" in the abstract, verbal sense is required. Thus, the same abilities that are involved in IQ predictors of success are used as criteria of success, producing self-fulfilling prophecy.

I would suggest that a more basic factor in maintaining the self-fulfilling prophecy is the generally low quality of instruction. The more confusing, inconsistent, and full of gaps that instruction is, the more the child has to figure out for himself. Thus, the child must be able to form abstractions, to generalize from scattered and incomplete evidence, even when the material to be learned is itself not of a very abstract nature. To ensure that IQ predicts achievement, it is necessary merely to teach badly. But educators have managed to make a virtue out of poor teaching by asserting that they want to make children think. The implication is that if the child cannot think, he has no business learning.

The term "thinking," as used above, ought perhaps to be severely qualified. It refers to the kinds of information processing that are called for on general intelligence tests. There are, of course, other sorts of mental activities that deserve to be called thinking—indeed, it would be difficult to justify assigning a very exalted place to the sort of thinking called for by intelligence tests. But for the present discussion, which deals exclusively with school learning, it will simplify the terminology a great deal if we understand the unqualified term "thinking" to refer to the kinds of thinking normally involved in mastering academic subjects.

The question Jensen has raised may then be given a more pointed form: Can we find ways for children to learn that require less thinking? It will immediately be seen that the main trend of curriculum reform in recent times has been in precisely the opposite direction—in the direction of requiring more thinking. Before considering whether this trend is a wise one, let us consider whether there are actually any viable alternatives.

An obvious alternative to thinking is rote memorization, and this is more or less the alternative that Jensen proposes. His own re-

* Unless otherwise designated, page numbers refer to pages in Jensen (1969).

search has indicated that, among lower-class children at least, substantial numbers can be found who have adequate associative learning ability even though they have low IQs. Accordingly, he has proposed that educational methods might be developed that make greater use of this ability and less use of what I am referring to as thinking.

The prospect of an instructional program based entirely on rote learning is not an attractive one by any account. This is not the necessary alternative, however. In the first place, the problem is not that children of below-average IQ cannot think, but that they may not be able to think well enough to cope with instruction as it is presently administered. In the second place, even the most rudimentary of scholastic skills, such as oral reading and arithmetic computation, require transformations of some complexity on the presented information, and hence, some amount of thinking. Thus the implied alternative to existing methods of instruction would not be one that eliminated thinking but one that kept it within more attainable limits of difficulty.

On the face of it, programmed instruction would appear to be the embodiment of this alternative—particularly programmed instruction that follows the principle of small steps and the minimizing of error rate. From the evidence, the most that could be claimed is that, on occasion, programmed instruction has managed to produce learning that is less dependent on IQ than the learning that takes place through conventional instruction (Tuel, 1966). It is difficult to draw any far-reaching conclusions from these results, however, since programmed instruction is a medium that differs from ordinary instruction in a number of ways, and, moreover, does not necessarily entail any change in the basic strategy by which subject matter is organized for teaching.

The work that I have been associated with over the past six years, in the design of instructional programs for young disadvantaged children, has, on the other hand, been consciously aimed at reducing the conceptual and problem-solving difficulties of school learning. The approach we have taken does not have any magic key to it. It has been a matter of trying to locate the underlying sources of difficulty in grasping various concepts and operations, and then trying to devise ways to overcome them. Expositions of the general approach and specific applications of it can be found in Engelmann (1969a), Engelmann (1969b), Bereiter and Engelmann (1966). Results reported by Engelmann in this volume indicate the striking achievements in subject-matter learning that have been obtained with this approach. How much can be accomplished through such an approach in raising achievement over the full span of school years and what overall value this might have to the individual learner remain to be seen. As they stand, however, the results lend strong support to Jensen's faith that basic scholastic skills can be put in the reach of children who lack the attributes needed to master them under conventional approaches.

While much remains unknown about the broader effects of reducing the thinking burden in learning, there are two consequences that are foreseeable enough to be worth considering at this time. One is that such instructional reforms cannot be expected to reduce the range of individual differences in achievement, nor, in the long run, to make achievement less dependent upon IQ. What such reforms can be expected to do, instead, is simply to raise the general level of achievement, while conceivably even increasing the spread of individual differences. Teach every child to read or to do multiplication, and you have raised the general level of attainment in these subjects; but the reasonable expectation, confirmed by results to date, is that "bright" children will still be able to read better and to solve more complex problems involving multiplication. By virtue of these superior skills, the "bright" children may be expected, in consequence, to show even greater attainment in more advanced mathematics and in what they learn through reading. Clearly, then, if our concern is purely with a child's attainment relative to his peers, we gain nothing much by improving instruction.

If, however, there is some value to be attached to absolute levels of attainment, then there is much to be hoped for from instructional reform. If, under method A, the lowest child in the class does not learn to read, while under method B he does, then he is better off under method B, even if he is still the lowest child in the class and just as far below the mean of his class as he would have been under method A. To the extent that scholastic skills have some value outside of school itself, one does children a service by increasing the general level of their attainment in them, regardless of individual differences; and if scholastic skills have no direct or indirect external value, they ought simply to be eliminated from the curriculum. Hence, what is worth teaching ought to be taught in ways that put it within the reach of the largest number of children, which means simply making it as easy as possible to learn. Any educational philosophy that insists on putting unnecessary thinking difficulties into the path of learning should be rejected as discriminatory.

Having made such a self-obvious pronouncement, however, it is necessary to look at its social implications. These implications can perhaps best be brought out by making them graphic. The kinds of children we have worked with have been ones who, by the usual sociological and psychometric predictors, would be expected to be, at best, plodding and unprepossessing students. In the course of our work with them, however, they consistently took on many of the characteristics of gifted or academically talented children—they were alert, confident, often to the point of being cocksure, proud of their ability, and eager for new challenges. This is not what would have been expected from the view that learning is a degrading form of drudgery, but it makes sense in light of the fact that their whole school experience consisted of a series of challenges successfully met. However, the challenges were challenges to learning rather than to thinking, and by the same token, what the children were good at and proud of was learning rather than thinking. Their confidence was easy to shatter, and their experiences after they left our program often managed to do this in short order. All that was needed to discourage them was to put them into a situation where they had to think at a higher level in order to succeed, and the ordinary school situation is well designed for that purpose.

One may speculate, however, as to what might happen if the entire school program were designed so as to permit continual progress in learning without excessive demands on thinking ability. One can imagine that under such a program children of below-average IQ could remain eager and self-confident learners throughout their school careers, emerging with a substantial fund of knowledge and skill and a readiness to go on learning in later life.

Presented in this way, the picture is a glowing one that almost anyone would endorse. But the other side of the picture is that such children would not necessarily have any deep understanding of what they were doing and might be severely limited in their ability to apply knowledge in new situations. We have to ask ourselves whether, in the prevailing climate of value, such children would indeed be accorded the respect without which their eagerness and self-confidence could never survive.

My view of the prevailing climate of value is largely based on the reactions of educators and child developmentalists to our program, and so it may be excessively jaundiced. But from this experience I conclude that the prevailing view is that high-achieving children who cannot think very well are impostors who ought to be exposed and their mentors denounced. Learning is viewed as a subhuman activity, more appropriate to rats or pigeons; among humans, only what is acquired through thinking is deemed of any value. The response of one noted educator is typical. Viewing a kindergarten class engaged in solving algebraic equations, he was impressed and inquired as to how this phenomenon had been brought about. When he heard the story, he announced, like someone uncovering a fraud, "Why, you taught them how to do it!" The almost universal response of visiting educationists or child developmentalists has been to ask "Could they do such-and-such?" (or, if they are bolder, to quiz the children themselves) until they have managed to convince themselves that the children are, after all, still stupid.

Although these reactions may have been exaggerated by other situational factors, I have not been able to discover anything in current educational writing that suggests people are willing to view learning, apart from thinking, as a dignifiable human accomplishment. This elitist view seems to be so widespread that it is held by a number of people who can lay no claim to superior thinking abilities themselves. People who cannot excel at thinking are allowed to dignify themselves through nonintellective accomplishments or personal virtue. And yet it seems clear that only a small minority of people are ever able to do much of the kind of thinking that is associated with intellectual disciplines. They may do a good deal of thinking in their daily lives, but it is thinking that owes practically nothing to school learning. So far as we know, schooling in modern societies does nothing to improve thinking abilities.

Schooling may do nothing to improve learning abilities either, but it can at least produce useful learning, including tool skills that permit further learning, and it could develop interest and confidence in learning. As technology assumes an increasingly prominent place in human activities, the need to figure things out becomes increasingly subordinated to the need to learn how to use equipment. Compare the old-fashioned automobile mechanic, who had to possess a good deal of diagnostic, problem-solving, and inventive skill, with the modern mechanic, whose main problem is to keep up with the increasing number of instruments, tools, procedures, and specifications that he must use.

It is all very well to speak, as Jensen does, of trying to discover ways of utilizing "the great and relatively untapped reservoirs of mental ability in the disadvantaged" (p. 117), but I see little prospect that any good will come of this, either for disadvantaged children or for the great mass of middle-class children of average or below-average IQ, so long as nonthinking routes to scholastic achievement are held in such low regard. A number of prejudices need to be eliminated. We need to eliminate the prejudice that simple learning is tedious and boring; it often is, just as what passes for thinking and problem-solving in our schools is often tedious and boring, but it doesn't have to be. (Learning to swim, for instance, is straightforward learning that involves practically no thinking, yet only under the worst conditions of instruction is it boring.) We need to eliminate the prejudice that managing simple learning is a pedestrian task and that only the promotion of thinking requires pedagogical art. (Those who think it is might try such a seemingly simple task as teaching an average class of second-graders how to tell time.) Above all, however, we need to eliminate the prejudice that thinking ability is the one true mark of scholastic achievement, all other marks being in some wise inferior or spurious.

At the root of this prejudice seems to be the illusion that schools teach children to think. There is no question that the kind of thinking exemplified in the academic disciplines is one of man's highest achievements. Teachers will always be pleased to observe signs of such thinking in their students, and they do right to encourage it. The trouble begins when teachers claim credit for having produced such thinking, for it is then but a small step to conclude that the less capable academic thinkers have failed to learn what was taught. It seems to me that Jensen has provided the most effective antidote available for this illusion, by showing the strength of genetic influences on thinking ability and pointing out the lack of evidence that schooling has any effect on it.

What Does Heritability Tell Us About Teachability?

In the preceding section, individual differences in IQ were treated as an unalterable fact, while attention was directed to the possibility of improving the scholastic success of children despite it. That there will always be IQ differences of some magnitude probably can be treated as an unalterable fact. On the other hand, IQ scores themselves are known to be in some measure influenced by experience. Hunt (1961) presented evidence to show that they could be influenced a great deal. Jensen (1969) has presented a documented case for the contrary conclusion. I do not propose to enter into weighing of the evidence, but to focus upon a question of interpretation that seems to be a pivotal one, namely, the question of what can be inferred about modifiability of IQ from its heritability ratio. Since the issue is one of interpretation, I shall accept without discussion Jensen's estimate that the heritability of IQ is 80 percent for the white population of the United States at the present time.

A heritability ratio, as Hirsch (1969) has emphasized, tells us about the population on which it is calculated, and tells us nothing of a fundamental nature about the determinants of the trait in question. It tells us what proportion of the variability of a trait in a particular population can be attributed to genetic differences. In principle, the same trait could show heritability ranging from zero in one population to 100 percent in another, depending on gene frequencies and frequencies of relevant environmental conditions. Accordingly, Hirsch has declared that, "High or low heritability tells us absolutely nothing about how a given individual might have developed under conditions different from those in which he actually did develop" (Hirsch, 1969, p. 19). He labels as fallacious Jensen's (p. 59) inference that the lower heritability of scholastic achievement indicates that it is potentially more susceptible to improvement through environmental means than is intelligence.

There seems to be no question that Hirsch is correct in saying that heritability tells us nothing about the *potential* susceptibility of a trait to environmental influence. Tomorrow, somebody could stumble upon an environmental variation that would have much more influence on IQ than the normal run of genetic variations. Are we to conclude, therefore, that heritability estimates have no educational significance whatsoever?

I would say no for the simple reason that virtually all educational effort is concerned, not with the creation of new environmental conditions, but merely with the allocation of existing ones. We may take Head Start as an example. What it has consisted of is taking certain environmental conditions (nursery school experience, medical examinations, lunches, availability of playthings) which had been largely restricted to children of higher socioeconomic status and applying them to children of lower socioeconomic status. Thus, it amounts to changing the distribution of existing environmental conditions.

If heritability ratios can tell us anything of practical importance, they can tell us something about the expected results of altering the distribution of existing variants. In order for them to tell us any thing definite, however, it is necessary to introduce additional assumptions that go beyond what is given by a simple heritability ratio. To illustrate this, I shall introduce a simplifying assumption that appears, from available evidence at least, not to fly in the face of fact.

Let us assume that the effects of heredity and environment on IQ are independent, that is, that the contribution of genetic and environmental sources of variation is the same for all levels of IQ. According to Jensen's calculations, as interpreted by Light and Smith (1969), interaction between source and level accounts for only about 1 percent of the variation.

Ignoring this interaction, we may conceive of the observed distribution of IQ scores (approximately normal in shape with a standard deviation of 15 IQ points) as being due to the summation of two other normal distributions, one a distribution of genetic effects on IQ, and the other a distribution of environmental effects. With a heritability ratio of 80 percent, the distribution of genetic effects will have a standard deviation of approximately 14 IQ points, and the distribution of environmental effects will have a distribution of 7 IQ points.

One thing worth noting immediately is that the distribution of environmental effects does allow for substantial IQ differences due to environment. Each standard deviation of difference in environmental effectiveness should make a difference of 7 points in average IQ. Extreme environmental differences could have very large effects. Although the present representation is entirely abstract and says nothing about what the relevant environmental differences are or how they might be scaled, we might consider as one anchor point the extremely impoverished and restricted orphanage environment from which children were taken in the Skeels and Dye (1939) study. From what we know about conditions of learning and intellectual development, it seems reasonable to put such an environment at the bottom extreme of the distribution—at, say, four standard deviations below the mean.* The expected result of taking children out of such an environment and rearing them in an average one would be a gain of 4 times 7 or 28 IQ points. That is approximately what was found. Thus, the Skeels and Dye findings, which have often been cited as evidence contrary to the hypothesis of high heritability of IQ (c.f. Hunt, 1964), are in fact adequately accounted for within a model based upon a heritability ratio of 80 percent.

This model allows us to make some inferences about the effects to be expected from utopian environmental conditions. Suppose, for instance, that the mean quality of environments, as they affect IQ, were raised two standard deviations, while the standard deviation of environments was reduced by half. This would amount to compressing the whole distribution of environmental conditions into the top half of the existing distribution. Accordingly, the poorest conditions to be found would correspond to what is average for the population today, although the most favorable conditions would be no better than the most favorable conditions today. This is what could be expected from a major improvement in social welfare accomplished without the creation of any new environmental conditions. The expected result of such environmental improvement would be to increase the mean IQ of the population by 14 points. This would be a major change and undoubtedly one of far-reaching consequence. On the other hand, the effect of such environmental improvement on the spread of individual differences in intelligence would be negligible. The standard deviation of IQs would be reduced by only about one point.†

* It seems reasonable to assume that not only the distribution of observed IQ scores but also the distributions of hypothetical components of these scores are bounded. That is, just as IQ scores do not have an infinite range, but vary from approximately zero to 200, it may be supposed that genetic environment factors are limited at the upper end by the fact that they consist of a finite number of pluses and at the lower end, by the fact that if things are too bad, the organism will perish. Setting the limits at ± 4 SD, while arbitrary, produces results that are conformable to the observed range of IQs.

† These observations are all implicit in Jensen's own account and have been noted explicitly by other writers (e.g., Stinchcombe, 1969). Jensen has emphasized the small potential effect of environmental change on variance; his critics have emphasized the large potential effect on mean IQ. What I have tried to make clear is that these two conclusions are not at odds, and furthermore, that they presuppose only change in the allocation of existing environmental conditions. The possibilities opened up by the introduction of novel environmental

What a high heritability ratio implies, therefore, is that changes within the existing range of environmental conditions can have substantial effects on the mean level of IQ in the population but they are unlikely to have much effect on the spread of individual differences in IQ within that population. If one is concerned with relative standing of individuals within the population, the prospects for doing anything about this through existing educational means are thus not good. Even with a massive redistribution of environmental conditions, one would expect to find the lowest quarter of the IQ distribution to be about as far removed from the upper quarter as before. On the other hand, if one is willing to attach some absolute importance to IQ levels, the prospects are much brighter. A 14-point gain in mean IQ for the population would reduce the number of people having IQs below 70 by a factor of more than 10 and would increase the number of those with IQs above 130 by a factor of about 5. Thus, there should be a substantial decrease in the number of people who are incapable of holding their own in the world and a substantial increase in the number capable of doing the more demanding kinds of intellectual work.

A high heritability ratio for IQ should not discourage people from pursuing environmental improvement in education or any other area. The potential effects on IQ are great, although it still remains to discover the environmental variables capable of producing these effects. It is clear from the Coleman report (Coleman et al., 1966) and subsequent analyses of it (Jencks, 1969) that the kinds of tangible environmental improvements usually sought by school administrators are not very relevant to intellectual differences.

What Are the Prospects for Reducing the Spread of Individual Differences in Intelligence?

I shall leave aside for now the question of why anyone should be concerned about reducing the spread of individual differences in intelligence. The discussion of how such differences might be reduced will, I think, convert the question of why into the question of why not.

---

conditions are much broader and unpredictable, as is indicated in the following section of this paper.

If a given environmental improvement affects everyone the same, regardless of their genetic endowment, then it may produce some general improvement in IQ but it will not reduce individual differences. On the other hand, if a given environmental improvement is more beneficial to those who would ordinarily manifest a low IQ than to those who would ordinarily manifest a high one, then it will have the effect of reducing the spread of individual differences in IQ. This is a heredity-environment interaction (Jensen, 1969, pp. 39–41). Of course, the interaction can work the other way; the environmental improvement can be more beneficial to those more favored genetically, thus increasing individual differences. From the evidence cited by Jensen, it appears that what heredity-environment interaction has so far been demonstrated has been of this latter kind, leading to the expectation that general environmental improvement will increase individual differences in IQ.

There is, however, one dramatic instance of heredity-environment interaction that has had a compensatory or difference-reducing effect. This is the dietary treatment of phenylketonuria. Phenylketonuria, or PKU, is an inherited metabolic defect which, under ordinary circumstances, has a high probability of leading to severe mental retardation. Given a special diet, however, this probability is considerably reduced. The "environmental improvement" represented by this special diet, therefore, has the effect of raising the mean IQ of PKU children but is presumably not beneficial to other children. Thus, it reduces individual differences.

We may suppose that in the days before this special diet was discovered, PKU children experienced the normal range of dietary variations fond in the infant population at large. If these normal variations made any difference in IQ, it was probably to the same small degree as for other children. What was required to make a real compensatory difference was an environmental variation that lay outside the normal range and that would probably never have been stumbled upon by chance.

This is a point worth keeping in mind when we consider Hirsch's statement, quoted previously, that "high and low heritability tells us absolutely nothing about how a given individual might have developed under conditions different from those in which he actually did develop." We do not know how changed environmental conditions will affect an individual's development; but on the other hand, it is wishful thinking to suppose that, through conventional ameliorative efforts, we will ever stumble upon environmental variations that will interact with genetic factors in such a way as to produce dramatic compensatory effects. If such effects could be produced in this way, the means would probably already have been discovered centuries ago as a kind of folk psychological medicine.

A more reasonable supposition is that, although environmental conditions vary greatly within a society as heterogeneous as that of North America, the variations are not consistent enough to interact *strongly* with genotypes in determining IQ. Diets vary considerably, for instance, and some diets undoubtedly contain smaller amounts than others of phenylalinine (the critical substance in determining the effects of PKU). However, it is unlikely that any normally occurring variations in diet would be low enough in phenylalinine to produce a noticeable effect on PKU children. Similarly, we might suppose that if there are certain conditions of experience that could interact with genotypes to produce markedly higher intelligence among some kinds of children than are observed under other conditions, these special conditions would not occur in great enough concentration within the life history of individuals to reveal such effects.

This point may be made clearer by considering a conjectural example. Suppose that there are some children who could develop high intelligence except that they are lacking in the normal sort of sensitivity to cognitive conflict—they remain indifferent to contradictions and incongruities that influence other children and promote cognitive growth. Conceivably, such a deficit could be offset by having people in the child's environment continually confronting him with the discrepancies which he is inclined to ignore. Now a certain amount of this confrontation goes on in any child's experience, and some children encounter more of it than others. But perhaps no adults normally instigate the amount of confrontation that would be necessary to produce noticeable results. Moreover, since adults themselves are sensitive to response from the child, it may be that the child who is most in need of such confrontations fails to respond in ways that would encourage adults to provide them.

If strong interactions between experience and heredity are to be obtained, it is likely that they must be created experimentally rather than discovered through observation of existing variations in experience. Blind probes are likely to have a very low probability of success. I am not in a position to offer specific suggestions of promising leads, but it may be worth noting some of the *least* promising sources of leads. Investigation of the differences in experience of high and low IQ children is not very promising. It might indicate experiential factors that would have a generally beneficial effect on IQ, but high and low IQ groups are so heterogeneous that the possibility of unearthing specific heredity-environment interactions is slight. Even the search for generally beneficial kinds of experience is likely to go astray in this kind of investigation because one will be comparing differences in home background of children whose parents probably differ significantly in genotype, so that differences in home experience may merely reflect the same genetic differences that are also reflected in IQ.

Another unpromising source of leads is the normal course of child development, as represented, for instance, in Piagetian stage theory. One cannot expect to discover the sources of individual differences by investigating those developmental changes that are universal. A final unpromising source is the developmental deficits of young children. These generally turn out to be time lags in development and, in themselves, tell nothing about why the children may turn out to be handicapped in the long run.

More promising sources of leads that might be mentioned are the following: (1) Intellectual deficits remaining at maturity. These may indicate specific intellectual skills or operations that some children never acquire, but which might be specifically taught to long-range effect. (2) Deficiencies in experience-producing behavior (Hayes, 1962). If certain children fail to engage in such activities as question-asking, exploration, verbal recontruction of events (what Church, 1961, calls "thematization"), or solitary thought, these might have serious but preventable effects on learning to think. (3) Special abilities or interests of low IQ children. These might suggest assets that could be, but under normal educational conditions are not utilized in the development of general intelligence.

It is possible that no very strong interactive effects can be discovered. It is also possible that they might require such massive or radical deviations from normal child-rearing practices or might have such undesirable side effects that no one would wish to bring them about. On the other hand, we cannot know this in advance and we ought not to be so wedded to the status quo in child-rearing and education that we refuse to investigate. The sorts of practices that have evolved over the centuries are undoubtedly ones that work fairly well on the average. But it may well be that there are other ways, equally reasonable and economical, that would work much better for some children, and that these children are now being unnecessarily handicapped in intellectual development.

What I am proposing is not quite the same thing as adapting education to individual differences. The usual kind of adaptation assumes that individual differences cannot be altered and that the children's life chances can be improved by allowing them to develop in accordance with their unique characteristics. Adaptation to individual differences is generally not compensatory in effect, but, if anything augments differences. Seeking out specific heredity-environment interactions, on the other hand, is an effort to neutralize the effect of certain genetic characteristics which handicap children under standard environmental conditions. Thus, it is a compensatory effort intended to make children less the victims of their heredity.

What Are the Educational Implications of Possible Hereditary Differences in Intelligence Associated with Race and Social Class?

Lower-class children are known to score lower in IQ than middle-class children, and Negroes lower than whites. By far the most controversial part of Jensen's paper has been his suggestion that these observed group differences may be, in part, due to hereditary differences. Jensen does not claim to have proved that this is true, but only to have presented enough evidence to show that it is a reasonable possibility. So far as I know the numerous rejoinders to Jensen have never directly attacked this claim; that is, no one has tried to show that it is *not* a reasonable possibility that social class and racial differences in IQ are partly due to heredity. Accordingly, I shall proceed on the grounds that it is a reasonable possibility and ask what the educational implications are if it should be true.

The issue of racial differences in intelligence has commanded much more attention among educators than the issue of socioeconomic differences, largely, it would seem, because of its bearing on school desegregation. As I see it, however, the question of genetic differences in racial intelligence ought not to have any bearing on the issue of school desegregation, and the fact that it does indicates that the desegregation issue has been poorly drawn. Genetic considerations become pertinent only when one makes the claim that schools should be integrated because integration will be good for black children—i.e., it will eliminate their lag in IQ and scholastic achievement. This claim, however, has already been largely rejected, not on grounds that it is false, but on grounds that it is invidious. It is, when you get down to it, just as invidious as the counterclaim that schools should not be integrated because integration would be bad for white children. Both claims are, at bottom, similar in their environmentalist assumptions. The status of minority groups in relation to public institutions can pose exceedingly complex problems, as no one needs to be told, but one principle that seems clear to me is that policy decisions with respect to such problems should not be based on generalizations about the personal attributes of individuals composing those groups. Such generalizations, no matter how benignly intended, almost invariably turn out to be invidious and unfair to "minorities within the minority." If such generalizations are ruled out of policy consideration, issues will not be drawn in such a way that questions of genetic difference assume importance—and that, it seems to me, is the surest way to avoid the idealistic contrafactuality that can lead to paternalistic do-goodism on one hand or to *apartheid* on the other.

There has been a good deal of "all or nothing" thinking on both sides of the issue of genetic group differences. On the one side, environmentalists seem to be taking the position that, so long as group differences in IQ *could* be explained by environmental factors, it follows that they are in fact solely due to such causes. Thus, Light and Smith (1969) have made use of hypothetical interaction effects to show that the Negro-white difference in IQ could be accounted for by environmental factors even within the limits of Jensen's estimate of heritability components. This is a significant point to have demonstrated,* but it should be recognized that Light and Smith have not demonstrated that genetic differences are

* The demonstration is, however, rather fanciful, for it seems to require that racial discrimination have the effect not merely of putting black people into generally unfavorable environmental conditions but of putting individuals into the particular environments that are worst for their particular genotypes. One result of such bizarre allocation would be greatly to increase the variance of IQ scores among blacks as compared to that among whites (a point brought to my attention by William Shockley, personal communication), whereas in fact the variance of IQs for blacks is less than that for whites.

not involved, or even that an environmentalist explanation is the more likely one.

On the other side, one may discern a type of all-or-none thinking which holds that if there are genetically determined IQ differences between racial or social class groups, then efforts to raise the IQs of such groups through education are doomed to failure. Jensen himself has not made any such bald assertion, but is easy enough to infer it from the general structure of his argument, which begins by asserting that compensatory education has apparently failed to boost the IQs of disadvantaged groups, then goes on to discuss genetic influence on IQ and the possibility that group differences have a genetic basis, and ends by proposing that educational treatment of disadvantaged groups should concern itself with abilities other than general intelligence. The message would seem to be, "Forget about raising the IQ of lower-class and Negro children. It can't be done."

However, there is no question that lower-class and Negro children are disadvantaged environmentally. The possibility that they may also be "disadvantaged" genetically does not rule out the possibility that a large part of the IQ deficit shown by these groups is due to environmental factors and can thus be remedied by environmental means. Even if compensatory education has failed to produce demonstrable effects, it is going too far to excuse this failure on genetic grounds. We know that environmental changes can have an effect, as shown, for instance, in the studies of Negroes migrating to Philadelphia from the South (Lee, 1951). The last thing I think Jensen or any other genetically oriented behavioral scientist would want to see happen is for his ideas to be used as an excuse for not doing anything. There have been plenty of such excuses in the past. "Cultural deprivation" itself has been used as an excuse for giving up on disadvantaged children—misused, we should say; and there is danger that the notion of hereditary differences will be misused in the same way.

There are, however, two points in Jensen's discussion that have pessimistic indications for educational efforts to raise the IQs of disadvantaged children. One of these is the idea of environment as a threshold variable and the other is the indication that educational variables might be among the least potent of environmental variables for influencing IQ. The threshold idea implies that, above a certain minimum level, further improvements in environment make no difference in IQ. If the environments of so-called disadvantaged groups are already above this minimum level, then no effects can be expected from environmental improvement of any kind. Jensen does not provide strong evidence for his threshold hypothesis. It appears that he has constructed it mainly to account for the evidence of large IQ gains through environmental change in the Skeels and Dye (1939) study. However, as we have seen in the second section of this paper, these gains can be adequately accounted for by the hypothesis that environmental effects are additive across the whole range of environments. The other argument he employs is that he has found "no report of a group of children being given permanently superior IQs by means of environmental manipulations" (p. 60). There is, however, some evidence of this from work done in Israel, where Kibbutz-reared Oriental Jews show above-average IQs while those outside the Kibbutzim show IQs substantially below average (cited in Bloom, 1969).

Although there is reason to be skeptical of these results because of unknown selection factors (only a small minority of Oriental Jews get into Kibbutzim, and those who do may be exceptional in some way), it is nevertheless evidence that would have to be ruled out before one could be confident in resorting to the threshold hypothesis. To me, a more reasonable and cautious hypothesis is that effectiveness of the environment in fostering IQ is not linearly related to gross indicators of socioeconomic status. As one moves from a slum environment to an ordinary working-class environment, the difference in IQ-related effect may be much greater than it is when one moves from a working-class environment to that of a middle-income suburb. Accordingly, if one is dealing with slum children, the more obvious sorts of environmental improvement—or those that would normally occur with increased prosperity—can be expected to have some effect on IQ, whereas if one is dealing with middle-class children, the kinds of environmental improvements that would be effective in raising IQ may be much less obvious and more difficult to bring about. Another way of putting it is that such environmental variables as income, quality of housing, and supply of reading material in the home may indeed be threshold variables, which make no difference above a certain level, but other, more primary variables (to which these are only indirectly related) may be significant at all levels.

The practical implication of this nonlinearity notion is one with which I think few people would disagree, but it is also a fairly pessimistic one. It is that environmental improvement is likely to show diminishing returns in its effect upon IQ or any related variables. Alleviating the more obvious social ills that go with poverty may produce some notable gains in IQ among the formerly afflicted, but it cannot then be expected that further social improvements will automatically result in further IQ gains, and environmental improvements capable of producing such gains will take some deeper research to discover.

Jensen's other point, that education seems to be among the less potent environmental variables in affecting IQ, is of course only pessimistic in its implications for educators. In the larger social context, it would be a good thing if medical care and nutrition proved to be more potent variables, because the chances of doing something directly about them are much greater. We at least have a clearer idea of what constitutes good medical care and nutrition than we have of what constitutes effective education. However, it seems to me that there is much greater reason to suppose that physical factors in the environment function as threshold variables than that experiential factors do. Physical factors influence the condition of the neurological system, and by analogy with other biological systems we might assume that there are certain minimum requirements of nutriments and so forth that need to be met, and that beyond that, further inputs are useless. Experience, on the other hand, determines what information is stored in the system, and there is no reason to suppose that this could not be improved upon indefinitely. So, again, it might be that, for severely impoverished children, physical improvements might make a decided difference, but for additional gains in intelligence or for gains among children whose physical conditions are already above the minimum level, experiential factors might be the only ones that could make a difference.

All of this is highly conjectural, but it suggests a potentially fruitful direction for educational research. Perhaps what educational research should be doing is not concentrating upon improving the lot of severely impoverished children, whose main problems may be of a nonexperiential nature and ones that can only be alleviated by more general kinds of environmental change, but should rather be concentrating upon finding ways of improving the intellectual abilities of children whose physical conditions are already in good order. These would be children from the great mass of "Middle America" whose environments are physically adequate but not rich in intellectual stimulation. The justification for doing research on them would be that it could allow us to identify experiential factors that are separate from the physical factors that are also being dealt with in poverty programs, and might thus make it possible for us to design educational programs for poorer children that would be beneficial over and above the general effects of eradicating poverty.

In raising the possibility that supposed cultural differences may in fact have genetic roots, Jensen has brought forth a point that casts profound doubt on a large range of research studies. I refer to studies that attempt to account for the behavior of children by examining the behavior of their parents toward them. Practically every psychological attribute of children, from creativity to bed-wetting, has been treated to a search for child-rearing antecedents, and the results of such studies have naturally been of considerable interest to educators, for they would seem to get at the causal experiential variables which educators in their turn might attempt to manipulate. Causal inferences from such studies must presume that the behaviors in question are environmentally determined. An alternate hypothesis is that the behaviors are genetically determined and that when correlations are found between parent behavior and subsequent child behavior, these correlations are due to genetic similarity of parents and children and reflect manifestations of the same inherited traits.

The genetic hypothesis has found considerable support in several areas where the hypothesis of experiential causation had previously held sway. Rimland (1964) has advanced such a hypothesis to account for the characteristics of parents of autistic children, rejecting experiential causation on grounds that the symptoms of infantile autism appear too early in life to have been influenced by parent behavior. Heston (1970), after substantiating the heritability of schizophrenia by evidence similar to that used by Jensen to demonstrate the heritability of IQ, went on to define a schizoid syndrome, shared by schizophrenics and their nonschizophrenic relatives alike, which accounts for parent-child relationships that had previously been interpreted as causal. In the case of intellectual abilities, it is easy to see how the same basic traits, which are manifested by children in their test performance, could be manifested by parents in how they talk to their children, teach them, discipline them, in their attitudes toward education, in what sorts of educational materials they have around the house, and so on. While one can always make a reasonable case for the causal influence of such parent characteristics, it is just as reasonable, even if less appealing to the educator, to treat them as evidence of genetic resemblance.

The practical upshot of this consideration is that findings on child-rearing antecedents of behavior must be regarded as ambiguous. At best, they can suggest causal hypotheses that must be tested experimentally before they can be held with confidence.

The most profound and far-reaching implication of the hereditarian view of ethnic and other group differences is one that I don't believe has been touched on by Jensen, although it has been developed somewhat by Eckland (1967).

The implication is that culture may not only influence the phenotypic expression of inherited traits but may also, through selection pressures, influence the distribution of genes within cultural groups, and that cultural differences may represent not only adaptations to environmental conditions but also to distributional genetic differences between groups, with the result that culture and heredity are intimately interconnected.

If true, this would not be an argument for separatism, but it would add weight to arguments for pluralism, for allowing cultures to maintain their distinct identities, and for not applying uniform standards of treatment and expectation, as has been done in public education. It could still be true that any given individual could adapt to any culture and might even be better suited to a culture different from the one into which he was born, while at the same time it was true that each culture was, on the whole, the best one for its members and should therefore not be willy-nilly absorbed into some other.

Although pluralism seems to be on its way to becoming a "good word" in education, the schools have not taken any but the most trivial steps toward achieving it (Green, 1969). The trend is toward individualization within heterogeneous pupil groups, an arrangement which among other things increases the opportunities and influences that might lead a child to deviate from the norms of his cultural group. Pluralism begins with a recognition of the right of groups to protect their members from such influences. The line between separatism or segregation and pluralism is a difficult one to establish in practice, as the current dilemmas in the education of black Americans reveals only too well, but it appears essential that the line be drawn in some rational and generalizable way. Thus far, education has had to deal largely with cultural groups defined on long-standing ethnic or religious lines. The future may well see the emergence of self-selected, experimental subcultures that cut across traditional lines and that have as a primary rather than a secondary distinction their rejection of the prevailing culture represented in the schools. These self-selected subcultures may well reflect, more strongly than subcultures of the past, complex differences in genetic disposition.

Conclusion

One apparently reasonable stance is that the educator need not concern himself with genetics because, in the first place, he is constrained to working with environmental variables and must therefore do the best he can with them, regardless of their relative potency compared to genetic variables; and because, in the second place, education deals with individual children of unknown genetic potential, so that normative data on genetic differences have no application. These are valid points with respect to the work of the teacher in the classroom, for whom genetic principles are most likely to function only as an after-the-fact excuse for educational failures.

At the level of policy, however, education deals with populations rather than with individuals, and it is at this level that genetics becomes potentially relevant. In this paper, I have tried to indicate some ways in which genetic considerations can be relevant to educational policy. The mere fact of individual differences in intelligence should encourage us to look for alternative methods of achieving educational objectives that do not rely so heavily upon the abilities represented by IQ. The apparently high heritability of IQ should influence our expectations as to what may be accomplished through allocation of existing environmental variants: reallocation may produce substantial gains in mean IQ but should not be expected to produce much alteration in the spread of individual differences. The idea of specific heredity-environment interactions suggests the possibility of producing substantial environmental effects on individual differences in intelligence, but it appears that we are a long way from knowing how to produce such effects.

On the matter of social and racial differences, it is probably safe to say that the educational policy maker need not concern himself with the question of whether these differences have a genetic basis.

It is necessary to avoid both the oversimplification that says if there are genetic group differences nothing can be accomplished through educational improvement and the oversimplification that says if group differences in IQ are environmentally caused they can be eliminated by conventional social amelioration. The possibility that cultural differences are related to heredity, however, adds force to the need for schools to come to grips with the problem of providing for cultural pluralism without separatism or segregation. This may well be the major policy problem facing public education in our time.

References

BEREITER, C., and ENGELMAN, S. 1966. *Teaching Disadvantaged Children in the Preschool.* Englewood Cliffs, N.J.: Prentice-Hall.

BLOOM, B. S. 1969. Letter to the Editor. *Harvard Educational Review* 39:419–421.

CHURCH, J. 1961. *Language and the Discovery of Reality.* New York: Random House.

COLEMAN, J. S., et al. 1966. *Equality of Educational Opportunity.* Washington, D.C.: U.S. Department of Health, Education, and Welfare.

ECKLAND, B. K. 1967. "Genetics and Sociology: A Reconsideration." *American Sociological Review* 32:173–194.

ENGELMANN, S. 1969b. *Preventing Failure in the Primary Grades.* Chicago: Science Research Associates.

ENGELMANN, S. 1969a. *Conceptual Learning.* San Rafael, California: Dimensions Publishing Co.

GREEN, T. F. 1969. "Schools and Communities: A Look Forward." *Harvard Educational Review* 39:221–252.

HAYES, K. J. 1962. "Genes, Drives, and Intelligence." *Psychological Reports* 10, no. 2:299–342. (Monogr. suppl. no. 2–VIO).

HESTON, L. L. 1970. "The Genetics of Schizophrenic and Schizoid Disease." *Science* 167:249–256.

HIRSCH, J. 1969. "Behavior-Genetic Analysis and Its Biosocial Consequences." Paper presented at University of Illinois, Conference on Contributions to Intelligence, November 14, 1969.

HUNT, J. MCV. 1961. *Intelligence and Experience.* New York: Ronald Press.

———. 1964. "The Psychological Basis for Using Pre-School Enrichment as an Antidote for Cultural Deprivation." *Merrill-Palmer Quarterly* 10:209–248.

JENCKS, C. 1969. "A Reappraisal of the Most Controversial Educational Document of Our Time." *New York Times Magazine,* August 10, pp. 12–13.

JENSEN, A. R. 1969. "How Much Can We Boost IQ and Scholastic Achievement?" *Harvard Educational Review* 39:1–123.

LEE, E. S. 1951. "Negro Intelligence and Selective Migration: A Philadelphia Test of the Klineberg Hypothesis." *American Sociological Review* 16:227–233.

LIGHT, R. J., and SMITH, P. V. 1969. "Social Allocation Models and Intelligence." *Harvard Educational Review* 39:484–510.

RIMLAND, B. 1964. *Infantile Autism*. New York: Appleton-Century-Crofts.

ROBINSON, H. B., and ROBINSON, N. M. 1965. *The Mentally Retarded Child: A Psychological Approach*. New York: McGraw-Hill Book Co.

SKEELS, H. M., and DYE, H. B. 1939. "A Study of the Effects of Differential Stimulation on Mentally Retarded Children." *Proceedings of the American Association of Mental Deficiency* 44:114–136.

STINCHCOMBE, A. L. 1969. "Environment: The Cumulation of Effects Is Yet to Be Understood." *Harvard Educational Review* 39:511–522.

TUEL, J. K. 1966. "The Relationship of Intelligence and Achievement Variables in Programmed Instruction." *California Journal of Educational Research* 17:68–72.

VERNON, P. E. 1960. *Intelligence and Attainment Tests*. New York: Philosophical Library.

PART IV

IQ, Heritability, Inequality

IQ, Heritability, and Inequality

N. J. Block and Gerald Dworkin

PERHAPS NO ISSUE IN THE HISTORY OF SCIENCE PRESENTS SUCH A COMplex mixture of conceptual, methodological, biological, psychological, ethical, political, and sociological questions as the controversy over whether intelligence has a substantial genetic component. The discussion and debate, which dates back to the publication of Galton's *Hereditary Genius* in 1869, was brought most prominently to public notice one hundred years later by the publication of Arthur Jensen's article "How Much Can We Boost IQ and Scholastic Achievement?"[1] in the *Harvard Educational Review*. Jensen has estimated that since then over 120 articles have been provoked by his article alone. Among the articles which have called attention to social and political implications of the controversy, one of the most discussed is Richard Herrnstein's "IQ."[2] Since then, Herrnstein has expanded his article into a book, *IQ in the Meritocracy*,[3] and Jensen has published two books. One of them, *Educability and Group Differences*,[4] argues that between 50 and 75 percent of black-white IQ differences are due to genetic differences.

The present essay is intended to fall into the tradition of philosophy as the critical examination of fundamental assumptions of organized bodies of knowledge. It is divided into two parts. The first discusses the conceptual and empirical issues raised by IQ testing. The second discusses the concept of heritability; the social, political, and educational implications of the thesis that IQ has high heritability; and the morality of research and publication in the area of genetic racial differences. Although we have tried to write without rancor, to substitute light for heat wherever possible,

Notes for this article begin on p. 521.

and to be fair to those we criticize, it will be (and should be) apparent that we do not write with magisterial impartiality out of a purely disinterested search for knowledge. We write, at least partly, because we feel strongly that ideas matter, that in Bradley's words "ink and paper can cut the throats of men and the sound of a breath may shake the world."

Part I

> . . . Psychological metatheory has remained seriously underdeveloped. With a few important exceptions, its history during the second quarter of this century has been an attempt to work out a variety of behaviorism that would satisfy the constraints imposed on psychological explanation by an acceptance and application of empiricist (and particularly operationalist) views of general scientific method. The better known accounts of psychological explanation have thus often failed to reflect the most important movement in current philosophy of science: the attempt to determine the consequences of rejecting key features of the empiricist program. Verificationism as an account of meaning; conventionalism as an account of theoretical constructs; sharp distinction between the observational and inferential language of theories; uncritical reliance upon the analytic-synthetic distinction—all these have recently come into question among philosophers of science who have realized that these doctrines are by no means indispensable to characterizations of scientific explanation and confirmation and that philosophical accounts that exploit them may in fact seriously distort the realities of scientific practice. Yet it is upon precisely these views that much of the implicit and explicit metatheory of American experimental psychology appears to rest.
>
> —J. A. FODOR[5]

Our aim here is to argue that it does not seem likely that IQ tests measure mainly intelligence. The evidence suggests that there is no good reason to believe IQ tests *do* measure mainly intelligence and that a number of other quantities (such as sociocultural background and personality-motivational-temperamental factors) appear to have as good a claim to be measured to some degree by IQ tests as intelligence does. Of course, some of these other quantities undoubtedly affect intelligence itself, but we are pointing out that they probably have substantial effect on IQ independently of affecting intelligence. We argue, in addition, that bad philosophy of science has played an important role in the construction of IQ tests and has served to keep many psychologists and educators from seeing the inadequacies of IQ tests as measures of intelligence.

Part I of this essay is organized around three questions:

1. Can we reasonably ask whether IQ tests measure intelligence?[6] (sections I and II)
2. Assuming we can, how good are the arguments that IQ tests do measure intelligence? (section III–VI)
3. If IQ tests do not measure intelligence, what do they measure? (sections VII and VIII)

We concede at the outset that "intelligence" is a vague term, often used to refer to different things in different cultures and even used differently by a single person on different occasions. For example, if national differences in dictionary definitions are any guide, mental speed is more important to the English concept of intelligence than to the American. English dictionaries often mention quickness of wit, while American dictionaries rarely do. Indeed, this sort of cultural difference indicates one respect in which any attempt to measure intelligence must be culturally biased: what abilities one attempts to measure depend on what abilities are taken to be part of intelligence. Thus, even if it were the case that IQ tests measure the ability or abilities that are generally taken to constitute intelligence in this culture, one might argue against IQ tests on the ground that the ability or cluster of abilities counted as intelligence in this culture are not the abilities that ought to count as intelligence. Our main concern here will be with the question: Is what intelligence is taken to be in our culture measured by IQ tests?

It is oversimple, however, to suppose that there is any one way "intelligence" is generally understood, even in our culture. Some think of intelligence as a general cognitive capacity which is employed in any task with a cognitive or intellectual element, while others think of intelligence as some sort of cluster of more specific cognitive capacities, e.g., mathematical or mechanical ability. Our arguments are intended to apply to either understanding of it, except in section VIII, which is devoted to the issue of "general intelligence."

Another common source of misunderstanding can be avoided if we keep in mind the distinction between a person's genetic potential for intelligence and his actual intelligence. A person with a high genetic potential for intelligence may sustain extreme brain damage—e.g., through a birth accident—and thus end up with low actual intelligence. On the other hand, a person with genes which would have resulted in low actual intelligence in an average environment for the development of intelligence can achieve higher actual intelligence by being raised in a more favorable environment. We will always be talking about actual ("phenotypic") intelligence and not genetic potential ("genotypic") intelligence.

It is sometimes said that the current controversy about IQ does not depend on any claims about what IQ is but only on the certainly true claim that whatever IQ is, it correlates with various measures of success. For example, Carl Bereiter comments: "IQ is important, not because it measures 'intelligence,' but because it predicts something important—school achievement."[7] However, the question whether IQ tests do measure intelligence is crucial to many of the current controversies. On the basis of claims of genotypic IQ differences, Jensen has suggested that education for the low-IQ disadvantaged (especially blacks) should be directed toward associative (memory) learning instead of conceptual and cognitive learning,[8] and Shockley has suggested that persons with below average IQ be encouraged to undergo sterilization—the lower the IQ, the greater the encouragement.[9] We doubt whether such suggestions could be taken seriously if IQ tests were known to measure not intelligence but, say, mainly sociocultural background and tendency toward aggression. For Jensen's suggestion depends on the assumption that IQ is closely related to the capacity to engage in conceptual learning, and Shockley's suggestion presupposes that IQ is of overwhelming social value.

Herrnstein has advanced an argument (discussed in detail in Part 2 of this essay) which says in essence: IQ is somewhat heritable; success requires IQ; therefore, success reflects inherited differences. Supporters of this argument sometimes say that it does not depend on any claim about what IQ is, but only on the claim that whatever IQ is, it is both heritable and correlated with success. But the second premise *does* depend on what IQ is. Herrnstein's claim that success requires IQ is based on the correlation of IQ with success, or more precisely, on the distribution of IQ within socioeconomic-status groups: the higher the socioeconomic-status group, the narrower the spread of IQs within it. Thus, there is a smaller proportion of auditors with low IQs than truck drivers with high IQs. From the distribution of auditors' IQs, Herrnstein (*IQM*, p.

123) concludes that "half of mankind (roughly the half below IQ 100) is not eligible for auditing." (Even though 10 percent of auditors have IQs below 88!) But such statistical facts cannot be taken to show that IQ determines *eligibility* for job category without definite assumptions about what IQ *is*. For example, even if the number of hairs in a person's nose correlated with success in the same way IQ does, no one would be entitled to conclude that a certain level of nose-hair numerosity is a requirement or a condition of eligibility for any level of success. Nose-hair numerosity may be epiphenomenal as far as success is concerned. Later we shall consider evidence that to a surprising extent, IQ is such an epiphenomenon. For now, our point is simply that one must have some idea of what IQ is in order to judge whether existing data about the distribution of IQ in SES groups support a claim for a causal relation between IQ and success.

I. Operationalist Doctrines Behind IQ Testing

In this section, we will discuss a cluster of operationalist methodological doctrines which have strongly influenced the construction of IQ tests[10] and which have been widely used to defend IQ tests against criticism. We do not mean to suggest that any one person holds *all* the views we are attacking, nor even that most psychologists hold most of them. Many psychometricians disparage IQ tests in favor of separate tests of special abilities; others disparage current mental-ability tests altogether. Even among psychometricians who more or less accept IQ tests as measuring some sort of a general cognitive ability, many, including some of the authors of the leading textbooks, often disparage use of words like "intelligence" and "capacity" in connection with them.[11] On the other hand, many of the contributors to the current controversy over IQ and heritability (including Jensen and Herrnstein) seem to hold most of the views we attack; further, we suspect that many if not most psychometricians hold at least some of these views, though perhaps in a more or less attenuated form.[12]

Operational Definition. We shall now discuss the first of the operationalist doctrines mentioned above: that one can avoid the whole question of what IQ tests measure simply by *defining* "intelligence" as what IQ tests measure. As Jensen puts it:

> The notion is sometimes expressed that psychologists have misaimed with their intelligence tests. Although the tests may predict scholastic performance, it is said, they do not *really* measure intelligence—as if somehow the "real thing" has eluded measurement and perhaps always will. But this is a misconception. We *can* measure intelligence. As the late Professor Edwin G. Boring pointed out, intelligence, by definition, is what intelligence tests measure. The trouble comes only when we attribute more to "intelligence" and to our measurements of it than do the psychologists who use the concept in its proper sense.[13]

What little plausibility Boring's dictum has issues from a simple ambiguity in phrases of the form "what *x*'s measure." Consider an analogous definition of "temperature" as what thermometers measure. One sense of "what thermometers measure" is: what thermometers read or say, i.e., the number (in the units in which the thermometer is calibrated) which the thermometer indicates when the object to be tested is placed in contact with the thermometer. Now in this sense of the phrase "what thermometers measure," the operationalist claim that temperature is by definition what thermometers measure, commits its proponents to absurdities. They would have to say, for example, that thermometers could not possibly be improved on or that no device could measure temperature more accurately.[14] The analogous absurdity invited by those who define "intelligence" as what IQ tests measure (in this sense of the phrase) is that nothing could measure intelligence better than an IQ test. These definitions would be acceptable if what was meant by "thermometer" or "IQ test" were an *ideal* thermometer or IQ test. But then one could ask how closely a real IQ test approximates an ideal one, and this is the sort of question the operationalist is trying to avoid.

There is, however, an obvious alternative sense of "what thermometers measure," namely, what thermometers are *supposed* to read or say. In this second sense of the phrase, defining "temperature" as what thermometers measure is harmless enough, though the definition would perhaps be more useful as a definition of "thermometer" than of "temperature." But notice that it is the first sense of "what IQ tests measure" that Jensen and Herrnstein require, not the second sense.[15] For in the second sense, even if one accepts the definition of "intelligence" as what IQ tests measure, one can reasonably ask whether IQ tests *do* say what they are supposed to say. But this was just the sort of question the "operational definition" was intended to forestall. The operationalist dilemma here is that one of the interpretations of Boring's definition is *absurd* while the other is *useless* for their purposes.

One could also understand a definition of "intelligence" as what IQ tests measure as fixing not the meaning of "intelligence" but rather its reference.[16] That is, the definition would fix the reference of "intelligence" as that quantity which standard IQ tests succeed in measuring, if indeed they succeed in measuring anything. Of course, this interpretation is of no more use to the operationalists than the other ones. For even if one accepts this definition, one can reasonably ask whether IQ tests do succeed in measuring anything, and whether, if they do, it is the quantity which people have insofar as they are smart and lack insofar as they are stupid. In other words, fixing the reference of "intelligence" just raises the question whether the quantity referred to as "intelligence" by those who accept the definition is the same quantity referred to as "intelligence" by those who do not accept it. Once again, this is just the type of question that the operational definition was meant to forestall. Incidentally, it is unlikely that operationalists have had this type of definition in mind, since they seem to understand definition as fixing meaning, not reference.[17]

Since operationalism, at least in so blatant a form, is a thoroughly discredited doctrine among philosophers, even among most of its former proponents,[18] we will not detail its inadequacies here.[19]

Is Theory Unnecessary? Another strand in this operationalist methodology is the view that measurement of intelligence can proceed independently of the construction of a theory of intelligence. What should such a theory do? It should explain the causal role of intelligence in phenomena in which intelligence has a causal role. For example, it should explain how intelligence affects learning, problem-solving, understanding, discovering, explaining, and so on. Also, it should explain how factors which affect intelligence do so. A good theory should say something about what intelligence is and what people who differ in intelligence differ in (information-processing capacity? memory?), though this latter task would presumably be a by-product of the former tasks. Of course, these are empirically based suggestions; it could turn out that no single true theory could explain all or even most of the phenomena we now take to involve intelligence.

While the view is widely held that measurement of intelligence does not require theory, it is rarely stated explicitly. Indications that Jensen and Herrnstein advocate an atheoretical doctrine are to be found in remarks such as: "Disagreements and arguments can perhaps be forestalled if we take an operational stance. First of all, this means that probably the most important fact about intelligence is that we can measure it. Intelligence, like electricity, is easier to measure than to define" (*Gen. & Ed.*, p. 72).

> Binet invented the modern intelligence test without saying what intelligence is. At first he was trying to sort out the mental defectives; later he was trying to rate all the children—defective, average, or superior. Some rough and ready notion of intelligence lurked in the background—having to do with mental alertness, comprehension, speed, and so on—but he was not forced to defend an abstract definition in order to sell the ideas of his test to the world. Instead, he could point to how well the test worked. Rarely did a bright child, as judged by the adults around him, score poorly, and rarely did a poor scorer seem otherwise bright. Occasionally a child would do worse than expected on the test because a teacher had confused obedience with brightness, or better than expected when rebelliousness had been mistaken for stupidity, but in general most children ended up about where they were expected to. . . . Once, when he was speculating about the nature of intelligence, Binet mentioned the attributes of directedness, comprehension, inventiveness, and critical capacity, which he thought may vary somewhat independently from person to person. Usually, however, he was too busy with his practical goals to dwell on hypotheses.[20]

It seems psychometricians are supposed to have managed to measure intelligence by simply finding IQ tests which correlate with intelligence, thereby sidestepping the problem of constructing theories of intelligence. (Correlation is explained in the Appendix.) An analogy is: you notice that when it rains you develop an ache in your knee. Its strength is proportional to the amount of rainfall that has accumulated at that time. You then realize you can measure[21] rainfall merely by attending to the severity of the pain in your knee. There is a fatal flaw involved in supposing this picture applies to measuring intelligence: you cannot measure intelligence by finding items which correlate with it unless you already have a way of measuring it.

In the case of measuring rainfall, you *do* have another measurement—merely looking at the height of water in a beaker. Even this measurement is not utterly atheoretical since one has to have a beaker of the correct shape and one has to place it in the appropriate place. The knowledge which allows one to do these things correctly is, of course, embedded in common sense. Although measurement usually requires theory, sometimes, as in the rainfall case, prescientific measurement is possible. In such cases, correlation with the prescientific measurement can license some new (perhaps more convenient) type of measurement.

People sometimes suppose that personal judgments of intelligence can play the role of the prescientific measurement in justifying IQ tests. But no one familiar with the problems of judging intelligence should take this seriously. One major difficulty can be expressed as a dilemma: If the judge has just met the subject, various well-known interview effects, such as the halo effect, may bias him.[22] The halo effect is the tendency judges have of evaluating some characteristics of a person on the basis of the impression the person has made with respect to other characteristics. On the other hand, if the judge knows the subject well, his relationship with the subject and his knowledge of the subject's background may bias him.

The use of judgments of intelligence in validation of IQ tests is now in general disrepute among psychometricians. Terman (who developed the Stanford-Binet Test) and McNemar state flatly: "In the early days of the development of group tests of mental ability an attempt was made to validate them by correlating scores with teachers' marks. As has been pointed out many times in the intervening years, this procedure is unsatisfactory because of the serious shortcomings in teachers' judgments of mental ability."[23]

Much of the psychometric negativism about judgments of intelligence is a response to the unreliability of these judgments. It is worth noting, however, that even if judgments of intelligence were reliable (in the sense that people tended to agree), the reliable judgments would not be necessarily *accurate*. Theory would be required to show that they were accurate. Of course, theory construction in this area might *start* with judgments of intelligence, just as thermodynamics started with judgments of temperature. But to say this is not to imply that these judgments are much good as validators.

In most kinds of cases, if one can measure a quantity, one has some sort of theory of it, and if one has a theory of it, one *can* frame a definition, at least in one respectable sense of "definition," simply by exhibiting its links with other notions of the theory. Thus measurement usually *presupposes* definition, or at least definability. Jensen's comparison of intelligence with electricity is instructive. What would a scientist be measuring if he measured electricity independently of theories about it? Electrical potential (measured in volts)? Electric current (amps)? Electric power (watts)? Electrical work (joules)? Electrical charge (coulombs)? Jensen says intelligence, like electricity, is easier to measure than to define. But the fact is that "electricity," like any serious theoretical term is easy to define, if one knows the theory. For example, "electricity" can be defined as the motion of electric charges, and its various aspects can be measured on the basis of the theory which licenses the definition (along with auxiliary hypotheses about the measuring apparatus).

But it would be a mistake to get caught up in what is really a pseudoissue: Which comes first, measurement or theory? Measurement and theory must progress together; each depends on the other. (Though, of course, usually measurement will be accurate only insofar as the theory is true.) Current scientific measurement must be based on current scientific theory, but current theory is only developed with the help of measurements based on past scientific theory. Ultimately, this scientific process is based on prescientific observation entwined with prescientific conceptions.

Many psychometricians attempt to justify IQ tests by comparing the development of such tests with the development of the thermometer, an instrument whose history does seem at first glance quite similar to the history of IQ tests. The picture some psychometricians seem to have in mind is that people could (or even did) measure temperature in the absence of any theory of temperature, simply by noticing (on the basis of a lucky guess, perhaps) the fact that the height of a column of mercury in a closed tube corresponds to temperature. Similarly, they suppose, one can simply *observe* that answering certain types of questions (correctly) correlates highly with intelligence. So one can measure both temperature and intelligence without theories, simply on the basis of observed correlations. But without *some* fragment of a theory of temperature (more precisely, of heat phenomena—a theory is needed even to know the proper domain of a phenomenon), it would be no more reasonable in ascertaining the temperature in a room to investigate the level of mercury in a tube in the room than to investigate the average height of people in the room. Of course, one might investigate the level of mercury by chance. But even then, without some sort of theory about heat phenomena, one would have no chance of determining the functional relation between temperature and mercury-column height. For all one would have to go on is *perceived* temperature. Even under ideal conditions (e.g. constant humidity, no wind) the relationship between temperature and perceived temperature is nonlinear; and with no theory, one would have no way of knowing what conditions are ideal. But without knowing that such variables as humidity, wind velocity, amount of exercise, illness (causing fever, previous experiences, and many other physical, psychological, and physiological factors affect one's perception of the temperature of one's surroundings, one would have no way at all of telling even whether there is a functional relation between temperature and mercury-column height.

Further, the distinction between temperature and perceived temperature which we have been presupposing is not a pretheoretically obvious one. A piece of wood and a piece of steel both at 50 degrees feel as though they have different temperatures, because the steel conducts heat so much better. Indeed, though devices which can perhaps be called thermometers were produced from the beginning of the seventeenth century, the theoretical level was apparently sufficiently low that, as Thomas Kuhn points out,

> it was totally unclear what the thermometer measured. Its readings obviously depended upon the "degree of heat" but apparently in immensely complex ways. "Degree of heat" had for a long time been defined by the senses, and senses responded quite differently to bodies which produced the same thermometric readings. Before the thermometer could become unequivocally a laboratory instrument rather than an experimental subject, thermometric reading had to be seen as the direct measure of "degree of heat," and sensation had simultaneously to be viewed as a complex and equivocal phenomenon dependent upon a number of different parameters.[24]

But perhaps we are attacking a straw man. For even if some psychometricians took an atheoretical approach, the development of IQ tests was not—and could not be—totally atheoretical. Further, while some elements of a theory of heat phenomena were required for the development of the thermometer, probably we would never have developed sophisticated modern theories of heat without *already having* the thermometer. Thus, it might reasonably be suggested that IQ tests measure intelligence about as well as early thermometers measured temperature, and further, that IQ tests are the condition for the possibility of theories of intelligence which will someday allow more nearly perfect tests of intelligence. We shall argue that while the modern IQ test may be, in some respects, analogous to some early thermometers, the development of the IQ test and the thermometer are crucially different. Later, we will give some reasons for doubting that IQ tests are the condition for the possibility of theories of intelligence.

The first modern thermometerlike device was probably made by Galileo at the end of the sixteenth century. (Philo of Byzantium apparently made one like it about 1,600 years earlier.) It was simply an open tube with a water-filled bulb at one end that was inverted in a dish of water. Changes in the temperature of the air in the bulb caused the air to expand or contract, thus changing the level of water. The open dish made the device as much a barometer as a thermometer, but this was not realized until the middle of the next century. In our view, this "air thermoscope" is the thermometer which most closely corresponds to standard IQ tests, although it may have been a better indicator of temperature than the IQ test is of intelligence. Just as standard IQ tests probably measure a number of other qualities to about the same degree as intelligence, the early air thermoscope readings reflected air pressure, idiosyncratic behavior of water, irregularities in the bore of the tube, and other quantities as well as temperature.

The subsequent history of the thermometer provides a good example of the way the development of measuring apparatus depends on—and is also the condition for—the development of theory. A crude device based on theories embedded in common sense provided the data for more sophisticated theories, and these theories in turn allowed the construction of more sophisticated measuring devices.

Once the responsiveness of the air thermoscope to air pressure was understood, closed thermometers were introduced. Some investigators tried to select two temperatures as fixed points, e.g., the temperature of melting snow, and the temperature of the human body or the boiling point of water. They would then divide the distance in between into a number of equal segments. This procedure gave rise to a number of serious theoretical difficulties.

First, investigators did not know whether their "fixed points" were really fixed. (Water, in fact, does not always freeze and boil at fixed temperatures.) Solving this sort of problem awaited theories of the relation between temperature, freezing, and other variables such as pressure. Second, they did not know whether the liquids they used (water, oil, spirit of wine) expanded linearly with temperature. The volume of water, for example, is not even an always increasing function of temperature; water contracts with decreasing temperature until about 39 degrees F at which time it starts to expand with decreasing temperature. Early experimentalists soon discovered that thermometers set to agree at a minimum and a maximum did not agree at points in between, because different liquids had different curves of expansion with temperature. They were further hampered because different samples of spirit of wine (a commonly used thermometric liquid) were often chemically different, and because they did not take account of the fact that tubes made of different substances (glass, brass, silver, and so on) changed differently with temperature as well. Another theoretical difficulty was their assumption that a thermometer placed out of the sun would register the temperature of the air. But thermometers in air under ordinary circumstances are much more responsive to the temperature of surrounding *objects* than the air temperature because, in ordinary circumstances, heat exchange in air is far more a matter of radiation than convection or conduction.

The early makers of thermometers had a good deal of plain luck. For example, it turned out by chance that many of the common liquids available to them expanded with temperature in a rather similar manner; in addition, this expansion happened to be fairly linear with temperature within the temperature limits they experienced.[25] Because the development of the thermometer depended on both luck and serious theory, it can provide no solace for defenders of IQ tests. For they have no serious theory; and further, without serious theory, how are we to know whether their luck has been as good as that of the early thermometer makers?

The history of the IQ test is quite different from the history of the thermometer. IQ testing as a field has developed to a surprising degree as the atheoreticism discussed above would have us believe a science *should* develop. Binet's original tests were constructed almost entirely without benefit of explicit theorizing.[26] The relative atheoreticity of Binet's tests would have been all right if the tests had been regarded simply as the first stage in a process which would someday yield a real understanding of intelligence on which measurements could be based. But the fact is that Binet's tests were the starting point of a *technological*, not a scientific process.

The Stanford-Binet is currently the most influential IQ test; other test makers often exhibit correlations with the Stanford-Binet by way of "validating" their tests. The first Standard-Binet, constructed in 1916 by Lewis Terman and others, was basically an adaptation (with extension and refinement) of Binet's tests.[27] There have been two subsequent revisions, in 1937 and 1960. But as Anastasi points out in *Psychological Testing*, "*Continuity* in the functions measured in the 1916, 1937, and 1960 scales was ensured by retaining in each version only those items that correlated satisfactorily with mental age on the preceding form" (p. 204). In the case of the 1960 revision, the most "successful" items in the two 1937 tests were simply collapsed into a single test.

We do not want to claim that there have been no changes with even the slightest theoretical basis since Binet. Binet reported his results as "mental ages"; only afterwards was the practice introduced of dividing mental age by chronological age to yield an intelligence quotient. (In 1960, this method was dropped from the Stanford-Binet for technical reasons.) Binet's tests were mainly verbal; nowadays, standard IQ tests are often substantially nonverbal. Binet's tests were administered individually; many modern tests are administered in groups and scored mechanically. But in spite of these and other developments, modern IQ tests do not differ in any essential scientific respect from Binet's tests. What Goodenough said in *Mental Testing* (New York, 1949, p. 57) is still applicable: "As we look back over the road along which we have come it is apparent at once that improvements in testing procedures have been made for the most part on an empirical rather than a theoretical basis." The English psychologist H. J. Eysenck, who has argued the importance of heritability of IQ in much the same way as Jensen has in the United States, rather candidly remarks:

> It is often believed that intelligence tests are developed and constructed according to a rationale deriving from some scientific theory. . . . In actual fact . . . intelligence tests are not based on any very sound scientific principles, and there is not a great deal of agreement among experts regarding the nature of intelligence. . . . Because the intelligence tests, originally constructed in the early years of this century, did such a good job when applied to various

practical problems, psychologists interested in the subject tended to become technologists eager to exploit and improve these tools, rather than scientists eager to carry out the requisite fundamental research, most of which still remains to be done. Society, of course, always interested in the immediate application of technological advances and disinterested in pure research, must bear its share of the responsibility for this unfortunate state of affairs.[28]

There has been some theoretical work on intelligence, especially recently: Cattell's theory of fluid versus crystallized intelligence, Guilford's theory of abilities, Cronbach's idea of a spectrum of ability tests, and Jensen's theory of Level I and Level II abilities. But theoretical work has had little effect on IQ tests. David Wechsler, the maker of one of the best-known IQ tests, put it this way: "Notwithstanding their theoretical views, authors of intelligence scales tend to make use of the same sort of tasks and items. Procedures may vary, but the tests themselves do not differ very much."[29]

The Operationalist Picture of Measurement. We shall now examine the operationalist approach to measurement in more detail. Its central tenet is that measurement is a matter of stipulation. The picture seems to be that the way scientists measure a quantity, be it intelligence or temperature, is to look for a reliable reproducible operation whose results more or less correlate with variables one might expect intelligence (or temperature) to correlate with, and then simply stipulate that the reliable reproducible operation is a measure of intelligence (or temperature). Evocations of this picture, like other aspects of this methodology, are seldom entirely explicit; they are often accompanied by the intimation that the only alternative is to become involved in the semantic morass of analyzing the ordinary concept of intelligence, a task no doubt better left to philosophers. Jensen says, for example:

If we are to study intelligence, we are ahead if we can measure it. Our measure is the IQ, obtained on tests which meet certain standards, one of which is a high *g* loading when factor analyzed among other tests. To object to this procedure by arguing that the IQ cannot be regarded as being interchangeable with intelligence, or that intelligence cannot really be measured, or that IQ is not the same as intelligence is to get bogged down in a semantic morass. It is equivalent to arguing that a column of mercury in a glass tube cannot be regarded as synonymous with temperature or that temperature cannot really be measured with a thermometer. *If the*

> *measurements are reliable and reproducible, and the operations by which they are obtained can be objectively agreed upon,* this is all that need be required for them to qualify as proper scientific data.[30]

Herrnstein's methodological views also seem to involve something like this operationalist picture. He says:

> Writing in the *New Republic* in 1923, Professor Edwin Boring of Harvard said, "Intelligence as a measurable capacity must at the start be defined as the capacity to do well in an intelligence test. Intelligence is what the tests test." Once we agree on a test, that is doubtless true, just as well-defined physical concepts like force and work can be identified with the instruments that measure them. But how do we get to that kind of agreement on the defining instrument? At some point, it becomes a matter of prior definition, arising in theory or common knowledge. For example, we would reject any intelligence test that totally discounted verbal ability or logical power, but how about athletic prowess or manual dexterity or the ability to carry a tune or qualities of heart and character? More data are not the final answer, for at bottom, subjective judgment must decide what we want the measure of intelligence to measure. So it is for all scales of measurement—physical as well as psychological. The idea of measuring length, weight or time comes first; the instrument comes thereafter. And the instrument must satisfy common expectations as well as be reliable and practical. . . . As for what intelligence "really is," the concept still has ragged edges where convenience and sheer intuition set boundaries that will no doubt change from time to time.[31]

This sort of methodology fails in three distinct ways. First, too many quantities satisfy common expectations. Thus, if the common expectations about intelligence are merely that it correlates moderately with various forms of success and with judgments of intelligence, more than likely, many composites of qualities other than intelligence would satisfy those common expectations. Second, some of our common expectations about what intelligence would correlate with may well be wrong. A scientifically acceptable test of intelligence, if we had one, might fail to satisfy some current common expectations. Correlation and causation often do not go together. Variables neither of which causally influences the other can be highly correlated, and even if one variable completely causally determines another, the two may have zero correlation. Operationalists may say they require only that some rather than all of the common expectations be satisfied. But this would make the quantity measured even more underdetermined, thus intensifying the first difficulty described above. Further, without theory, how is the operationalist supposed to decide which common expectations to reject? Third, the quantities that are susceptible to very reliable reproducible measurement at a prescientific stage of inquiry may *not* be the quantities one set out to measure. If we do not now have the scientific capability of measuring intelligence in a reliable and reproducible way, to require that our methods of assessing intelligence satisfy reliability and reproducibility conditions may well be to preclude the possibility that our methods of assessing intelligence actually assess intelligence.

Thus, the operationalist constraints on measurement may be simultaneously too weak and too strong. They are too weak to rule out measurement of the wrong quantities and may be so strong as to direct measurement away from the right quantities. What makes these operationalist constraints immediately suspect is that it is hard to see how one could justify them from the point of view of truth in measurement. That is, methodological constraints on measurement ought to be justifiable on the grounds that measurements satisfying the constraints are more likely to yield adequate measurements of what the measurer intends to measure than measurements not satisfying the constraints. But one should want a measuring device to satisfy common expectations only insofar as one believes common expectations are correct, and one should not demand a greater degree of reliability and reproducibility than the state of knowledge allows.

On the other hand, it is easy to see how these methodological constraints can be "justified" on other grounds. A measuring device which fails to satisfy common expectations and whose results are not reliable and reproducible is hardly likely to help the profession which proffers it to the public to flourish—especially if the device also fails to have any good scientific rationale.

Further, this operationalist methodology is not that actually deployed in the physical sciences, though there are some misguided physical scientists who think otherwise. Measurement in the physical sciences is, in general, based on objective theory construction, not subjective judgment (though subjective judgments can, of course, be data).

A theory in the physical sciences talks about the quantities it talks about because talk of those quantities is required to explain the phenomena the theory is supposed to explain. The same ought to be true in the social sciences. For example, Newtonian mechanics talks about kinetic energy because kinetic energy is involved in various laws of physics which must be adverted to in order to explain phenomena in mechanics. A physicist who wanted to measure the kinetic energy of a bouncing ball would use the theory of mechanics, not some "agreement on the defining instrument."

If the version of operationalism we are discussing were right, seventeenth-century thermometer makers simply could have agreed to the stipulation that temperature is what a closed tube filled with wine and marked off in arbitrary segments says it is. Of course, this would have been an inconvenient definition, but mainly because it is false that such a device measures temperature, and it is in general inconvenient to believe falsehoods. Operationalism goes wrong in construing as a linguistic stipulation a theoretic inference that a particular interaction between a thing and a device is a measurement. As Hilary Putnam puts it:

> We know that this object (the meter) measures electrical charge, *not* because we have adopted a "convention" or a "definition" of electrical charge in terms of meter readings, but because we have accepted a body of theory that includes a *description of the meter itself in the language of the scientific theory.* And *it follows from the theory,* including this description, that the meter measures electrical charge (approximately, and under suitable circumstances). The operationalist view disagrees with the actual procedure of science by replacing a probabilistic inference within a theory by a nonprobabilistic inference based on an unexplained linguistic stipulation.[32]

On the other hand, operationalism probably would not be so widely accepted if it did not contain some elements of truth. For example, Herrnstein and Jensen are surely right insofar as they are saying that one should replace prescientific concepts by more precise scientific concepts instead of doing conceptual analysis of the prescientific concepts. Seventeenth-century physicists did not waste their time figuring out exactly what their neighbors meant by words like "force"; rather, they constructed theories, finding lawlike relations which allowed them to replace such notions with more precise versions.

Operationalism in psychometrics is also right in viewing the speculations of the 1920s and 1930s about the nature of intelligence as having little lasting scientific merit. On the other hand, it is wrong to suppose the alternatives are exhausted by analysis of the concept of intelligence, speculation about the nature of intelligence, and subjective judgment made in a virtual theoretical vacuum. There is another possibility: develop a theory of intelligence and construct measurement procedures on the basis of the theory. As is often the case in an infant science, it is not very clear *how* to go about developing a theory of intelligence. We suspect that an understanding of intelligence awaits the at least partial success of the program of cognitive psychology: investigating the psychological processes which occur in tasks involving cognitive activity. It is likely that, only when we understand cognitive processes much better than we do now, shall we be able to see what sort of differences there are in cognitive processes, how they arise, and how they give rise to what we prescientifically describe as differences in intelligence. Some prescientific descriptions turn out to be roughly correct; others turn out to be misguided. What will turn out to be the case with our ability ascriptions we cannot now know.

When cognitive processes are well understood, we may decide not to retain the term "intelligence" at all, for it may turn out that intelligence phenomena do not constitute a "natural kind," that no single theory of a class of cognitive phenomena can explain most of the phenomena we would want a theory of intelligence to explain. Or, like contemporary geneticists with respect to the term "gene," we may be faced with choosing which of a number of things to attach the term to.

Take the concept of hardness, for example.[33] If one material can be used to scratch, abrade, indent, drill, and saw another, but not vice versa, our prescientific concept of hardness would dictate that the first is harder. But as it happens, these criteria involve various combinations of physical magnitudes and thus do not always occur together. The resistance of a piece of material to abrasion, for example, depends heavily on the properties of the surface as well as on the kind of material. Resistance to indentation depends mainly on plasticity, while resistance to scratching depends on both plasticity and fracture characteristics. "Hardness phenomena" are apparently not a natural kind. In the light of such findings, it is not totally clear how we should use the word "hardness." Perhaps we should speak of "scratch hardness," "indentation hardness," "abrasion hardness," and so on, and simply drop "hardness." Or perhaps we should use "hardness" to denote some sort of average of these quantities. When we understand as much about intelligence as we now do about hardness we may be faced by a similar choice. However, such choices are often largely matters of stipulation, and insofar as they are, they are usually of little scientific interest.

It might be objected that the theory-building process we envision is a long one. Is it reasonable to give up a useful instrument like the IQ test until this process is further along? Suppose thermodynamics had been much harder than it was. Suppose that for 200 years, Galileo's knowledge of thermodynamics could not be significantly improved on. Should people have refrained from using Galileo's air thermoscope as a poor but useful measure of temperature? Of course not. But neither should they have pretended that a finely honed supersophisticated air thermoscope is anything more than an air thermoscope. We do not object to IQ tests being *used*, but rather to their use under descriptions like "intelligence tests"—descriptions they are not known to satisfy.

II. Misleading Use of the Word "Intelligence"

Many psychologists seem to have the impression that IQ has roughly the same relation to the ordinary notion of intelligence that Newton's concept of mass had to common seventeenth-century notions of mass. At the same time, many of them (Jensen and Herrnstein included) actually use "smart," "stupid," "bright," "dull," "intelligent," "unintelligent," in stylistic variation with "high/low IQ." They thereby transfer to IQ all the emotional and conceptual associations the reader has to intelligence. All this after we are disarmed by being told that after all, "intelligence" is just a word like any other word, and how a scientist chooses to use it is just a matter of stipulation. Thus, they manage to have their cake and eat it too—that is, they have their new "scientific" concept of intelligence at the same time retaining the social importance of the old "prescientific" concept of intelligence.

Given their practice of using "IQ" interchangeably with "intelligence," "smartness," "brightness," and so forth, it is somewhat surprising to find these psychologists deploring public misunderstanding of the psychometric use of the term "intelligence." These disclaimers are typically accompanied by descriptions of IQ tests and their construction so that the reader can see what psychometricians mean by "intelligence" (following Boring's dictum that intelligence is what IQ tests test). For example, in the handbook to the current Stanford-Binet IQ Test—called *Stanford-Binet INTELLIGENCE Scale*[34]—Terman and Merrill actually say that "a brief resume of standardization procedures and reexamination of the rationale underlying their use seems to be indicated . . . because in current popular usage IQ is so often used as a synonym for intelligence, unrealistic expectations have come to be attached to *intelligence* tests" (p. 7, emphasis added). In a similar vein, after a brief description of IQ tests, Jensen says, "The trouble comes only when we attribute more to 'intelligence' and to our measurements of it than do the psychologists who use the concept in its proper sense" (*Gen. & Ed.*, p. 76). But the habit of using "IQ" interchangeably with "intelligence" and related words has the effect of *inculcating* precisely the misunderstanding which the authors seem to deplore: namely, that IQ tests measure how smart or intelligent a person is in the *ordinary senses* of these terms.

The psychometric practice of using "IQ," "intelligence," and so on, interchangeably apparently reflects a feeling that "intelligence" is basically a technical term of psychometrics. Thus Jensen says: "The term 'intelligence' as used by psychologists, is itself of fairly recent origin. Having been introduced as a technical term in psychology near the turn of the century, it has since filtered down into common parlance, and therefore some restriction and clarification of the term as it will be used in the following discussion is called for" (*Gen & Ed.*, p. 72). Of course, restrictions and clarifications made by Jensen on page 72 tend to be forgotten by page 203, especially when the intervening pages contain countless uses of "bright," "dull," etc., in the ordinary relation to "intelligent."

Eysenck says that the modern use of "intelligence" derives from Herbert Spencer who revived it in his *Principles of Psychology* in 1870.

> Newton met the same problem when in his *Principia Mathematica* he dealt with Mass; he points out there that what the physicist means by this concept and what the man in the street means by it are two very different things. He adds that it is a "vulgar error" to confuse the two, but he does not advocate referring to his "mass" by a simple letter *m*. The term intelligence originated, as we have seen, in departments of psychology and philosophy; it is by no means clear why we should give up using it because it is suggested that the man in the street uses it in rather a different sense. . . .[35]

But Newton's use of "mass" differs from psychometric uses of "intelligence" in the crucial respect that the social role of Newtonian mechanics was not based on physicists' leading people to believe physics uses "mass" in one of the everyday senses of the term. Further, Eysenck's philology is suspect. The *OED* lists many uses of "intelligence" and "intelligent" in their present senses from the fifteenth century onward. But philology is not our concern. The point is that one cannot have it *both* ways. One cannot use "intelligence" to refer to what people ordinarily refer to when they use words like "smart," and then also use it as a technical term stipulated to refer to whatever it may be that IQ tests test. The current social role of IQ tests is based on the presupposition that these two usages of "intelligence" have the same reference. But this presupposition is insistently ignored by too many of those whose responsibility it is to either justify or reject it.

One argument seems to involve the claim that the word "intelligence" should be understood as meaning a disposition or tendency toward occupational success in our type of society. The view is that intelligence is a pure disposition; to assert a person has high intelligence is to say or imply nothing about what characteristics he has in virtue of which he is likely to succeed, but rather only that he is likely to succeed. O. D. Duncan seems to have this sort of view in mind when he says " 'intelligence' is a socially defined quality and this social definition is not essentially different from that of achievement or status in the occupational sphere . . . what we now *mean* by intelligence is something like the probability of acceptable performance (given the opportunity) in occupations varying in social status."[36]

But if words are to mean anything at all, to say a person is intelligent is to say something about his *cognitive* or intellectual aspect. A test can correlate with success—and thus its results represent a disposition to succeed—even if it does not measure anything cognitive at all. For example, a measure of the socioeconomic status of a child's *parents* correlates as highly with the child's occupational status as his IQ does.[37] Clearly, whether a test which correlates with success measures intelligence depends on *why* it correlates with success. Chomsky suggests that in our society, "Wealth and power tend to accrue to those who are ruthless, cunning, avaricious, self-seeking, lacking in sympathy and compassion, subservient to authority, willing to abandon principle for material gain, and so on."[38] Surely a test which measured only such characteristics would not be an *intelligence* test, even if it did correlate highly with success. In any case, it is clear that test makers do not take this "pure disposition" view seriously in practice. There are some supposedly nonintellective tests which predict grades about as well as IQ does.[39]

For example, H. G. Gough[40] devised a test which contains about sixty true-false items of which the following are samples. "Correct" answers are in parentheses.

> I get sort of annoyed with writers who go out of their way to use strange and unusual words. (F)
> I am not afraid of picking up a disease or germs from doorknobs. (T)
> I daydream very little. (F)
> My parents have often disapproved of my friends. (F)
> I have had no difficulty in starting or holding my urine. (T)

The test also contains a few items whose connection with ability seems less tenuous, e.g., "I was a slow learner in school (F)." Scores on this test correlated around .5 with IQ for high-school students, and .42 for college undergraduates and graduate students. The test predicted school grades for girls slightly better than IQ, though for boys IQ was better. Though this test was devised in 1953, its type of item has never been incorporated in any serious IQ test we know of, even though the predictive value of IQ tests could clearly be increased by so doing.

If makers of IQ and aptitude tests were really interested only in obtaining scores which correlate as highly as possible with measures of success, they would certainly add personality-motivation-temperament and other nonintellective items to their tests. Indeed, the best "test" from the point of view of prediction would also include school grades[41] and measures of parental socioeconomic status. The fact that IQ tests do not include such items indicates that test makers (and their customers) find it implausible that such tests would measure intelligence.

III. The Logic of Correlational Validation of IQ Tests

Thus far, we have, in effect, been arguing that it is important to *ask* whether IQ tests measure intelligence. In the next three sections, we shift to a critical examination of arguments that IQ tests *do* measure intelligence, i.e., arguments about IQ-test validation.

Psychologists who believe IQ tests measure intelligence—operationalists and nonoperationalists alike—base their belief in one way or another mainly on the correlations of IQ with various measures of success. For example, Eysenck argues: "There is no doubt that success in life defined either in terms of income or of social prestige (and thus defined essentially in terms of what the man in the street thinks) correlates quite well with IQ (though not perfectly, as we have seen); *hence* IQ does clearly measure much the same sort of thing as the man in the street means by intelligence."[42] In this section, we argue that these correlations are useless for validation of IQ tests.

First, what sorts of success does IQ correlate with? The highest correlations are with measures of school success. The average correlation of IQ with highest grade completed in school (abbreviated "schooling") is .68 and correlation coefficients of IQ with tests of general scholastic achievement are often higher.[43] Correlations of IQ with grades vary a great deal, but usually fall within the range of .4 to .6.[44] IQ correlates about .5 with occupational status (a measure based on average educational level and income of persons in a given occupational category) but only about .35 with income.

In the next section, we shall argue that each of these correlations is useless for validation of IQ tests for one or another reason. For the moment, however, let us take these correlations at face value, that is, as representing causal influence of IQ on characteristics which are both important and different from IQ. For purposes of discussion, we shall ignore the particular values of the correlations and simply speak of the .5 correlation of IQ with success.

How is the .5 correlation of IQ with success supposed to show that IQ tests measure intelligence? One argument (A) is

(1) Success is a measure of intelligence.
(2) IQ correlates with success.
(3) Therefore, IQ is a measure of intelligence.

The logic of this argument is represented in figure 1. The only known quantity is the correlation of IQ with success. In order to justify a conclusion about the correlation of IQ with intelligence, one must obtain some knowledge of the size of the correlation between success and intelligence. One's estimate of this correlation will affect both the degree to which the premises appear to make the conclusion probable and also the size of the inferred IQ-intelligence correlation.

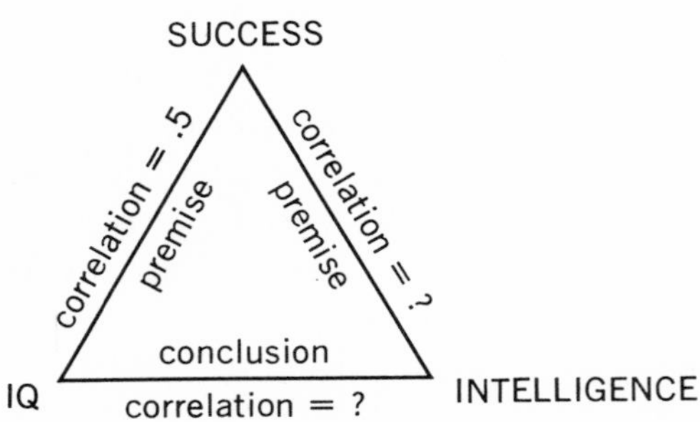

Figure 1

The most obvious problem with the argument is that no one knows what the correlation of intelligence with success is. Indeed, no one *could* know this without some way of measuring intelligence justified by some *other* argument. Thus argument (A) can, at best, play a secondary role in validating IQ tests. Nonetheless, since arguments that IQ tests measure intelligence which have any force at all are hard to come by, it will be worth examining (A) in detail. We should notice first that insofar as the correlation of intelligence with success is assumed to be low, the argument gives us little reason to infer any substantial correlation between IQ and intelligence. So if the argument is to get off the gound, it must be assumed that success has a substantial correlation with intelligence.

For the sake of argument, let us suppose success is a perfect measure of intelligence, i.e., that the correlation of intelligence (as it would be measured by an ideally perfect intelligence test) with success is 1.0. Then, since IQ correlates .5 with success, it follows (deductively) that the correlation of IQ with intelligence is .5 (*see* fig. 2). This would make IQ tests a rather poor measure of intelligence, for it would mean that *at most* 25 percent of the variance in IQ scores is due to variation in intelligence (*see* Appendix). That is, even if success is a perfect measure of intelligence, quantities other than intelligence would account for at least 75 percent of IQ variation. So making premise (1) of (A) as strong as possible makes the argument as strong as possible (i.e., deductively valid) but only at the cost of making the conclusion rather weak.

But how strong need the conclusion be? That is, how high must the correlation of IQ with intelligence be for it to be legitimate to

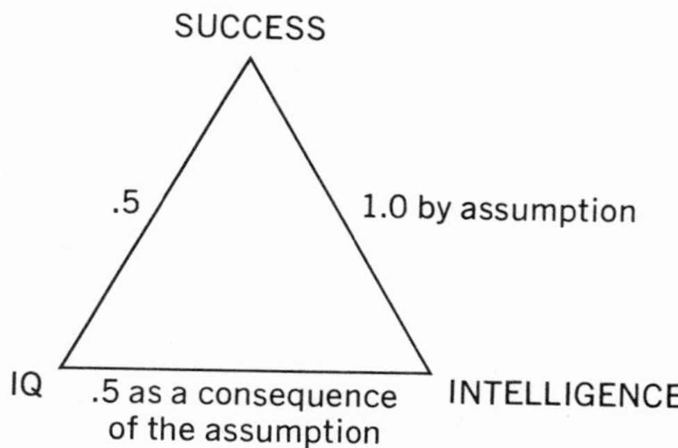

Figure 2

say IQ tests measure intelligence? For IQ tests to measure intelligence, we would require that IQ scores reflect mainly intelligence, e.g., that somewhere in the vicinity of, say, three-quarters of the variance in intelligence can be predicted from IQ. This would correspond to a correlation between IQ and intelligence of .87.[45]

We have seen that if we interpret premise (1) as strongly as possible, we are stuck with a weak conclusion. We will now show that if the conclusion of (A) is interpreted very strongly, the premises can no longer provide much reason for believing it. Let us suppose, for the sake of argument, that the conclusion of (A) is as strong as it could be: that IQ is a perfect measure of intelligence, that is, that the correlation of IQ with intelligence is 1.0. Now, if intelligence correlates 1.0 with IQ, and IQ correlates .5 with success, it follows that intelligence correlates .5 with success (*see* fig. 3). But then a person who propounds argument (A) and also accepts its conclusion is in the position of arguing that IQ is intelligence on the ground that IQ correlates .5 with something (success) which itself correlates only .5 with intelligence. That is, he is arguing that $x = y$ on the ground that x and y both correlate .5 with z. But it is mathematically possible for x, y, and two other variables, u and v, each to correlate .5 with z even though the intercorrelations among x, y, u, and v are all *zero*. Thus, in the absence of additional information, the fact that x and y each correlate .5 with z is weak evidence that x and y are highly correlated with one another. Correlation is not transitive. Perhaps an example will make this point clearer. Often, one disease will mimic the symptoms of another, quite different disease—e.g., both stomach

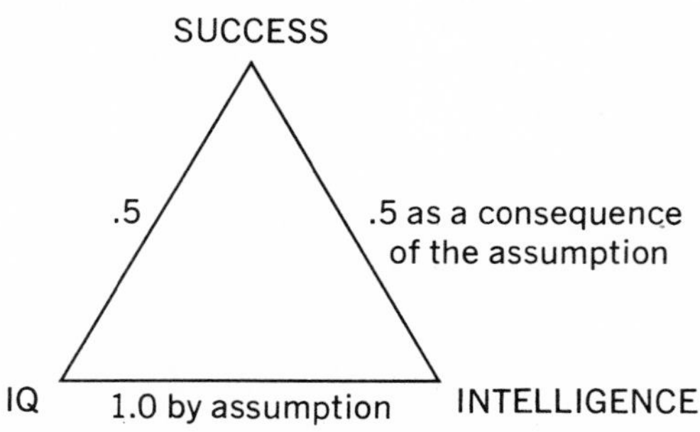

Figure 3

and heart disease often produce chest pains. Suppose that stomach disease and heart disease are each correlated .5 with pain in the chest. This may be the case even if stomach disease and heart disease hardly even occur together. Analogously, it could be that both IQ and intelligence contribute to variation in success even if IQ and intelligence are themselves mainly uncorrelated.[46]

Thus the proponent of argument (A) finds himself in a mild dilemma: insofar as he supposes success is a perfect measure of intelligence, his premises give him good reason to believe that IQ is related to intelligence, but only at the cost of decreasing the size of the inferred relation between IQ and intelligence. On the other hand, insofar as he supposes success is a weak measure of intelligence in order to allow a strong relation between intelligence and IQ, he finds the premises give him little reason to believe the strong conclusion.

Of course, nothing we have said conflicts with the fact that a moderately high correlation of intelligence with success (if such turns out to be the case) would provide a *good* reason for believing there is *some* (i.e. more than zero) relation between IQ and intelligence, and also *some* reason for believing there is a *moderate* relation between IQ and intelligence.

If the correlation of success with intelligence were known to be low, we could not infer any correlation between IQ and intelligence. If the correlation of success with intelligence were known to be high, we could conclude with certainty that there is a moderate correlation (no more, no less) of IQ with intelligence. Such a moderate correlation would be too low to justify the claim that IQ tests measure intelligence. If the correlation of success with intelligence were known to be moderate, we would have only a weak reason to believe there is a high enough correlation between IQ and intelligence to justify the claim that IQ tests measure intelligence, though at the same time we would have a good reason for believing there is (at least) a small correlation between IQ and intelligence. Thus, no information about the correlation of intelligence with success would make argument (A) a good argument that IQ tests measure intelligence.

Another argument (B) that IQ tests measure intelligence is of the "inference to the best explanation" variety:

> It seems reasonable that intelligence would correlate about .5 with success; after all, we all know that many factors other than intelligence such as luck, looks, family, personality, etc., contribute to success. Thus the best explanation of the fact that IQ correlates .5 with success is that IQ tests measure intelligence. Hence it is likely that IQ tests measure intelligence.

This argument would have some force, if there were any reason to believe that there are no *other* qualities or amalgams of qualities which are substantially independent of intelligence and which might be expected to have a .5 correlation with success. But if the conclusion of (B) (understood as: IQ tests are perfect measures of intelligence) is *true*, then there *must* be a number of such qualities. For if the conclusion is true, then intelligence correlates only .5 with success, and thus *at least three-quarters* of the variance in success is due to qualities other than intelligence. Moreover, many of the qualities which appear to contribute to success (personality, motivation, temperament, sociocultural background, school achievement) are also plausibly measured to some degree by IQ tests.

With argument (B) as with argument (A), insofar as IQ tests are supposed to be good measures of intelligence, the *argument* that they are good measures loses its force. One might reasonably ask why it is that, despite the weakness of available arguments supporting correlational validation of IQ tests, so many people nevertheless appear to suppose otherwise. We doubt there is a single explanation of this. The operationalist methodological views discussed above have played a major role in preventing clear thinking on this issue. In addition, some of those who insist IQ is intelligence appear to use the word "intelligence" to mean something like "probability of occupational success." Others (usually non-psychometricians) are overimpressed with the .5 correlation of IQ with success because they think this correlation can be regarded as a confirmed prediction of the theory underlying IQ tests as measures of intelligence. The implicit theory behind IQ testing is, in essence, that differences in the skills and knowledge demonstrated on certain tasks will be due mainly to differences in capacity to acquire knowledge and skills. But, as we shall argue in section VII, these differences in test performance seem equally likely to be due to differences in sociocultural background or personality-motivational-cultural factors. And these factors could also be expected to correlate with success. A second major flaw in the "confirmed prediction" reasoning is that the .5 correlation of IQ with success cannot be viewed as *confirmation* of anything because the correlations with success have simply been built into the tests. (How this happened will be explained later.)

It is worth noting that the logic of confirmation in this case is quite similar to the logic of the arguments discussed above. Even if we did have independent reason for believing IQ tests measure intelligence, and even if the correlation between IQ and success could be viewed as a prediction, still the .5 correlation of IQ with success would not necessarily be impressive. For if a test correlated 1.0 with the types of success we have been talking about, that would show the test is *not* a perfect measure of intelligence. We all *know* on the basis of personal experience that many factors other than intelligence contribute to success. A perfect intelligence test would have whatever correlation with success intelligence *actually* has. A correlation *greater* than this value, whatever it is, would count against a test as surely as a correlation *less* than this value. Since we do not have an acceptable theory of intelligence, or of school or occupational success for that matter, we simply do not know what correlation intelligence actually has with school and occupational success. So even if we did have independent evidence for believing IQ tests measure intelligence, and even if the correlations of IQ with success could be viewed as predictions, we would not know how strongly the obtained correlations of IQ with success support the claim that IQ is intelligence.

Of course, it *could* turn out that .5 as opposed to .3 or .7 is an impressive correlation. This would depend on (a) whether the actual correlation of intelligence with success could be expected to be .5, (b) how many *other* quantities or amalgams of quantities which IQ tests might measure could be expected to correlate .5 with success, and (c) the degree to which these other quantities or amalgams are uncorrelated with intelligence.

IV. The Correlations of IQ with Success Cannot Be Taken at Face Value

Thus far, we have, in effect, taken the correlations of IQ with various measures of success at face value, that is, as representing causal influences of IQ on characteristics which are both important and different from IQ. We have argued that even taken at face value, these correlations cannot be used to validate IQ tests. We shall now argue that the correlations of IQ with these measures of success fail in one way or another to be useful for validating IQ tests—even if the arguments of the last section are ignored.

Correlation of IQ with Achievement Scores. The correlations of IQ with scores on standardized tests of general scholastic achievement are very high—but this is little more impressive than the high correlations of IQ tests with one another. The fact is that most tests of general scholastic achievement fare pretty well in satisfying the criteria which have in practice been adopted for group IQ tests.

This point is recognized by most of the leading psychometrics textbooks. For example, Anastasi says that

> it is questionable whether broadly oriented achievement tests are contributing any information about individuals that could not be obtained from intelligence or other aptitude tests. Although the various available instruments may all serve useful purposes, there seems to be little justification for retaining different labels for them. In fact calling one an aptitude and another an achievement test may lead to confusion and to the misuse of test results. [*Psychological Testing,* p. 393]

Many psychometricians have explicitly abandoned the distinction between achievement and aptitude tests in favor of a continuum or spectrum of educational loading, i.e., the extent to which the tests tap specific school instruction. Both Cronbach and Anastasi, for example, rank standard IQ tests close to general achievement tests in their respective spectra.[47]

The explanation of this similarity of IQ and achievement tests is not hard to find. The rationale behind most IQ-test items is roughly that differences in the degree to which individuals have acquired information and intellectual skills are due mainly to differences in intelligence. So IQ tests attempt to assess a broad range of information and intellectual skills. But this is just what general achievement tests attempt to do.

IQ and Grades. While IQ correlates from .4 to .6 with grades, there is reason to believe that insofar as there are qualities in common to doing well both on IQ tests and in school, they have little importance outside the "charmed circle" of IQ tests and schools.[48] While studies show that number of years in school is highly correlated with life success, studies of the relation between various measures of life success and *grades* consistently show little or no relationship, even for college grades. In a review article, D. P. Hoyt concludes: "The bulk of the studies in the area of teaching, business, scientific research, and medicine show no significant relationship between college grades and vocational success."[49] It may surprise those of us who teach in colleges that the same appears to hold even for performance in highly intellectual jobs like scientific research. For example, an excellent study by Taylor, Smith, and Ghiselin indicates that superior performance in scientific research has no significant relation to college grades.[50] What is the explanation for this counterintuitive fact? We would not be surprised if there were a common emphasis on relatively unimportant skills both in our schools and in the tests which are designed to predict success in them.

Of course, we are not denying that intelligence itself may have little to do with success within occupations (even intellectual ones) and much more to do with success in school. Were there evidence of this, the correlation of IQ with grades might be good reason to think that IQ tests measure intelligence. However, such evidence probably awaits an independently validated measure of intelligence.

Correlations of IQ with Proficiency within Occupations. Recent reviews have shown, as Jensen observes, "surprisingly low correlations between a wide variety of intelligence tests and actual proficiency on the job. Such correlations average about .20 to .25, and thus predict only four or five percent of the variance in work proficiency."[51] McClelland argues persuasively that even these "surprisingly low" correlations are inflated (*see* n. 48). Supervisors' ratings are often used as a measure of job proficiency, and these are sometimes biased by the supervisor's knowledge of the subject's education, social class, or even his IQ; or by his verbal skills in talking about his work.

Since both IQ and grades fail to correlate well with occupational proficiency, it would seem that either intelligence is not very involved in occupational proficiency, or else neither IQ nor grades have much to do with intelligence.

The Correlation of IQ with Schooling. The correlation of *adult* IQ with highest grade of school completed (schooling) is substantially higher than the correlation of *childhood* IQ with schooling. This is not surprising since the tests are in large part tests of school learning. One would expect (or at least hope) learning to be correlated with schooling. Thus the correlation of childhood IQ with schooling is of more interest from the point of view of validation. Jencks estimates a correlation of .58 between a sixth-to-eighth-grade IQ and schooling. We shall argue that this correlation is mainly a result of one or another sort of artifact, and is thus nonvalidatory (i.e., is useless for validation). This is important, not because of its independent interest, but because it will help us to show that the correlation of IQ with occupational status is also probably nonvalidatory.

Artifact 1: Score Causation. While our belief is not anchored firmly in data, we think it is likely that a substantial portion of the correlation between IQ and schooling is due to the effect of people *knowing* the *score itself* during a student's stay in school. We shall call this type of artifact "score causation." The most obvious type of score causation is admissions decisions based on aptitude or IQ scores. Because aptitude scores are good predictors of freshman grades, they are the major factor by far in college admissions and are also important for admission to a wide variety of professional and graduate schools.[52] Obviously, insofar as people are admitted to college or graduate school *because* of administrators' knowledge of their scores, correlation between scores and attending college is nonvalidatory. On the other hand, we have no idea just how large an effect scores have on the quantity of schooling people receive. Perhaps they affect the quality more than the quantity.

Another plausible type of score causation depends on the seriousness with which teachers, parents, guidance counselors, and students themselves tend to take aptitude-test scores, especially in considering whether a student should drop out of or go on in school.

Another score-causation mechanism which *is* backed up by data involves the "tracking" system. About 85 percent of high-school districts put students in separate programs called "tracks," or separate sections of the same course on the basis of "ability" judged to a large extent by IQ and aptitude tests. Students placed in noncollege tracks are overwhelmingly less likely to go to college or even to want to go to college. Using a mathematical technique called "path analysis," Jencks concludes that "curriculum placement seems to explain between a third and a sixth of the relationship between test scores and a student's chances of entering college." (See *Inequality,* pp. 34, 145.) Finally, insofar as IQ affects grades, and grades affect schooling, other things being equal, IQ will affect schooling. (This form of argument must be employed with caution.) But if we are right in arguing that the correlation of IQ with grades is nonvalidatory, then the portion of the IQ-schooling correlation due to the effect of grades is also nonvalidatory.

Let us be clear about what this sort of point does *not* imply. First, insofar as the correlation between IQ and any sort of social reward is due to score causation, this does not show that the correlation represents a nonmeritocratic aspect of our social system. Insofar as IQ is a characteristic deserving of reward, reward mechanisms which in effect distribute rewards on the basis of IQ *scores* are meritocratic. Nonetheless, insofar as the correlation is due to score causation, it obviously cannot be used to validate IQ tests. Second, we are not saying we know that the correlation of IQ with schooling would be lower without score causation. Indeed we do not know this.[53] What we *are* saying is that to the extent that score causation exists, the great mass of data showing a large correlation of IQ with schooling cannot be used to validate IQ tests.

Artifact 2: Noncausal Contribution. We have just argued that much of the causal contribution of IQ to schooling is nonvalidatory. But even more importantly, it appears that about half of the correlation between IQ and schooling is *not causal at all,* but rather is due to the effect of variables which themselves affect both IQ and (independently) years in school.

Children with high IQs have a double advantage as far as staying in school is concerned. First, their IQ causes them to stay in school longer because of the artifact of score causation, and insofar as IQ tests do measure, to some degree, abilities useful in school (e.g., intelligence) through the effects of these abilities. Second, high-IQ children tend to have home backgrounds with economic and cultural characteristics that will tend to keep them in school. They are apt to have sufficient funds (so that quitting school to earn money is not imperative), they are apt to live in a place where their friends and neighbors go to college, and perhaps most importantly, their parents are apt to pressure them to go on in school.

Jencks used path analysis to divide the correlation of IQ with schooling into two parts: one, ascribable to the causal effect of IQ on schooling and the other to background factors associated with IQ which affect schooling independently of IQ. He estimates that two elementary school children born during World War II with IQs differing by 15 points would differ by 1.6 years of schooling. He then calculates that if the two children *also* had similar *economic* backgrounds, their expected educational difference would be 1.25 years. If they were raised in the same *family*, their estimated educational difference would be about .9 of a year. He concludes: "This suggests that the actual effect of cognitive skills [IQ] on educational attainment is around half of the observed correlation. The other half of the association arises because high test scores are related to having the right parents."[54]

If Jencks's analytic models are right, the available evidence suggests that the relationship between IQ and schooling is largely noncausal, and even the causal component may have a large artifactual element. Thus, the correlation of IQ and schooling is probably mainly nonvalidatory.

Correlation of IQ with Occupational Status. Mathematical techniques such as path analysis and linear regression analysis have shown that insofar as IQ affects occupational status, it is mainly because IQ affects schooling and schooling affects occupational status. That is, almost all the causal action of IQ on occupational status is *via* schooling. As Jensen puts it:

> Duncan's detailed analysis of the nature of the relationship between intelligence and occupational status led him to the conclusion that "the bulk of the influence of intelligence on occupation is indirect, via education."
>
> If the correlation of intelligence with education and of education with occupation is, in effect, "partialed out" the remaining "direct"

correlation between intelligence and occupation is almost negligible.[55]

Our argument that the correlation of IQ with occupational status is probably mainly nonvalidatory is simple. First, a substantial part of the causal effect of IQ on schooling is probably due to score causation. Since almost all of the causal effect of IQ on occupational status is via schooling, a substantial part of the causal component of the correlation of IQ with occupational status is probably nonvalidatory. Second, a large part of the effect of schooling on occupational status may be due to educational *credentials,* and not to the skills or knowledge acquired in school. Third, only about half the correlation of IQ with schooling is causal, and thus probably much of the correlation of IQ with occupational status is not causal. Indeed, a recent article by Bowles and Nelson shows that only about half the correlation of IQ with occupational status is due to the causal effect of IQ on occupational status.[56] But surely, insofar as the correlation of IQ with occupational status is not even causal, this correlation cannot be used to validate IQ tests. We conclude that the correlation of IQ with occupational status is probably mainly nonvalidatory.

We remarked earlier that Herrnstein used data about the distribution of IQs in occupational categories to justify such conclusions as "half of mankind (roughly the half below IQ 100) is not eligible for auditing" (*IQM*, p. 123). We are not exactly sure what Herrnstein means by "eligibility," but he clearly intends some sort of causal relation between IQ and job category: e.g., that someone with an IQ below 100 is likely to be mentally incapable of competing successfully for an auditor's job. The results discussed in this section expose the weakness of such inferences and the importance of the old saw that correlation does not entail causation. Not only is the causal contribution of IQ to occupational category much smaller than the distribution of IQ in job categories would suggest, but also, insofar as people with IQ below 100 are ineligible for auditing because of people's knowledge of their IQ, this ineligibility is clearly artifactual.

V. How Correlations with Success Are Built into IQ Tests

We have been criticizing arguments that IQ tests measure intelligence. We now pause to fill a gap in an earlier argument. We have said that the correlations of IQ with measures of success do not confirm the claim that IQ tests measure intelligence because these correlations have been built into the tests. How did this happen? In the late nineteenth and early twentieth centuries, there were many attempts to construct tests of intelligence using mainly motor and sensory discrimination tasks, e.g., detecting differences in weight, or reacting to a sound.[57] Though all the pre-Binet tests were failures, some of the individual items they deployed were not; such successful items (successful from the point of view of correlation with "criteria of intelligence") provided a pool of material which was used by Binet in constructing his tests.[58]

This historical process provides one mechanism by which correlations with success are built into IQ tests. Survival of the fittest operates in this case to insure that items and types of items which survive have substantial correlations with a number of measures of success. Moreover, the same process operates at the level of whole tests. The history of IQ testing right up to the present is littered with the corpses of tests which were dropped because they failed to correlate sufficiently with measures of success (e.g., the Cattell-Wissler tests and the Davis-Eels games). Extant tests are those which have passed the correlational hurdles.

The construction of an IQ test can be viewed as the result of two processes: first, the initial selection of candidate test items, and second, the sifting of these items. It appears that both processes make use of data concerning correlation with success, although precise details of the construction of most tests are hard to come by. In particular, it is hard to know a test maker's basis for selection of candidate items. Some respected commentators are of the opinion that a (or *the*) major factor involved in choice of test items is whether an item seems likely to correlate with success. Thus, Duncan says, "When psychologists came to propose operational counterparts to the notion of intelligence, or to devise measures thereof, they wittingly or unwittingly looked for indicators of capability to function in the system of key roles in the society. . . . Our argument tends to imply that a correlation between IQ and occupational achievement was more or less built into IQ tests by virtue of the psychologists' implicit acceptance of the social standards of the general populace" (quoted, apparently approvingly, in *Gen. & Ed.*, p. 82). Certainly, it is undeniable that test makers' choices of items for consideration are influenced by how well types of items on past tests have correlated with measures of success. Moreover, David

Wechsler, the author of the most highly respected adult IQ test, is quite open about having chosen his items on the basis of the successes of past tests.[59] Terman and Merrill describe the second revision of the Stanford-Binet as follows: "The search for suitable material yielded thousands of test items, some of them of unknown value and most of them of unknown difficulty. The first principle of sifting was to give preference, other things equal, to types of test items that experience had shown to yield high correlations with acceptable criteria of intelligence."[60]

The use of correlations with success in weeding out test items is a complex business. For example, a test maker might try out his tentative items on two groups: (1) professionals with college degrees who had done well on another IQ test and (2) manual laborers who had not finished high school and had done less well on another IQ test. If the former group averages much higher on the set of items as a whole, while the latter group averages much better on, say, item 37, other things equal the test constructor would reject or modify item 37. In general, the factors most often used in sifting items in children's tests are age discrimination, teacher's judgments, school grades, and more commonly, scores on other IQ tests. In the case of adults' tests, the factors seem to be educational and occupational level, judgments of supervisors (e.g., army officers and business executives), and probably most commonly, scores on other tests.

It may be objected, at this point, that our practice of lumping educational success and occupational status and income together under the heading of "success" is misleading. While correlation with occupation and income may play a major role in the selection of items, occupation plays only a small role in the sifting of items, and income per se probably plays no role at all. So (it may be argued) the correlations of IQ with occupation and income are not as clearly built-in as correlation with educational success.

Our reply is as follows. First, techniques such as linear-regression analysis and path analysis have shown to the satisfaction of virtually everyone in this field (including Jensen) that insofar as IQ causally affects occupational status, it does so by affecting educational level attained (see section IV). Educational attainment, in turn, is the factor with the largest demonstrable contribution to occupational status. Thus, other things equal, a test which has a correlation with educational attainment built into it will also have a correlation with occupational status built into it. But the correlation with educational attainment is itself built in: in part directly, as mentioned above; in part by building in correlations with other types of educational success which are intimately related to attainment; and in part, by a mechanism discussed in section VI.

Second, the *other* IQ tests used in sifting items are ones which have "survived" and thus correlate with occupational status and income. So the use of other IQ tests in sifting items contributes to building into the test correlations with occupational status and income. The correlation of IQ with income is much lower than the correlation of IQ with occupational status and educational attainment: .35 versus .5 and .7, though the independent effect of IQ on income is proportionately greater than the independent effect of IQ on occupation. Given the substantial correlations of occupational status and educational attainment with income, it would be difficult to design a test which correlated well with these variables but had a negligible correlation with income.

VI. Does Age Discrimination Add Anything to Validation Arguments?

There is a validation argument which is applied only to children's tests. It appeals to Binet's exploitation of the fact that as children age, their ability to perform a wide variety of intellectual tasks increases. For example, at one age, most Western children can copy a square but not a diamond; a year later, most can copy a diamond as well. Items on IQ tests for children under fifteen are still chosen partly on the basis of how well they discriminate among children of different ages. A good test item will be one which can be done by less than half the children at age n but most of the children at age $n + 1$. The first two versions of the Stanford-Binet calculated IQ for children as mental age divided by chronological age times 100. IQ is no longer calculated in exactly this way, but the idea is the same: A child's IQ is a matter of how advanced he is for his age on those tasks which distinguish most sharply among children of different ages—and also satisfy other psychometric criteria, like correlation with success and reliability.

Defenders of IQ tests often suppose that Binet's technological innovation embodies a great insight (*see IQM*, p. 67). But there are good reasons to doubt that age discrimination can add anything to the correlational validation arguments we have been criticizing. These tests were, and still are, constructed without the benefit of anything approximating a theory of intellectual development. Without a good theory of intellectual development, and we do not believe a good theory exists, we simply have no way of knowing which age trends in answering questions correctly reflect basic intellectual changes, and which age trends reflect aspects of maturation which are irrelevant to intelligence.

A group of British psychometricians developing a new IQ test freely admit: ". . . none of the existing scales are based on any recognized theory of intellectual development. . . . The placement of items at a given age level has depended on that item meeting statistical requirements rather than psychological criteria."[61]

Terman was well aware of the problem we are discussing. He used *mental* age gradients—as measured by previous IQ tests—to sift items. In *Measuring Intelligence* (p. 10), Terman and Merrill say:

> Increase in percents passing at successive chronological ages is indirect but not conclusive evidence of validity. Height for example increases with age, but is known to be practically uncorrelated with brightness. Increase in percents passing by mental age is better, but exclusive reliance upon this technique predetermines that the scale based upon this criterion will measure approximately the same functions as that used in selecting the mental age groups. In the present case, this was not objectionable, since the purpose of the revision was to provide scales closely comparable to the old with respect to the mental abilities tested.

What about the *first* Stanford-Binet, the one used to select the mental-age groups? In that test, Terman selected among items whose answers improved with age by using *Binet's* tests.[62] Binet's procedure, however, was to use criteria such as teachers' judgments, educational attainments, and so on. Thus even in the view of the constructors of these tests, age discrimination was only a necessary condition of inclusion.

But what reason is there to believe that adding age discrimination to the set of criteria used for selecting items makes the test any better as a test of intelligence? In our view, this question depends on two other questions. (1) What reason is there to believe that correlations with school success, teachers' judgments, and so on can be used to select, from among age changes in children's responses to test items, those age changes which reflect basic intellectual changes rather than intellectually irrelevant aspects of maturation? (2)

What reason is there to believe that intelligence corresponds to rate of basic intellectual development? We know of little reason to have much confidence in either supposition. While many people in this field seem to be aware of the problem posed by (1), we have seen no discussion of the problem posed by (2). Question (2) could turn out to be more serious because even if we did develop a theory of intellectual development, and even if it proved possible to construct a Binet-type of test using basic intellectual changes,[63] we might still have little reason to believe the test measured intelligence. Even an ideal Binet-type children's test presupposes that children who are both smart and slow in basic development and children who are both stupid and fast in basic development are rare. We know of no good reason to believe this is true.

VII. The Implicit Theory and What Is Wrong with It

Most items on standard IQ tests require some degree of knowledge or developed skills. For example, the Wechsler tests for both children and adults contain eleven subtests, among which are a subtest of miscellaneous information (e.g., Who wrote *Faust*?); a vocabulary subtest (e.g., What is the meaning of "microscope"?); a comprehension subtest (in which the subject has to explain, for example, the meaning of a proverb and answer, Why should people pay taxes?); an arithmetic subtest; and a subtest on which the subject has to compare the meanings of two words, e.g., "liberty" and "justice." Of standard IQ test tasks, only a few, e.g., rote memory for digits, seem to require virtually no knowledge or developed skills at all. The rationale apparently implicit in using this procedure seems to be that the knowledge and skills a person demonstrates at the time of testing mainly reflect something central to intelligence: his capacity to acquire knowledge and skills.

We do not mean to imply that test makers often have this theory in mind. Indeed, the rationale they usually give for including blatantly culture-loaded items is that scores on particular items of this sort correlate with the rest of the test, and thus can be seen to measure intelligence.[64] (Of course, one might equally well use that fact to argue that the rest of the test—which itself has no independent justification—does not measure intelligence either.)

The prima facie problems with the implicit theory are obvious enough. People with equal intelligence may differ in knowledge and skills because of differential access to cultural information due to differences in parental vocabulary and knowledge; presence of books, magazines, and newspapers in the home, and quality of local schools; differences in parental encouragement of intellectual and scholastic activities; desire to appear cultured or to do well in school, anxiety about culture or school, attentiveness, and so on.[65] Of course, many such factors may be only slightly independent of others. Further, these factors may affect intelligence as well as IQ. But insofar as IQ scores reflect differences in opportunity, interest, and so on, over and above differences in intelligence, the tests measure opportunity, interest, and so on, rather than intelligence.

Suppose you wanted to judge a person's talent for typing. It would not be enough to compare his speed and error rate on a standard passage with the scores of others. You would also want to know how much typing training and practice he had, the degree of interest and energy with which he approached the training, and how well he knows English. The most obvious fault of the kind of IQ-test item we are talking about is that it makes use of only the first type of information: present skills. We do not deny that an IQ score might be useful to a counselor in judging a person's mental capacities if the counselor already knows a great deal about the person's background, personality, and motivation. But in such a case, the counselor's detailed knowledge would probably already include the information an IQ test would provide. The great evil of IQ tests is that they are supposed to measure intelligence independently of such background information.

We have been arguing that people of equal intelligence probably differ in the knowledge and skills they demonstrate on IQ tests simply because they probably differ in knowledge and skills. A more insidious fault of IQ tests is that people with the same knowledge and skills (and equal intelligence) probably differ markedly in their ability to *demonstrate* this knowledge and these skills on an IQ test for reasons whose connection with intelligence is tenuous. People differ in their anxiety in the testing situation, their "test-wiseness," their desire to do well on the test, their response to the tester, their concentration, persistence, mental speed, distractability, tendency to guess, tendency to "give up" on hard questions, tendency to check over answers, and even familiarity with paper and pencil.[66]

In *The Inequality of Man* (p. 63), Eysenck says:

> ... the general outcome of the work done along these lines has been quite clear-cut. It appears that performance on IQ tests can be accounted for in terms of three major mechanisms: mental speed, error-checking, and persistence or continuance. These mechanisms are independent of each other, although this depends to some extent on the instructions given, the motivation present, and other extraneous variables which can be manipulated experimentally. Just as the atom can be analysed into protons and electrons, and many other constituents, so the IQ can be split into these three major components; for serious theoretical work, and I suspect for many practical purposes too, it would seem that we ought to think in terms of speed, error-checking, and continuance, rather than simply in terms of IQ.

Eysenck's claim, if true, would seem to us grounds for thinking IQ tests do not measure intelligence. We are not sure whether persistence and error-checking should count as personality-motivational-temperamental characteristics, but they certainly are not central to the ordinary concept of intelligence. It is common to say that a person is bright, though intellectually lazy and careless. Mental speed is more plausibly part of intelligence. But surely, some people are quick without being smart while others have enormous mental capacities though their thought processes are slow. In the year of publication of *On the Origin of Species*, Charles Darwin wrote a letter in which he said, "I suppose I am a very slow thinker." Darwin's son Francis remarks that Charles Darwin "used to say of himself that he was not quick enough to hold an argument with anyone, and I think this was true."[67]

Of course, it *could* turn out to be the case that the phenomena we think of as involving intelligence can all be explained in terms of persistence, error-checking, and mental speed, but if so, we should be in favor of dropping the word "intelligence" altogether. Further, the educational and social implications of group differences in persistence and so on would be quite different from the implications of group differences in a characteristic of the kind intelligence appears to be.

Thus far, we have pointed out that IQ tests measure qualities other than intelligence (though we have not said that they do not measure intelligence as well, to some extent). However, it is also plausible that IQ tests *fail* to measure important aspects of intelligence. The practical purpose of IQ tests—evaluating intelligence reliably in an hour—has been taken to require a large number of items, each of which can be dealt with quickly. The hope is that with a large number of items, accidental advantages on particular items will tend to cancel out over the test as a whole. This has led to the exclusion of problems which require deep sustained thought. A recent review of tests of the type of problem which requires the subject to "restructure the elements of the problem" concludes that this type of problem-solving has a very low correlation with IQ.[68] There are psychometric tests which are supposed to measure creativity. If they do, then IQ tests do not. In *The Psychology of Human Differences,* Tyler observes: "One interesting fact has turned up again and again in these and other studies. There is only a very low relationship between these creative abilities and intelligence as ordinarily measured" (p. 413). Wallach and Kogan administered a battery of "creativity" tests designed to minimize the kind of test anxiety present in the usual psychometric situation; they found tiny (less than .1) correlations between creativity and IQ tests.[69] While a high correlation does not show two tests measure the same thing, a low correlation shows they do not. Even if the Wallach and Kogan tests do not measure creativity, they have at least as much of a prima facie claim to measure an interesting aspect of intellectual ability as IQ tests do. The low correlation of the two types of tests suggests that, insofar as IQ tests measure intellectual abilities, they measure only some of them.

The sorts of criticism we have sketched in this section are by no means new. Indeed, many constructors of tests and some widely used textbooks agree with much of what we have said. For example, Binet wrote: "There are tests which require knowledge outside the intelligence of the child. To know his age, count his fingers, recite the days of the week indicate that he has learned these little facts from his parents or friends; we have thought well to suppress these three tests."[70] Nonetheless, the current Stanford-Binet asks children to name the days of the week, mainly because this task correlates highly with the test as a whole. Wechsler notes that "neurotic subjects do badly [on the digit-symbol substitution test, one of Wechsler's eleven subtests] because they have difficulty in concentrating and applying themselves for any length of time and because of their emotional reactivity to any task requiring persistent effort." He also says of his arithmetic subtest, "its merits are lessened by the fact that it is influenced by educational and occupational pursuit. Clerks, engineers, and businessmen usually do well on arithmetic tests while housewives, day laborers, and illiterates are often penalized by them." Of his picture-completion subtest (the subject has to say what is missing in a picture) Wechsler says: "But one must note again, that the ability of an individual to do this depends in a large measure upon his relative familiarity with the object with which he is presented, that is to say, upon the actual content of the picture."[71]

Anastasi says, "It is now widely recognized that an individual's performance on an aptitude test, in school, on the job, or in any other context, is significantly influenced by his achievement drive, his self-concept, his persistence and goal orientation, his value system, his freedom from handicapping emotional problems, and every other aspect of his so-called personality." Why do many people who know these criticisms accept IQ tests as measures of intelligence? Probably the most common reply to the charges we have sketched above is that the differences in people that underlie differences in IQ scores—differences in cultural background, parental encouragement, school quality, attitude toward school, attentiveness, speed, ability to concentrate, persistence, tendency to check for errors—are the very differences in people which determine differences in success in life. People who take IQ tests *have* developed minds, and the point of the tests is to assess those aspects of their developed minds which lead to success in the world. For example, Anastasi, one of the psychometricians most insistent on dropping labels such as "intelligence test," observes:

> . . . if we were to rule out cultural differentials from a test, we might thereby lower its validity against the criterion we are trying to predict. The same cultural differentials that impair an individual's test performance are likely to handicap him in school work, job performance, or whatever other subsequent achievement we are trying to predict.
>
> Tests are designed to show what an individual can do at a given point in time. They cannot tell us *why* he performs as he does. To answer that question we need to investigate his background, motivations, and other pertinent circumstances.[72]

If this reply is used to defend the claim that IQ is intelligence, it is totally inadequate—unless the defender deploys the word "intelligence" in a very peculiar way: e.g., to refer to a "pure disposition" to succeed. As we pointed out in section II, whether a measure which correlates with success is a measure of intelligence depends on *why* it correlates with success. For example, a measure of a child's parents' socioeconomic status is a measure of something which helps him to succeed in school and in life, and thereby correlates with success, but it is not thereby a measure of intelligence. Insofar as people of equal intelligence differ in IQ because they differ in cultural opportunity, anxiety, and persistence, IQ tests fail to measure intelligence—even if the quantities they do measure are important to success.

Part of the point we are making here is that there is a good prima facie case that IQ tests measure nonintellective characteristics to a large degree. Some psychometricians respond to this by claiming that there is no real difference between intellective and so-called nonintellective characteristics. Cronbach and Anastasi regard this view favorably, and Wechsler says that "temperamental or personality factors . . . must be recognized as important in all actual measures of intelligence," but he goes on to describe these temperamental and personality factors somewhat paradoxically as "the non-intellective factors in general intelligence." However, as we pointed out earlier, if words are to mean anything at all, "intelligence" must be used to refer to intellectual or cognitive characteristics as opposed to personality, motivational, and temperamental characteristics. Wechsler's arguments for "impulse, instinct and temperament as basic factors in general intelligence" seem to be as follows. (1) IQ tests measure intelligence by definition; they also measure personality and temperament; therefore, personality and temperament, are part of intelligence. (2) "Personality traits *enter into* the effectiveness of intelligent behavior, and, hence, into any global concept of intelligence itself."[73] The first argument involves the kind of operationalism we criticized earlier. The second argument seems to presuppose the very distinction it is intended to blur; further, it is hard to see what the "global" concept of intelligence could come to if not some sort of behaviorist concept of intelligence.

Whatever one thinks of Wechsler's arguments, it would be foolish to insist that the distinction between intellective attributes and personality-motivational-temperamental attributes could not turn out to be a distinction without a difference. Certainly, many characteristics do not fit neatly into either the intellective or the nonintellective category. The ability to understand general relativity, for example, is usually a matter both of cognitive abilities and of concentration and perseverance. On the other hand, to conclude from this that there is no difference between traits of intellect and personality would be a simple fallacy.

It is this sort of question which most strikingly reveals the primitive nature of current psychological theory in this area. Whether developed psychological theory will indicate that there is a real difference between qualities of intellect and qualities of personality, motivation, etc., is an open question. In either case, however, Wechsler's talk of "the non-intellective factors in general intelligence" is unacceptable. If there *is* a real difference, the tests cannot be claimed to be intelligence tests because they probably measure mainly other characteristics. On the other hand, if there is no difference, the tests cannot be claimed to be "intelligence" tests either. For the common usage of "intelligence" presupposes the very distinction which is claimed to be faulty. Psychometricians who do not believe there is any difference between intellective and "nonintellective" attributes should drop words such as "intelligence" altogether (indeed, some have).

Since differences in IQ scores probably reflect all sorts of differences other than intelligence, it could be said that IQ tests are *biased* against people with certain cultural backgrounds, personality characteristics, and so forth. Arguments in the psychometric literature about test bias often center around the question whether IQ tests are biased against certain groups on the grounds of socioeconomic status. But for present purposes, this would be a misplaced emphasis. Socioeconomic background is not the same as cultural background; and it may be that most of the variation in cultural background in this country (at least insofar as it influences IQ) is *within* socioeconomic groups. For example, virtually whatever it is that IQ tests measure, it is overwhelmingly plausible that the IQ of a child's parents would be a major component of aspects of the child's cultural environment which affect his IQ. Vocabulary subtests are usually the most predictive subtests of IQ tests, and information subtests are often second. But surely the *parents'* vocabulary and store of miscellaneous information would be an influence on those of the child—insofar as his cultural environment influences him at all. But *three-quarters* of the IQ variation among adults (and even more of the variation among children) is *within* socioeconomic groups.

We suspect that some psychologists might reply to our argument by saying that 75 percent of the variance in IQ is due to genetic variation. (This is the figure Jensen and Herrnstein usually mention as the best current estimate.) Then, at most 25 percent could be attributed to IQ tests measuring cultural background—and not even that, since some of the environmental component would be "used up" by such biological factors as womb environment, and by test error. We certainly concede that the heritability of IQ provides an upper bound on the degree to which the tests could simply be measures of cultural background; however, we doubt that this upper bound is uncomfortably low.

First, recent reviews by Jencks and Kamin[74] lead us to believe that any estimate of the heritability of IQ based on current data alone is a guess. Our guess would be that it is below 50 percent. Second, we are by no means suggesting IQ tests measure just cultural background. Our view is that IQ differences are probably due to differences in many qualities; some, like cultural background, are nonheritable. Others are possibly heritable; for example, personality, motivation, temperament, special abilities (e.g., verbal and perceptual), and even general cognitive ability, if there is such a thing. Indeed, the sort of study which is supposed to show that IQ is largely heritable is also supposed to show that personality-motivation characteristics are substantially heritable. So even if IQ is largely heritable, if personality-motivation tests measure what they are supposed to measure, and if IQ tests measure personality-motivation characteristics substantially (as we have argued), the heritability of IQ could be, in large part, a matter of the heritability of personality and motivation. Further, studies which claim to estimate the heritability of particular abilities indicate that spatial visualization, word fluency, and vocabulary have the highest heritabilities (in that order) of all abilities while reasoning ability—whose centrality to intelligence seems far greater than verbal and spatial ability—has low heritability.[75]

At this point, the objection may arise that we are just making a semantic point about the word "intelligence." Our position is that probably IQ tests measure mainly sociocultural background, personality-motivational-temperamental characteristics, and cognitive abilities (and perhaps other independent quantities). Such measures may well be justified for some of the current uses of IQ tests. But we must remind the reader of such suggestions as Shockley's—that low IQ persons be encouraged to undergo sterilization—and Jensen's—that education for the low IQ disadvantaged be directed toward associative learning instead of conceptual cognitive learning.[76]

In the context of current controversies over eugenics, support of poverty programs and educational policy, our criticisms of IQ tests can hardly be termed "semantic." Further, questions of the sort we are raising are important to a number of educational and social matters which are of interest somewhat independently of current controversies. They are also crucial to the issue of whether we want to raise IQ levels and, if so, how this can be done. For example, if IQ tests measure personality-motivational-temperamental characteristics substantially, perhaps something like an encounter group would be more effective than traditional cognitive exercises in changing IQ. Furthermore, if IQ is important in the learning process, then knowledge of what IQ tests measure may be important in designing educational methods for teaching children of different IQs. More importantly, if our criticisms of IQ tests are correct, they would seem to throw some doubt on common assumptions about the connection between IQ and learning. Also, whether IQ tests measure nonintellective characteristics, and if so, what kind of nonintellective characteristics they measure is relevant to whether social and economic rewards ought to be, in effect, distributed partly on the basis of IQ. Finally, our criticisms are relevant to those who have low IQs. Whether they are branded by themselves and others as *dumb* matters.[77]

Race and Class Bias. Are IQ tests biased toward race or class? First, what is class or race bias? The answer to this question depends somewhat on the questioner's purpose. If one is asking—as Jensen does—whether whites are on the average more intelligent than blacks, the relevant concept of bias is clear: IQ tests are race-biased just in case there is a pair of races such that the racial IQ difference is not the same as the racial intelligence difference. Let us imagine we have a perfect intelligence test scored on the same scale as an IQ test. Suppose for the sake of argument that Jensen is right that blacks are lower than whites in average IQ by 15 points. Then IQ tests are biased against blacks just in case whites are less intelligent than blacks or less than 15 points more intelligent.

Given the 15-point black-white IQ difference, what can we infer about the likely black-white intelligence difference? Without some assumption about the kind of relation that obtains between IQ and intelligence, we can infer nothing at all about black and white intelligence. Even if we assume the correlation of IQ with intelligence is, say, .5, we can infer nothing about the expected black-white intelligence difference without an assumption about the degree of linearity (*see* Appendix) of the IQ-intelligence relation. The correlation coefficient yields the best linear prediction, not the best prediction.

We know of no good reason to think linearity obtains. But since the assumption of linearity is favorable to IQ tests and is an assumption we suspect most psychometricians would be willing to make, we shall assume a roughly linear relation in what follows. Given linearity and the obvious fact that IQ tests are less than perfect as measures of intelligence, a rather surprising conclusion follows: If IQ tests are color-blind on an *individual* basis, they are likely to be biased against racial *groups*, such as blacks, which score below the population mean. The argument for this is slightly technical, but because of its importance, we shall sketch it below.

First, what are we supposing in supposing tests are color-blind on an individual basis? We mean the tests measure intelligence equally well in individuals irrespective of color. That is, IQ "predicts" intelligence in the same way for blacks as for whites; if we choose a black and a white randomly, and if they have the same IQ, their expected intelligence will be the same.

Suppose the correlation of IQ with intelligence is zero. Then IQ would show nothing about intelligence. So the expected intelligence of a randomly chosen person with an IQ of 85 will be the mean: 100. On the other hand, if the correlation of IQ with intelligence is 1.0. then IQ will predict intelligence exactly; so, the expected intelligence of a person with an IQ of 85 will be 85. If the correlation of IQ with intelligence is between 0 and 1.0, as it presumably is, then the expected intelligence of a person with an IQ of 85 is *between* 85 and 100. For example, if the correlation of IQ with intelligence is .5, and if all we know about a person is that his IQ is 85, his expected intelligence will be 92.5, halfway between 85 and 100. (Here we are assuming perfect linearity, but only for illustrative purposes.)

Thus, on the average, people with below-average IQs have their intelligence *underestimated* by IQ tests. Now, there is no reason to believe blacks have an average intelligence any lower than the IQ 85 population as a whole. Indeed, if blacks differ in intelligence from the IQ 85 population as a whole, they are probably more intelligent, not less (*see* the discussion of Mercer's results below). Hence, the expected intelligence of blacks is no less than that of a group of persons chosen at random which happens to have an average IQ of 85. It follows that the black intelligence expected on the basis of IQ is higher than 85. If the IQ-intelligence correlation is .5, the black-white intelligence gap expected on the basis of IQ is half the black-white IQ gap. And this is predicted on the assumption that IQ tests are color-blind to individuals. If indeed IQ tests are biased against blacks on an individual basis, the expected black-white intelligence gap would be still smaller, or even reversed.

One way to get an intuitive appreciation of this phenomenon is as follows. Since IQ tests are obviously not perfect measures of intelligence, differences in IQ reflect differences in characteristics other than intelligence. For purposes of exposition, suppose what IQ tests measure is intelligence plus one other quality: test-wiseness. Now if IQ tests are fair on an individual basis, a person with a below-average IQ, whatever his color, on the average will be below average in intelligence too. But he will *also* be below average in test-wiseness. People who are below average in test-wiseness have their intelligence underestimated by IQ tests relative to persons who are average or above in test-wiseness. So those who are below average in IQ lose out as a group, both fairly (because of their below-average intelligence) and unfairly (because of their below-average test-wiseness).

Many people find this sort of argument confusing because they notice that correlation is a symmetric relation. For example, there is a .5 correlation between the IQ of a mother and her offspring. One might argue, if a mother's IQ is 200, the expected IQ of her children is 150 (halfway between 200 and 100). But if a child has IQ 150, one could equally well infer that the expected IQ of his mother would be 125 (halfway between 150 and 100), not 200. This is quite correct; there is no paradox. If a randomly chosen child has an IQ of 150, the expected IQ of his mother is 125. If one chose a 150-IQ child on the basis of the child's mother having an IQ of 200, one would hardly have chosen the child randomly.[78]

Why is all this important? It is important because even if psychometricians have succeeded in making IQ tests color-blind on an individual basis, and even if the relation between IQ and intelligence is linear, given that IQ tests are less than perfect as measures of intelligence (as ought to be admitted by all parties), one can infer from their IQ that blacks and other such low-scoring ethnic and status groups are probably smarter on the average than IQ tests say they are. However, a test which is fair to blacks as a group would overpredict individual intelligence. A test cannot be fair both to low-scoring groups and to the individuals in these groups. The policy suggested by this point would seem to be different tests for different purposes.

In spite of the obvious importance of the point we have been making and the widespread awareness among psychometricians of Thorndike's formally analogous argument (see note 79), we have never seen our point made. The explanation is not hard to find. Because of the nonrealist perspectives (mainly operationalism and instrumentalism) prevalent in psychometrics, test bias is understood, not in terms of difference between IQ gap and intelligence gap, but rather in terms of differential prediction. For example, Anastasi says: "In the psychometric sense, test bias refers to overprediction or underprediction of criterion measures. If a test consistently underpredicts criterion performance for a given group, it shows unfair discrimination or 'bias' against this group."[79] But there is no necessary connection between the psychometric sense of bias and the sense we have been deploying, for why should we suppose the correlation of intelligence with measures of success is the same in different racial or status groups?

In any case, at least in the psychometric sense of "bias," it is now clear that IQ tests are biased against lower-class groups, especially lower-class blacks, at least in some respects.[80] Jane Mercer constructed a set of behavioral indicators of retardedness and administered it to a sample of predominantly lower-class blacks, Chicanos, and predominantly middle-class whites. She used an age-graded set of indicators, e.g., being able to tie one's shoe at age seven. For adults, the tests were such things as being able to travel alone or do one's own shopping, reading books or magazines, holding a job, and staying in school. Mercer's results were rather startling. Of blacks with IQ below 70, 91 percent passed it, 60 percent of Chicanos passed it, but *none* of the whites with IQs below 70 passed it.[81]

One conceptual confusion that often infects discussion of race and class bias is the confusion between bias and loading. A test is culture (or race or class) loaded in the clearest sense if it requires knowledge available only in a certain culture (or race or class). Thus a test could in principle be culture-loaded but not race- or class-biased if all races and classes who take the test share the same cultural access to the knowledge (*IQM*, p. 72). Further, a race-biased test need not in principle be culture-loaded. For example, brain-wave readings in persons responding to flashes of light have been claimed to correlate with IQ. If such a measure were used as an IQ test, it could be biased in the sense of ranking blacks below whites of equal intelligence, even though it is not culture-loaded.

Standard IQ tests are without any doubt highly culture-loaded.[82] In our view, they are all also clearly culture biased in that they require knowledge of, for example, literary, musical, and geographical facts which are differentially available to people with different sociocultural backgrounds. There are many reasons other than those mentioned for being wary of class and (especially) racial comparisons of IQ. First, it seems clear that there are substantial differences in racial and status groups in opportunity to acquire the kind of knowledge and skills useful on IQ tests. Second, even the best standard IQ tests contain items on which performance differences would seem due in part to differences in culturally specific values or practices. For example, the Stanford-Binet shows four-year-olds three pairs of pictures asking, Which one is prettier? In all cases, the "correct" picture has the classic Anglo-Saxon features, while the "incorrect" one has features common to other ethnic groups, e.g., a wide, flat nose and thick lips in two cases, and a hooked nose in the last case. Third, though the results on this are very inconsistent, it does seem as if disadvantaged children are affected more by features of the testing situation.[83]

Fourth, the intellectual "profiles"—the pattern of developed abilities—of different racial or status groups may be different. But what abilities are counted as part of intelligence is a cultural matter. Hence, even if a test designed to measure the abilities culture A counts as intelligence measures perfectly in both culture A and culture B, the test may in a sense be biased against culture B.

This point is strengthened by a look at the kinds of items that are actually included in IQ tests. Different groups are known to perform differently on different types of items. For example, men and Orientals do better on certain types of spatial items but worse on verbal items. There are no significant race and status differences on some nonverbal types of items, though there are substantial differences on others. (*See* Jensen, "Another Look at Culture Fair Testing," p. 98.) The balance of different types of items on standard IQ tests is extremely arbitrary. On the first Stanford-Binet, girls were judged to be higher than boys in IQ by 2 to 4 percent at every age;[84] in subsequent revisions of this and many other tests, the types of items were manipulated so as to try to bring about equal averages for males and females. Thus roughly equal IQ means for males and females is essentially a *convention*. This point is inconclusive, however, as far as class and race differences are concerned, since no one has ever succeeded in constructing a test which eliminates black-white differences or class differences *and* satisfies standard psychometric criteria (*Ed. & Grp. Diff.*, p. 300). This should not be too surprising, however, since one of these criteria is correlation with various measures of success—measures on which blacks and low-status whites are substantially lower.

VIII. General Intelligence and Internal Validity

Facts about the correlation of IQ-test items and subtests with one another and with the whole test—usually deploying a set of mathematical techniques known as factor analysis—are often taken to show that about 50 percent of the variance in standard IQ-test scores is due to a general intellectual capacity called "general intelligence," "general mental ability," "general ability," or just *g*; and this is often taken to constitute a validation of IQ tests.[85] In this section we shall argue, first, that the rationale of the construction of IQ tests (as measures of intelligence) actually *presupposes* the existence of general intelligence. Second, there is no consensus even among factor analysts about the existence of general intelligence. Third, arguments for general intelligence contain numerous widely recognized problematic aspects and, finally, a version of fictionalism—a view related to the operationalist views discussed earlier—has played a major role in facilitating belief in general intelligence.

IQ Tests Presuppose General Intelligence. One argument, that the rationale of the construction of IQ tests as measures of intelligence presupposes general intelligence, is stated by David Wechsler:

> . . . so far as measuring intelligence is concerned, these specific tasks are only means to an end. Their object is not to test a person's memory, judgment or reasoning ability, but to measure something which it is hoped will emerge from the sum total of the subject's performance, namely, his general intelligence. One of the greatest

> contributions of Binet was his intuitive assumption that in the selection of tests, it made little difference what sort of task you used, provided that in some way it was a measure of the child's general intelligence. This explains in part the large variety of tasks employed in the original Binet scale. It also accounts for the fact that certain types of items which were found useful at one age level were not necessarily employed at other age levels. More important than either of these details is the fact that for all practical purposes, the combining of a variety of tests into a single measure of intelligence, *ipso facto,* presupposes a certain functional unity or equivalence between them.[86]

There is another feature of IQ-test construction whose rationale even more clearly presupposes the existence of general intelligence: substantial correlations of items with the test as a whole are *built in by simply eliminating items and subtests which have low correlation with the test as a whole and including items which would otherwise be highly suspect* (e.g., Wechsler's information test) *because they have a high correlation with the test as a whole.* Terman and Merrill state flatly: "Tests that had low correlation with the total were dropped even though they were satisfactory in other respects" (*Stanford-Binet Intelligence Scale,* p. 33).

Accounts of the construction of the Wechsler and Stanford-Binet tests show this principle has had enormous influence. To see how this procedure presupposes the existence of general intelligence, suppose there is such an ability as general intelligence. Then a test containing a wide variety of intellectual tasks should reflect general intelligence to a fair degree. So items which correlate *less* well with the whole test must involve relatively little general intelligence. Hence, throwing out items which correlate less well with the whole test will make the test an even better measure of general intelligence.

On the other hand, suppose there is no such thing as general intelligence. Then the practice of throwing out items which correlate less well with the whole test will have the effect of making the test a less adequate test of intelligence. For if there is no general intelligence, the best one could expect from an intelligence test would be an average over a wide range of mental abilites. But throwing out items which do not correlate well with the whole test would *decrease* the range of abilities over which the test would average, thereby making the test less broad, and a less adequate measure of intelligence conceived of as an average over a wide range of abilities.[87]

Given that the rationale for construction (or using) IQ tests as measures of intelligence presupposes general intelligence, a person who rejects general intelligence and thinks of intelligence as some sort of a composite of a wide range of mental abilities should not accept IQ tests as measures of intelligence. Thus we find Herrnstein's position somewhat strained. He discusses a number of views, some of which accept general intelligence, others of which reject it, concluding:

> For now, we cannot say which, if any, of the foregoing approaches to mental measurement will prevail, but perhaps further consensus is unnecessary. The I.Q. cuts across the fine structure of the various theories coming up with what is a weighted average of a set of abilities. Any modern intelligence-test battery samples so broadly that most of the abilities get tapped. And since the abilities themselves tend to be intercorrelated, an omission here and there will have little effect. The high correlations among full test scores for different sorts of intelligence tests bear this out. When the task is to get a single number measuring a person's intellectual power, the I.Q. still does the job, even with the proliferation of theories and tests. [*IQM*, p. 106]

But it is hard to share Herrnstein's confidence that an IQ test "samples so broadly that most of the abilities get tapped" given that one of the major principles of test construction may have precisely the opposite effect. Further, Herrnstein's premises are suspect. He appeals to the claim that different sorts of intelligence tests correlate highly. But the simple fact is that a test would not be counted as an "intelligence test" if it did not correlate well with accepted "intelligence tests." Herrnstein also appeals to the claim that abilities tend to be intercorrelated. But Guilford examined over 7,000 intercorrelations among tests of an intellectual nature, and found about 24 percent of these were close to zero.[88] Finally, it seems that Herrnstein's view that IQ tests measure a broad range of abilities is probably a minority view even in psychometrics. Indeed, the narrowness of IQ tests is one of the major reasons many psychometricians have abandoned labels such as "intelligence test." For example, Anastasi says: "It is certainly true that the limited sample of cognitive functions included in standard intelligence tests is inconsistent with the global connotations of the test names. This is but one more reason for discarding the label 'intelligence test,' as some psychologists have been advocating for several decades."[89]

There Is No Consensus Among Psychometricians About General Intelligence. Practically since the beginning of IQ testing, there has been a disagreement between psychometricians who, following Spearman, believed in general intelligence and those who, following Thorndike and Thomson, believed there is no such thing as general intelligence, but rather many individual abilities.

Jensen places himself firmly in the former camp. Indeed, though Jensen often talks as if he identifies intelligence with IQ, he repeatedly states that when he uses the word "intelligence" he refers to general intelligence or *g* (the term he prefers). He defines *g* alternatively as "the factor common to all tests of complex problem solving" and as "the general factor common to standard tests of intelligence" (*Gen. & Ed.*, pp. 77, 88). On the latter definition, one might wonder just how general Jensen's *g* is, given the points we made above about the probable narrowness of standard IQ tests. In any case, Jensen clearly states that there are aspects of mental abilities not covered by *g*. Indeed, his position appears to be that while *g* is something like abstract reasoning ability, IQ is a better predictor because it involves both *g* and other characteristics required in school and in life; he attributes only about *half* the variance in IQ to *g* (*Gen. & Ed.*, p. 77). Since he carefully identifies intelligence with *g*, and since most of what he has to say about intelligence is based on *IQ data*, one might well wonder why he thinks he has succeeded in talking about *intelligence*.

Jensen gives us the impression that the existence of general ability is now a well-established fact which no competent psychometrician really acquainted with the data could deny. For example, he says: "Despite the numerous theoretical attacks on Spearman's basic notion of a general factor, *g* has stood like a Rock of Gibraltar in psychometrics, defying any attempt to construct a test of complex problem solving which includes it" (*Gen. & Ed.*, p. 77).

But the fact is that there is no consensus about this matter among competent psychometricians. Even psychometricians such as Burt and Butcher who share Jensen's views about the existence of *g* and the relation of *g* to other abilities are willing to admit that psychometricians they respect disagree. Burt observes: "Perhaps the most illuminating publications on this subject that have appeared during recent years are Professor Guilford's papers on the 'three faces of intellect' culminating in his book *The Nature of Human Intelligence . . . Guilford finds no evidence for general ability.*"[90] In another recently revised article, Burt says that *g* is rejected by most American writers.[91] Most textbooks agree that there is a difference of opinion on this issue, with British psychologists tending to favor *g* and American psychologists tending to reject it. In discussing this dispute, Butcher (a British supporter of *g*) says in *Human Intelligence* that "there is little doubt that on the whole, with a few conspicuous exceptions, theories of general intelligence as the prime mover, perhaps modified or 'perpetuated' by subsidiary influences, have flourished in Britain, whereas the generally preferred picture in the United States has been of multiple abilities of more or less equal status and influence" (p. 51).

Thus far, we have argued that the rationale of the construction of IQ tests (as tests of intelligence) presupposes the existence of general intelligence and that psychometric factor analysts themselves do not agree about the existence of general intelligence. But why should we accept a view which presupposes a technical claim on which even the technicians are divided? The same objection applies to "factorial validation"—roughly the argument that IQ tests can be seen to measure intelligence because they measure general intelligence. The argument of this section could stop here, but we think it will be worthwhile to look into the logic of factorial validation in greater detail.

Problematic Aspects of Factor-analytic Arguments. The techniques of factor analysis, developed mainly by Spearman and Thurstone, are simply mathematical methods of finding patterns in large numbers of correlation coefficients. If one has a set of fifty tests, there will be 1,225 coefficients of correlation of each test with each of the others. If these coefficients cluster in some very simple way, one could ascertain the pattern by inspection instead of factor analysis. However, if the coefficients cluster in complex ways as they usually do, factor analysis or some similar technique is useful in finding patterns.

Factor analysts would be the first to agree that the techniques of factor analysis *alone* are insufficient to establish any of the interesting conclusions that have been based on them: additional assumptions are needed. Indeed, some of the weaknesses in factor-analytic arguments can be illustrated by "internal validity" arguments which deploy some of these additional assumptions without involving factor analysis at all. For example, F. S. Freeman makes the following argument (in his psychometrics text) on the basis of the correlation coefficients for each of the 258 items of the Stanford-Binet (the version before the current one) with the whole of the test: "Of the 258 coefficients, 201, or very nearly 78% are .50 or higher. This fact, and the [factor-analytic] data presented . . . later . . . provide strong evidence that the Stanford-Binet scale measures 'general ability' by means of test items that have psychological processes in common to a high degree." Terman and Merrill give the same argument for the validity of the current Stanford-Binet in the handbook for the test.[92] The reasoning behind this argument would seem to be: Since the items correlate with the whole test, they must (insofar as they correlate with it) measure the same ability as the whole test; but what single ability could be measured by each of such a set of heterogeneous items *and* the whole test is not general intelligence?

This reasoning has a number of serious weaknesses. First, it argues for the validity of the test while at the same time it presupposes something about the validity of the individual items.[93] But many of the reasons we have given for skepticism about the validity of the test as a whole—e.g., that the test might equally well measure sociocultural background or nonintellective attributes—are also reasons for skepticism about the validity of the individual items and subtests. Second, it depends on the principle that insofar as two tests of abilities correlate, their correlation is due to a common ability measured by both tests. Let us call this the Correlation-Entails-Commonality Principle. We shall criticize it in detail later. Third, even if the individual items do measure a variety of abilities, and even if the Correlation-Entails-Commonality Principle were right, there would be no good reason to think the ability common to the items is anything deserving of the name "general intelligence." For as we pointed out above, the deliberate rejection of items with low correlations with the test as a whole simply presupposes the existence of general intelligence. If there is no such thing, even if the Correlation-Entails-Commonality Principle were right, the ability common to the items might be whatever specific ability was dominant in the original tests. This is certainly the view of many anti-*g* psychometricians. For example, Guilford says: "As it actually happens, a Stanford-Binet IQ or any IQ from a test whose components are predominantly verbal is a total score heavily dominated by the verbal-comprehension factor."[94]

Some of the additional problems in factor-analytic arguments for general intelligence are: (1) What sorts of clusters of correlations factor analysis yields depends on the initial choice of tests to factor-analyze. A test which according to a factor analysis based on one group of tests is measuring pure *g* may measure *g* to a small extent according to a factor analysis based on another group of tests (see *Gen. & Ed.*, p. 79). Guilford is responding chiefly to this fact when he says: "A 'general factor' found by whatever method is not invariant from one analysis to another, and hence fails to qualify as a unity, independent of research circumstances."[95] (2) What clusters of correlations factor analysis yields depends on the particular method of factor analysis deployed; a given set of correlation coefficients can in general be broken down into clusters in different ways. Different factor analysts deploy different techniques and thus arrive at different results.

Points (1) and (2) together are the basis for the charge that factor analysts to a large degree *assume* what they want to show in the first place. The factor analyst chooses the set of tests to factor analyze and he also makes the methodological choices of how to separate the set of correlations of these tests with one another into clusters. Thus, Herrnstein says of Burt's and Vernon's (and apparently Jensen's) hierarchical theory: "The ascending layers, essential to the idea of a hierarchy, emerge only from statistical procedures that, to some degree, assume their existence in the first place" (*IQM*, p. 97).

The Correlation–Entails–Commonality Principle. We observed above that factor-analytic arguments for general intelligence often are based on the Correlation-Entails-Commonality Principle (henceforth, the CEC Principle): Insofar as two tests of abilities correlate, this is due to a common ability measured by both tests. The obvious objection against the CEC Principle is that correlation is a necessary but not a sufficient condition for commonality. For example, imagine that a terrible tyrant decides to produce a correlation between jumping ability and speaking ability by picking people at random and damaging their tongues and their legs while picking others at random and giving them elocution and jumping lessons. Tests of jumping and speaking would correlate, but would not measure a common ability. Of course, one can *sometimes* infer commonality from correlation, e.g., in a case where commonality is the only plausible explanation of correlation. Our first point against the CEC Principle is that going from correlation to commonality *does* require argument.

Our second point is that in the case of psychometric tests which are supposed to measure particular abilities, there are *many* equally plausible hypotheses. Suppose (for the moment) that tests which are supposed to measure mainly intellectual abilities actually do. Nonetheless, each of the tests may sample a set of abilities, and the tests may intercorrelate in virtue of the overlap of these sets.[96] Another possibility is: whatever environmental conditions nurture one ability (e.g. mental and physical health, good nutrition, sociocultural and educational advantages, intellectual encouragement) also nurture a number of abilities. Conversely, whatever conditions hinder the development of one ability (e.g, disease, neurosis, being hit on the head or poisoned, malnutrition, sociocultural and educational disadvantages) hinder the development of a number of abilities. Personality-motivation traits like curiosity and achievement orientation may nurture a number of abilities—or their absence may hinder a number of abilities. Cronbach remarks:

> Factor analysts often speak as if they were discovering natural dimensions that reflect the nature of the nervous system. Correlations and factors, however, merely express the way performances covary in the culture from which the sample is drawn. The usual high correlation between verbal and numerical abilities is due in part to the fact that persons who remain in school are trained on both types of content. If a culture were to treat map-making as a fundamental subject, then map-making proficiency, it is suspected, would correlate highly with verbal and numerical attainments.[97]

Another hypothesis as to why abilities might correlate is that people who excel in one ability may tend to marry people who excel in another ability. Insofar as abilities turn out to be heritable, they might correlate for genetic reasons.

We have been supposing intellectual ability tests actually do measure mainly intellectual abilities. But as we pointed out in the last section, many test makers and widely accepted psychometrics texts agree that standard tests measure personality-motivational-temperamental characteristics to a fair degree. But insofar as characteristics such as concentration, persistence, and tendency to check for errors help, they are likely to help in many (or even all) subtests, thus contributing to the intercorrelations among subtests.

In the light of the number and plausibility of the alternative hypotheses, the widespread tendency among believers in general intelligence to take correlation as a sufficient condition of commonality calls for some explanation.

This tendency may be in part due to a fallacy based on the extended use of "measurement" common in psychometrics. If a test score *correlates* with *x*, the test is said to measure *x*. Thus if a test measuring arithmetical ability correlates with a test measuring boxing ability, it does indeed follow that there is an ability which both tests measure—namely arithmetical ability itself and in addition, boxing ability itself. What does not follow is that there is a single common ability of which both arithmetical and boxing ability are facets.

However, methodologically sophisticated psychometricians are not guilty of this fallacy; many of them subscribe to a version of fictionalism with respect to abilities (though not under that name) which is closely related to the operationalism discussed earlier. Fictionalism is the doctrine that the use of theoretical terms (e.g., "electron," "gene") in science should be understood as a way of speaking in which scientists engage in order to facilitate prediction and control, but that such terms are not to be construed as actually referring to things in the world. The mistake of taking theoretical terms at face value is described as "reifying." Fictionalism with respect to abilities is this doctrine applied to ability terms. Fictionalists usually come in two varieties: those who say electrons, for example, do not exist, and those who say electrons do exist but that statements about electrons should be analyzed in terms of what we might naïvely think electrons are the causes: meter readings, cloud-chamber tracks, and the like. Fictionalism with respect to states, events, and capacities is more likely to be of the latter variety. Thus fictionalism with respect to mental states, events, and capacities often amounts to a version of behaviorism: Statements about mental events, processes, and capacities are analyzed in terms of what we would naïvely think of as their behavioral effects.

Psychometricians have debated about fictionalism since the beginning of factor analysis. Spearman tended to take a realist position, thinking of general intelligence as "amount of general mental energy" while modern realists sometimes suggest general intelligence is something like information-processing capacity. Jensen appears to identify general intelligence with something like capacity for abstract reasoning and problem-solving. However, the fictionalist position now appears dominant, especially among those who believe in general intelligence.

Cronbach's textbook gives a clear statement of a version of the fictionalist position with respect to abilities (though not under that description):

> Looking at a collection of scores such as the Wechsler subtests, the psychologist must ask: just how many different abilities are present? *The word "ability" in such a question refers to performances, all of which correlate highly with one another, and which as a group are distinct from (have low correlations with) performances that do not belong to the group.* To take a specific example, Wechsler Vocabulary items call for recall of word meanings, and Wechsler Similarities items call for verbal comparison of concepts. Are these measures of the same ability? Or do some people consistently do well on one but not the other? [The reader should note the alternatives presented here.] For a group of adolescents, we have these data:
>
> Form-to-form correlation of Vocabulary on same day = 0.90
> Form-to-form correlation of Similarities on same day = 0.80
> Correlation of Vocabulary and Similarities = 0.52
>
> The two tests evidently overlap. About 52 percent of either test can be regarded as representing a shared ability or "common factor." Twenty percent of the Similarities variance is due to form-to-form variation. This leaves 28 percent that must be due to some distinct ability that is not tested by Vocabulary. Likewise, 38 percent of Vocabulary is due to an ability not involved in Similarities. There is a common factor of verbal facility or reasoning, but each test also involves something extra. Hence, the two tests do involve distinct abilities.
>
> Factor analysis works along these general lines, starting from correlations. Binet applied such reasoning when he decided that his tests, all having a substantial relation to each other, must reflect a pervasive general intelligence.[98]

Cronbach deploys the CEC Principle here in concluding that all of the variance common to the two tests is due to a single ability measured by each test. He makes the connection between the CEC Principle and behaviorism clear in specifying that he takes the word "ability" to refer to performances. It should be obvious why this position is incompatible with realism about abilities. We mentioned above a number of reasons other than common abilities why two tests might correlate. For example, vocabulary and similarities subtests may correlate in part because people who check their answers over carefully may tend to do better on both. Cronbach's position with respect to such an objection is that checking for errors can be viewed as a component of general intelligence—though it is equally correctly viewed as a component of each task-specific ability.[99]

What this fictionalism with respect to abilities amounts to is a decision to use the word "ability" in such a way that it becomes simply a stipulation that the variance which statistical terminology describes as "common" to two tests in virtue of their correlation is due to a common ability. We should remind the reader that "common variance" implies nothing at all about an underlying unity. If the thickness of ice at the North Pole varies with the seasons so as to correlate .7 with the varying number of platypuses in Albanian zoos, the variance in common to ice thickness and platypus numerosity is precisely 70 percent *whatever* the mechanism of correlation.

Our opposition to fictionalism with respect to abilities should be obvious enough by now.[100] In our view, the same arguments which have been widely acknowledged (by philosophers at least) to defeat fictionalism in physics, chemistry, genetics, and the like, also serve in psychology.[101] Since we cannot enumerate them here, we shall have to be satisfied with pointing out that arguments for general intelligence rest to a large degree on a position in philosophy of science which all parties must at least recognize as highly debatable.[102]

Fictionalism in psychometrics would be less harmful if its role were more widely acknowledged. A careful reader of Cronbach's test would probably not be misled, but Jensen's writings are another matter. Jensen often commits himself explicitly to fictionalism. For example, he says: "We should not reify *g* as an entity of course, since it is only a hypothetical construct intended to explain covariation among tests. It is a hypothetical source of variance (individual differences) in test scores" (*Gen. & Ed.*, p. 77; *see also* p. 79). Further, Jensen makes heavy use of the claim that certain current tests measure *g* to a large extent, a claim which is based squarely on the CEC Principle, and thus presupposes fictionalism. Nonetheless, one cannot take Jensen's commitment to fictionalism totally seriously, since he also says:

> The term "intelligence" should be reserved for the rather specific meaning I have assigned to it, namely the general factor common

> to standard tests of intelligence. Any one verbal definition of this factor is really inadequate, but, if we must define it in so many words, it is probably best thought of as a capacity for abstract reasoning and problem solving.
>
> Intelligence fully meets the usual scientific criteria for being regarded as an aspect of *objective reality,* just as much as do *atoms, genes and electromagnetic fields.* Intelligence has indeed been singled out as especially important by the educational and occupational demands prevailing in all industrial societies, but it is nevertheless a *biological reality and not just a figment of social convention.* . . . But keep in mind that it is this technical meaning of "intelligence" to which the term specifically refers throughout the present article. [*Gen. & Ed.*, p. 88, emphasis added]

Jensen apparently contradicts himself, possibly because he simply holds contradictory views. Or perhaps, at some level, he takes a fictionalist view of *all* "theoretical entities" including atoms, genes, and electromagnetic fields.

Whichever explanation is correct, the effect of Jensen's peculiar stance is that he manages to gain the *use* of claims about general intelligence which presuppose fictionalism while at the same time giving the appearance of talking about real abilities in which groups are supposed to differ. His recommendation that education for lower-class, lower-IQ children be oriented toward associative (memory) learning rather than conceptual cognitive learning is based on the claim that these children are deficient not in a fiction, but in a real thing: general intelligence, something best thought of as "a capacity for abstract reasoning and problem solving." But the arguments used to justify the claim that the tests measure general intelligence depend on understanding general intelligence as just such a fiction. Here, as in the case of the operational definition of intelligence we discussed at the outset, Jensen tries to have his cake and eat it too.

Part II

I. Heritability

What Is Heritability? NO ONE KNOWS HOW TO DETERMINE THE DEGREE to which an individual's characteristics are genetically determined. Indeed, it is not at all clear that it makes sense to speak of doing so. Any characteristic which is influenced by a person's genes is also influenced by his environment, because genetic mechanisms operate differently in different environments and cease to operate at all in some environments. Each of us would have been half as tall as he is or would have had half as many limbs if he had had either an appropriate genetic defect *or* an environmental disadvantage, e.g., a womb environment affected by thalidomide. But even if it makes no sense to compare the genetic and environmental contributions to an individual's characteristics, it does make sense to compare the effects of the existing genetic and environmental differences in a group of persons.

The heritability of a characteristic in a population is the proportion of the characteristic's variation in that population which is due to genetic differences. The measure of variation is variance, the average of the squares of deviations from the population average. Thus heritability is the variance caused by genetic differences divided by the total variance. The following features of heritability are crucial to understanding its implications.

(1) Like infant mortality or divorce rate, heritability is a population statistic. It depends on the genetic and environmental variation present in a given set of persons at a given time. For this reason, heritability calculations made in one population can be extrapolated to another population only in very special circumstances. (We shall return to this point.) Consider, for example, the following two populations: (A) Jews who live in New York or

Miami Beach and (B) people in New York, excluding blacks. Although the total variance of skin color in A is probably about the same as in B, most of the variance in A is probably due to environmental (sun exposure) differences while most of the variance in B is probably due to genetic differences. Hence the heritability of skin color is probably low in A but high in B.[1]

The skin-color example shows that heritability can vary from place to place; but it also can vary temporally. For example, the heritability of having TB was (probably) once rather high, whereas now it is fairly low.[2] The reason is that contracting TB depends on two factors: contact with the germ, and constitutional susceptibility (probably highly heritable). When conditions were such that the germs were everywhere, whether a person contracted TB or not depended mainly on his constitutional susceptibility. Now, however, because TB germs are more localized in slums, whether a person contracts TB depends more on such factors as where he lives. So the heritability of contracting TB has changed even though the physiological events which occur in a person who contracts TB are the same as they always were.[3]

(2) One cannot meaningfully speak of the heritability of a characteristic of an individual person. Heritability calculations can be used to make estimates of an individual's genotype (see *Gen. & Ed.*, p. 116) or of genotypic differences between individuals, but because of its statistical nature the reasoning involved can be misleading. For example, since height is highly heritable, if all one knows about persons A and B is that A is three feet taller, one would be justified in estimating A's genotypic height to be greater than B's genotypic height. But in fact, A and B might be identical twins, in which case the phenotypic difference would be entirely environmental, or A might be genotypically shorter; for example, his phenotypic advantage might be due to pituitary hormone injections.[4]

(3) We must distinguish between a characteristic's having a genetic basis and its being heritable. Genes play a causal role in the development of height. In this sense, we will say that height has a genetic basis. In addition, height currently has high heritability in the United States. But imagine a population in which *everyone* has the same genes, i.e., in which everyone is a one-egg twin of everyone else. In such a population, height would still have a genetic basis, but its heritability would be *zero*, because there would be no genetically caused differences in height. The point is that, even if

genes play an important part in the development of a characteristic, if there is little genetically caused variation, the heritability of the characteristic can be quite low. *It is the causation of differences in a characteristic, not causal influence on the characteristic itself which is relevant to heritability.* For example, ask yourself: What is the heritability of having two legs among humans alive today? The answer is that the heritability of having two legs is probably near zero. When a person has a number of legs other than two, this is usually a result of one or another sort of environmental mishap such as an automobile accident or a defect in womb environment. Probably, little of the variation in two-leggedness is due to genetic variation, so the heritability is probably low.[5]

Insofar as it seems odd that two-leggedness has low heritability, this is due probably to a tendency to conflate heritability with genetic basis. Genetic mechanisms are involved in the development of two-leggedness, so two-leggedness has a genetic basis. Of course, in the same sense, two-leggedness *also* has an *environmental* basis: environmental mechanisms are also involved. As we pointed out, a characteristic can have a genetic basis and still have a heritability of zero—when the genes for the characteristic are homogeneous in the population or when different genes have the same phenotypic effects—while at the same time, environmental differences do have a differential effect. Having a genetic basis is not a sufficient condition of nonzero heritability, but it is a necessary condition.

(4) Consider the following fallacious reasoning: IQ is genetically determined among whites; blacks may differ from whites in *some* biological respects, but it would be far-fetched to suppose that blacks and whites are such different kinds of organisms that what is genetically determined in whites is not genetically determined in blacks; so IQ must be genetically determined in blacks too; hence the black-white IQ difference must be due to a black-white genetic difference.

But there is no respectable sense of "genetically determined" (least of all heritable) in which this reasoning is valid. The heritability of IQ may be high among whites though low among blacks. More importantly, the heritability of a characteristic within each of two groups can be 1.0, even though the difference between the two groups is entirely environmentally caused.[6] Imagine that we take two handfuls of seeds from a genetically heterogeneous sack. We carefully prepare two homogeneous nutrient solutions. One is normal; the other lacks essential nutrients and trace elements. We grow the two handfuls in the two homogeneous nutrient solutions with homogeneous lighting, temperature, humidity and so on. Since each lot has perfectly uniform environmental conditions, the heritability of height in each lot will be 1.0. But there will be a large difference in the average height of the two lots, a difference ascribable entirely to the environmental difference in nutrients.

How Is Heritability Estimated? Imagine that human embryos at conception could be selected at random from a population, then duplicated and reared separately in environments chosen at random from a given population.[7] Any IQ differences between such genetically identical individuals would have to be due to environmental differences. Given certain assumptions,[8] insofar as the correlation between such twins' IQs approaches the correlation between the IQs of persons picked at random (zero), we can conclude that genetic differences do not contribute to IQ variation. Insofar as the correlation between the twins' IQs approaches 1.0, environmental differences do not contribute to IQ variation. Indeed, the correlation between the IQs of such twins estimates the heritability of IQ in the twins' population.

Actual studies of one-egg twins reared apart differ grossly from this ideal procedure. One-egg twins share prenatal and some postnatal environments, twins may be different from other persons in relevant ways, and children given up for adoption cannot be supposed to be a random sample from the population. In addition, the placement of children in adoptive homes is not random, but is affected by the tendency of adoption agencies to try to match natural and adoptive parents on the basis of social and economic variables.

We have been discussing genetically identical individuals reared apart; now, consider genetically unrelated individuals reared together. Imagine that embryos could be sorted randomly into pairs and reared together in the same households, also picked at random. If both children in a pair reared together had, in effect, the same environmental influences with respect to IQ (which they do not), one would expect the correlation between members of the pairs to estimate one minus H, where H is the heritability of IQ. Allowing that children in the same household have different environments, insofar as H is very low, one would expect the correlation between these children to approach that of natural siblings reared together. On the other hand, insofar as H is high, the correlation should approach zero. Again, actual studies differ from idealized studies in important respects.

Even the idealized procedures mentioned above can be expected to overestimate H in the whole population because H in the whole population is probably quite a bit lower than H in any single age group.[9] The average IQ in the United States has increased markedly in the last sixty years (i.e., since the first IQ tests). Tests given to World War I and II soldiers showed that the average World War II soldier would have scored in the eighty-third percentile of the World War I group. Most investigators agree that this improvement is due mainly to increases in average education and other environmental changes.[10] Thus the large differences in average IQ between the old and the young probably are mainly environmentally caused. But almost all standard methods of estimating heritability have the effect of *excluding* this "intergenerational" environmentally caused IQ variance. For example, since twins are the same age, methods using only differences between twins cannot reflect general trends toward increasing IQ. Further, there is good reason to believe H in adults is lower than in children. Newton Morton and his colleagues at the Population Genetics Laboratory at the University of Hawaii have constructed a model which estimates heritabilities for both children and adults. They estimate H at .75 (plus or minus .06) for children, but only .12 (plus or minus .15) for adults.[11]

How seriously should one take current estimates of H? We agree with what is, in effect, the arguments of Jensen and Herrnstein: There are so many lines of evidence for substantial H (though we would restrict our conclusion to children in the United States and England) that, in spite of substantial bias in most of them, it is unlikely that all are misleading. On the other hand, Jensen and Herrnstein give the grossly misleading impression that there is a mass of data pointing reliably and unequivocally in the direction of high H. Any attempt to estimate the value of H on the basis of current data more precisely than saying it is substantial is, in our view, a guess. If we had to guess, we would guess H even in children is below .5.

Estimates of H are attempts to find the best fit to a jumble of data points. But given the character of the data, best-fit estimates seem to us to be of little value. In his attack on the whole data base for estimates of heritability of IQ, Leon Kamin has documented serious problems in all the major lines of evidence, including a number of bizarre irregularities in some of the most crucial studies.[12] The following is representative of the degree of seriousness of the problems Kamin describes. Newman, Freeman, and Holzinger examined nineteen pairs of separated one-egg twins. Their twins were obtained in response to nationwide newspaper and radio advertisements. Because of the expense of transportation and lodging (the study was conducted during the Depression), they made sure all the volunteers they accepted were one-egg twins by rejecting potential subjects who indicated in response to mailed questions that they were in some ways quite different from one another.[13] But since similarity of twins was taken to be evidence for heritability, rejection of volunteer twins on grounds of dissimilarity can be expected to have seriously biased the results. This and every other estimate of the heritability of IQ we know of either relies on highly flawed data or depends on assumptions so arbitrary that their main justification would seem to be that without some such assumptions one could not arrive at any heritability estimate at all.[14]

The Obscurity of Heritability. Let us leave the question of estimating heritability, and return to an examination of the concept. Heritability is the proportion of the variance which is genetically caused. But what is genetically caused variance? Consider the following sources of variance. (1) A child with an advantage in genotypic intelligence may magnify that advantage by shaping his own environment so as to provide intellectual stimulation for himself, e.g., by reading books, solving intellectual problems, and seeking out learning situations (see *Gen. & Ed.*, p. 110). (2) A child with an advantage in genotypic intelligence may be more interesting to talk to and thus may get more attention from his parents than he otherwise would, thereby nurturing his intelligence further. Jencks points out that "this cycle is likely to be repeated at school. The child who starts off with a small genetic advantage may learn quickly, receive encouragement, and learn more. The child who starts off with a small genetic disadvantage may learn more slowly, be discouraged by the teacher, and stop trying to learn at all. Small genetic differences may therefore end up producing big environmental differences." Big environmental differences, caused by genetic differences, may have a large effect on intelligence. (3) Intelligent parents tend to give their children a double advantage by providing them with higher genotypic intelligence *and* with better than average environments for the development of intelligence. Similarly, parents with lower than average intelligence give their children a double disadvantage. Unlike (1) and (2), this phenomenon is not a matter of a causal effect of a child's genes on his environment, but rather, a matter of the correlation of genetic and environmental advantages. Jencks estimates that about 20 percent of the variance in IQ is due to this phenomenon, though other investigators have made lower estimates.[15]

Sources of variance which (like 1, 2, and 3 above) result from anything other than a random distribution of environments for each genotype are usually grouped together under the heading "covariance." Standard methods of estimating heritability of IQ usually count covariance as part of the genetic component of variance, though Jencks and others have removed variance ascribable to the double advantage–double disadvantage phenomenon (3) from the genetic component. These phenomena illustrate obscurities in what is to count as genetically caused variance and thus show corresponding obscurities in the notion of heritability.

The problem posed by the double advantage–double disadvantage phenomenon is easily resolved. Those who include the variance due to it in the genetic component are using one definition of "heritability"; those who exclude it are using another (though it is hard to see what the interest of the former definition would be). Phenomenon 2 (and to a lesser extent, phenomenon 1) poses a more serious difficulty.

Prima facie, one would suppose that genes affect intelligence by some sort of internal biochemical process. A process by which genes affect a person's intelligence by affecting the behavior of other persons toward him is quite different. But the methods used to assess the heritability of IQ automatically count variance produced by genetic variation as genetically caused variance even if it is also environmentally caused (as illustrated above in 1 and 2).

We shall introduce an important though somewhat pragmatic distinction between direct and indirect causation of a phenotypic characteristic by a gene (or causation of a phenotypic difference between individuals by a genetic difference between those individuals). A gene produces a specific product, a polypeptide or a protein. We shall call the effect of a gene on a phenotypic characteristic a *direct* genetic effect just in case the gene affects the characteristic by means of an internal biochemical process initiated by its product. A gene affects a characteristic *indirectly* when it produces a direct effect which in turn produces or affects a *feature of the environment* (including the immediate environment) which itself affects the characteristic.

For example, consider ways in which genetic variation can cause variation in height. Differences in genes produce polypeptides or proteins which in turn control secretion of pituitary hormone—which, in conjunction with diet and other factors, affects height—this would be direct causation. The following is an (imaginary) example of indirect causation. Imagine a population in which, for religious reasons, all red-haired children (but not other children) are given a near-starvation diet. Then, since such a diet can affect height, differences in hair-color genes would indirectly cause differences in height.[16] Standard methods of computing heritability would count the variance in height produced by this phenomenon as genetically caused variance. Insofar as a rationale might be adduced for this assignment, it might be that the variation in hair-color genes causes the variation in diet, so the genetic variation is causally prior. But what this rationale misses is that the variation in diet is *also* caused by an environmental condition: discrimination based on hair color. The genetic variation together with the environmental condition causes the environmental (dietary) variation; neither the genetic variation nor the environmental condition is causally prior to the other. To see how methods of estimating heritability automatically attribute to genes the effect of variance jointly caused by genes and environment, consider how this phenomenon would affect a study of one-egg twins reared apart. A pair of red-haired twins *both* will be ill-nourished in their respective adoptive homes, while a pair of blonde twins will both be adequately nourished. Thus the pervasive discrimination will contribute to the correlation between the twins' heights, and thus contribute to the assessed heritability of height in the population.

There are several different kinds of indirect genetic causation. For example, not all indirect genetic effects are mainly external. When a child's genotypic intelligence results in his reading books and otherwise stimulating himself intellectually, much of the subsequent magnification of his initial advantage may be a result of internal processes. This sort of effect, in which an initial advantage in a meritorious characteristic is magnified, might be called a *meritocratic* indirect genetic effect. Another distinction is that between indirect genetic effects, in which the causal path from gene to characteristic passes through the behavior of *other* persons (as in the example of discrimination against red-haired children), and indirect genetic effects, in which the causal path does not do so (as in the self-stimulation example mentioned above).

The last distinction is of special moral interest. It is often unjust for people to act so as to contribute to an indirect genetic effect, even when the effect is also meritocratic. For example, it is arguable that it is unjust for parents and teachers to respond to children in a way that helps the smart get smarter and the stupid get stupider, though the analogous phenomenon with respect to, say, athletic ability seems innocuous. While it seems plausible that indirect genetic effects contribute substantially to the heritability of IQ, so little is known about the genetic mechanisms underlying the development of IQ that no estimate of the importance of indirect genetic effects can be made.[17]

The Abstract Nature of Heritability. We have just pointed out that the definition of "heritability" as the proportion of variance caused by genetic differences is obscure because environmental differences also contribute to some (perhaps most) of the genetically caused variance in IQ. It is natural to suggest that this obscurity can be removed by stipulation. That is, for some purposes the relevant statistic might be the "direct heritability," the proportion of variance directly caused by genetic variation, while for other purposes "ordinary heritability" might be the relevant statistic. This seems reasonable enough. As we pointed out with respect to the double advantage–double disadvantage phenomenon, different definitions of heritability are in use.[18] There is, however, one difficulty with the stipulation: no one knows *how* to separate the variance due to indirect genetic effects from the variance due to direct genetic effects, at least within the constraints on human experimentation. Such a separation would involve investigation of the details of the mechanisms by which genes affect psychological characteristics, a task which is well beyond present knowledge.

On the other hand, the methodology behind the calculation of heritability[19] is simple from a conceptual point of view, though difficult practically. It involves examining correlations between individuals of different degrees of biological relatedness in natural and adoptive families.

The heritability of a characteristic tells us the degree to which variation in the characteristic is caused by genetic differences, but the heritability does not tell us *how* the causal mechanisms work, nor the degree to which environmental mechanisms play a role. This kind of superficiality (abstraction from causal mechanisms) makes heritability a much less interesting notion than one might otherwise suppose. It might be true that if the heritability of IQ is high, then even if everyone in the United States were given an ordinary middle-class environment, the variance in IQ would not be very different from what it is now. But the public policy interest of this would depend very much on the type of genetic mechanisms involved in the production of IQ variance. For example, on the (probably false) assumption that lower-IQ populations are substantially outbreeding higher-IQ populations (and assuming IQ is highly heritable), Shockley has argued that lower-IQ persons be encouraged to undergo sterilization. However, suppose (for illustrative purposes) we were to find out that IQ is indeed highly heritable, but solely because adults react to children primarily on the basis of their heritable looks. Presumably, in such circumstances, it would be more reasonable to investigate ways of changing child-rearing practices, and less reasonable to consider eugenic programs.

Heritability has the virtues of its vices. We can calculate the heritability of a characteristic without knowing anything about the causal mechanisms involved in the development of the characteristic. Indeed, we can calculate the heritability of IQ without knowing what IQ *is*. The vice of this virtue, however, is that heritability may be of little sociopolitical interest.

The Norm of Reaction. We shall now explain in detail why the fact that heritability is a population statistic depending on distributions at a particular time and place makes predictions based on current heritability calculations problematic. R. C. Lewontin points out that the real objects of study in population genetics for both programmatic and theoretical purposes are the norms of reaction for traits.[20] A person's phenotypic height, for example, is a function of both his environment and his genotype. The norm of reaction for height is (by definition) the function which maps environment and genotype into phenotypic height. A representation of a (not atypical) norm of reaction for an hypothetical phenotypic variable P, environmental variable E, and two genotypes G_1 and G_2 is given in Figure 1.

Consider a population in which G_1 and G_2 are the only genotypes, and whose norm of reaction is correctly represented by

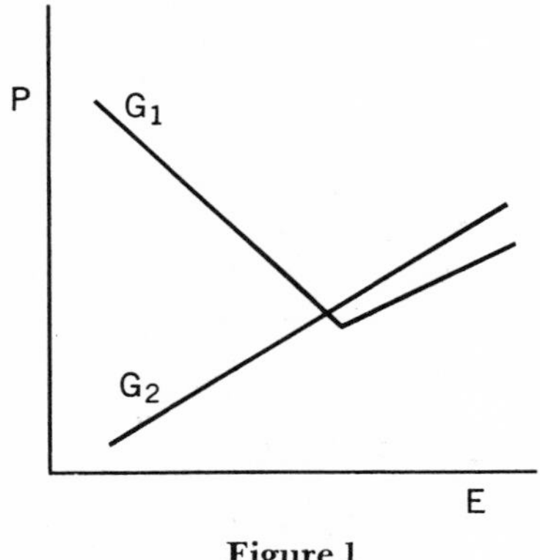

Figure 1

Figure 1. What is the heritability of P? That depends on the proportions of individuals with G_1 and G_2 in the population and the distribution of the environmental variable E among those individuals. For example, suppose there are roughly equal numbers of persons with G_1 and G_2, and suppose there is a moderate degree of variation in E. Consider two different distributions of E. (A) most of the persons have high values of E (and are thus clustered on the right-hand side of the E scale). Then the amount of genetically caused variation in P will be small; for on the right-hand side of the E scale, the P differences between the genotypes are small. So the heritability of P will be low. (B) Most of the persons have low values of E (and are thus clustered on the left-hand side of the E scale). Then the amount of genetically caused variation in P will be large; for on the left-hand side of the E scale, the P differences between the two genotypes are large. So the heritability of P will be high. In general, the amounts of environmentally caused variance and genetically caused variance (and thus the heritability) depend on the distributions of genotypes and environments, and on the norm of reaction.[21]

Jensen and Herrnstein emphasize that one consequence of high heritability is that giving everyone uniform improved environments will not reduce the total variance very much. But that depends on *what* environment you give everyone, and what the norms of reaction are. In the above example, a population clustered narrowly on the left side of the environmental scale would have high heritability (assuming roughly equal numbers of G_1 and G_2), but giving everyone in that population a single environment on the right side of the E scale would reduce the total variance enormously.

The assumption made by those who conclude that improving the environment for everyone to the same level would have little effect on IQ variance is that the curves representing the norm of reaction for different IQ genotypes are roughly parallel, at least within the range of environments we are likely to encounter.[22] Thus they suppose that for IQ, unlike many human traits, a given environmental change produces the same phenotypic effect for every genotype. There is a bit of evidence for this assumption, based on the data from the four studies of identical twins reared apart.[23] But these studies involve only 122 pairs of twins and the range of environments of the adoptive homes, and probably the genotypes of the twins as well, was very narrow compared to the range in the population as a whole (at least in three of the studies). Further, Sir Cyril Burt's study, which has the widest range of environments and is also the largest (fifty-three pairs of twins), has been shown by Leon Kamin to be riddled with inconsistencies and anomalies. In a long article on Burt's work, Jensen has conceded that much of Burt's data (apparently including the study of twins) cannot be relied upon for theoretical purposes.[24]

These four studies provide only a modicum of support for the claim of parallel curves (representing the norm of reaction for different IQ genotypes) and only for the types of environment most common in these studies. Thus it may be that giving everyone one of these particular types of environments will not reduce IQ variance much. This speculation is of little use, however, because we have little data about environments in these four studies, and even if data were available, we do not know enough about what aspects of environment affect IQ to draw any conclusions.

Most of the discussion of the support in population genetics for Jensen's and Herrnstein's claims has focused on the evidence for heritability of IQ. This is a misplaced emphasis, since only with the use of currently unjustified assumptions about the norm of reaction can heritability estimates tell us anything useful about the effects of providing uniform improved environments.

II. The Significance of Heritability

Three claims have been made about the significance of heritability estimates. The first is that they have important consequences for the provision of educational resources and for the design of intervention strategies aimed at raising IQ scores and educational achievement. In the current debate, this claim originates with Jensen's article in the *Harvard Educational Review*. The second claim, put forward most prominently by Herrnstein, is that the heritability of IQ has important implications for the patterns of class structure and social stratification of most contemporary societies. Knowledge of heritability, it is claimed, tells us something important about the ways in which our society is going to develop. The third claim concerns the possible utility of heritability estimates in avoiding dysgenic trends.[25] We shall discuss each of these claims in turn.

Intervention Strategies. It is clear from many studies on the distribution of IQ scores in the American population that there are subgroups, mainly the poor and the black, whose mean performance is well below average. The same pattern holds for scores on standard tests of academic achievement which are quite similar to IQ tests. We argued in Part I that IQ scores reflect many qualities, probably including academic knowledge and skills, qualities of personality and motivation conducive to achievement, and various abilities such as verbal, spatial, and reasoning ability. Assuming that it is desirable to improve the IQ scores of such groups, is it the case that estimates of the heritability of IQ (H) will contribute to our ability to make such changes? What do we know about the possibilities for environmental alteration of IQ if we know that H is high, low, or medium for a given population?

As we pointed out in the last section, the consequences of a given value of H for environmental intervention depend on the norm of reaction. In spite of the paucity of the evidence, discussions have tended to suppose that the curves representing the norm of reaction for different genotypes are roughly parallel, that is, that IQ is in effect a roughly linear function of genotype and environment. In what follows, we will go along with this assumption, not because we think it is true, but rather because the consequences of various values of H even under the assumption of linearity have not been properly understood.

It is essential to bear in mind that H is defined and measured with respect to *existing* environmental and genetic variation in a given population. Therefore, whatever H tells us is relative to that population and the existing sources of variation. It tells us nothing whatever about populations that differ from the initial one in genetic and environmental variation.

Failure to understand this results in the equation of high heritability of a trait with the idea that the trait is fixed and immutable. People often suppose that if H is high for some trait there is relatively little we can do to reduce differences among individuals by environmental manipulation. The converse—if H is low, there is relatively more we can do to reduce variance by manipulation of the environment—is equally inaccurate. Both claims are false because H tells us only what the relative importance of *existing* environmental variation is to phenotypic variation. If H is low, then we know that the existing environmental variations play a relatively large part in producing the existing phenotypic variance among individuals. If H is high, then the existing environmental variations within the population in which H is calculated are a relatively small factor in causing the phenotypic variance. Nothing follows about the possibility of reducing differences in a trait when new sources of environmental variation are introduced which do not currently cause much of the population variance.

Thus a high H for IQ does not preclude the possibility that novel intervention projects can produce large gains in IQ. One way to find out whether that possibility exists is to attempt to do it. One such attempt is currently being made by Heber and is reported as having considerable success. His Milwaukee Project consists essentially in removing children from the home for most of their waking hours and placing them in a highly structured learning environment with a low pupil-teacher ratio. In addition, the mothers—all of whom have IQs of 75 or less and are severely deprived economically—have been given job training and education in homemaking and child-rearing skills. At the time of the latest report, the original infants were four years of age and were maintaining a 28-point IQ difference relative to a control group based on random assignment from the original sample.[26] What we do not know about such intervention programs is whether such gains will be maintained over a long term. All the evidence points to the likelihood of their being washed out within a few years after intervention stops—if the family is not actively involved in the intervention process. This is what the widespread failure of Head Start programs shows.[27] But more recent intervention programs, such as Phyllis Levenstein's Verbal Interaction Project, focusing on both parents and children, have shown that initial successes can be maintained.[28]

So far, we have argued that a high H for IQ is perfectly compatible with making great changes in individual phenotypes by altering the existing range of environmental conditions, but it is *equally* important to see that high H is *also* compatible with making significant changes in mean IQ by offering all individuals the "best" (assuming we knew what this was) of the conditions currently available. This will have the most significant effects at the lower end of the IQ scale for children who are in the most deprived environments. Skodak and Skeels found in their study of children of low socioeconomic status adopted into homes of higher socioeconomic status that the children's IQs were an average of 12 points higher than indicated by predictions based on the IQ of the natural mothers.[29]

Further evidence that innovative strategies for severely deprived children can result in great and enduring changes in IQ comes from Skeels' follow-up study over thirty years of two groups of mentally retarded, institutionalized children who constituted an experimental and control group. The average IQ of the children and their mothers was under 70. At the age of two, thirteen children were placed in the care of mentally retarded female inmates in a state institution. The control group remained in the orphanage. During the experimental period, which lasted about eighteen months, the experimental group showed a mean rise in IQ of 28 points, from 64 to 92, whereas the control group dropped 26 points. Upon completion of the experiment, it was possible for eleven of the experimental children to be legally adopted; two and a half years later, they showed a further 9-point rise to a mean IQ of 101. Thirty years later, all of the original thirteen children were self-supporting, all but two had completed high school, and four had had one or more years of college. In the control group all were either dead or still institutionalized.[30] Neither of these results refute a finding of high heritability for IQ. The gains involved are within the "reaction range" of phenotypic IQ if—to take Jensen's most extreme estimate—20 percent of the variance is environmental.

The amount of change in mean IQ to be expected fom changing the allocation of the existing environmental sources of variation is relatively insensitive to the exact value of H. Only if E (the component of variance due to environmental variation) is very low—the difference between E = 0 and E = .1 makes a difference of almost 5 points in mean IQ change for every standard deviation of environmental change—is the exact value very important.

Given all this, is there any practical interest in the assessment of the precise value of heritability at all? The answer given by Jensen is that such estimates help suggest what kinds of educational intervention strategies are most likely to succeed. But to succeed in doing what? There are two distinct questions at issue. How much can one improve average performance in a given population? How much can one reduce the spread of differences within the given population?

A high heritability for IQ would show that adopting low-IQ children into environmentally advantaged homes will not do much good in reducing average differences. But, as Skodak and Skeels show, it will do a lot of good for some, namely, those whose IQs are low because their environments are very disadvantageous with respect to IQ. High heritability would further imply that improving the environmental conditions of all families, so that the worst conditions to be found were equivalent to the average found in the population today, would again not substantially reduce the total variance.[31] All this depends on the assumption, mentioned earlier, about the norm of reaction.

Nonetheless, all of this is perfectly compatible with supposing that novel environmental manipulations can greatly reduce variance. Novel environments are those which do not occur at all or occur only on a limited basis in the existing range of environments. To use the standard example, consider phenylketonuria. This is an inherited metabolic defect—an inability to convert phenylalanine into tyrosine—which leads in most cases to severe mental retardation. By providing a phenylalanine-free diet to children with this condition, we not only raise the mean IQ of such children but, because such a diet does not improve the IQ of normal children, we reduce individual differences as well. As this example shows, novel environments need not be something out of science fiction; they are merely conditions not commonly found in the range of current environments.

With respect to the second question, namely, how to improve mean phenotypic performance, heritability estimates could provide some indication of the efficacy of existing environments. If a trait has a low heritabilty in a population, then it is probable that we can improve mean performance simply by providing better environments than those which now exist. If H is high but the population has largely unfavorable environments, then simply providing better environments will probably work. If H is high and the environments are largely favorable, then one has to look for novel sources of environmental variation. With respect to IQ, even assuming a heritability of .8, Bereiter calculates a mean increase of 14 points "could be expected from a major improvement in social welfare accomplished without the creation of any new environmental conditions."[32] Because of the statistical properties of a normal distribution curve, this would imply that the number of people with IQs below 70—a figure sometimes used as a defining criterion of mental retardation—would decrease by a factor of more than 10, and those with IQs above 130 would increase by a factor of about 5.

Thus heritability estimates for a trait tell us something about the possibilities for change in variance and mean within existing sources of environmental variation, although knowledge about changes in the mean is not affected greatly by widely varying estimates of H. For the potential changes arising from new environmental sources of variation, heritability estimates, by definition, are useless.

Dysgenics. The relationship between IQ and differential fertility rates has been discussed by geneticists since Galton. The observation that social class is positively related to mean IQ scores and that parental IQ seems to be negatively correlated with fertility (lower-IQ parents having, on average, larger families than higher-IQ parents) gave rise to predictions of declining "intelligence" and the decline of civilization. Failure to observe a decline in measured IQ over time was explained by two famous studies in the 1960s which showed that previous studies were all flawed by failure to consider those individuals who failed to procreate.[33] While it is true that those individuals with IQs below 70 tend to have larger families than those with IQs above 130, it is also true that 30 percent of such persons have no children as compared to 10 percent of those above 130. If one considers completed fertility, then recent studies indicate a slightly positive correlation between IQ and completed fertility. Because such results pertain to particular populations at a fixed time interval, generalizations are not easy to establish.

The relevance of heritability to the issue of dysgenics is the following. If one wanted to improve by selective breeding the value of a trait in a given population, then the response of the characteristic to a given level of selection of parents would be proportional to the narrow heritability. The relevance of this fact to human populations is restricted by both theoretical and practical limitations. In theory, the actual gains may differ from those predicted because of interactions between genes and the environment.

In practice, the prospects of compulsory selective breeding are both unwelcome and unlikely. And any voluntary program would have to be on a very large scale to achieve any significant results. For example, if every male above IQ 140 donated sperm to be used with participating females selected randomly with respect to IQ, and as many as 10 percent of all females participated, the expected change in mean IQ in the first generation, assuming H is .50, would be 1.1 points. In later generations, the mean gain would decrease and eventually level off. If selection were also carried out on females, the gains, at most, would double.

As Cavalli-Sforza and Bodmer, to whom the above discussion is due, point out: "If we could alter environmentally intelligence to the same extent as stature has been, we could obtain an amount of change that would only be possible with 100 percent participation of the females in the eutelegenesis program."[34] Prediction of the IQ of succeeding generations based on possible differential fertility of various subpopulations also depends on a knowledge of the narrow heritability of IQ. But given the arguments in part I of this essay about the questionable nature of IQ tests as measures of intelligence, given the lack of evidence for a fertility asymmetry between low- and high-IQ persons, given the difficulty in carrying out eugenic programs and the obvious dangers of such policies, the significance of heritability estimates for eugenic social policy is severely limited.

Sociological Implications. We have seen what relevance the heritability of IQ has to decisions about the allocation of educational resources as well as about the design of intervention programs and eugenic programs. Is the heritability of IQ relevant to *other* social issues? R. J. Herrnstein, in effect, answers "emphatically, yes" in his much discussed article in the *Atlantic Monthly* and his recent book, *IQ in the Meritocracy*. The heart of Herrnstein's claim is his "syllogism":

1 If differences in mental abilities are inherited, and

2 If success requires those abilities, and

3 If earnings and prestige depend on success,

4 Then social standing (which reflects earnings and prestige) will be based to some extent on inherited differences among people.[35]

From the syllogism Herrnstein derives a number of corollaries which will make the syllogism "more relevant to the future than to the past or present" (*IQM*, p. 198). In particular, he argues that as society becomes more egalitarian the tendency toward a hereditary caste system based on the genetic transmission of IQ will increase.

We shall argue, in this section, that the syllogism as given is not a valid argument, that one can reconstruct the syllogism to make it valid; but that the premises of this valid argument are either false, not known to be true, or if true of society as it is now, they need not continue to be true of it in the future. To simplify the discussion, we shall demonstrate the invalidity of a somewhat truncated version of the syllogism advanced by Herrnstein. Once we see that the shorter version is invalid, it will be obvious that the original version is invalid too. Herrnstein says: "First, what is my argument? Concisely stated, it is that, (A) since people inherit their mental capacities (as indexed for example, in intelligence tests) to some extent, and (B) since success in our society calls for those mental capacities, therefore, (C) it follows that success in our society reflects inherited differences between people." Herrnstein adds "that the 2nd premise . . . means specifically that mental competence is necessary but not sufficient for getting 'ahead.' "[36] Thus we shall read "calls for" in premise (B) as "requires."

To see that this argument is invalid, consider the following counterexample. Imagine a society in which (as in our society) height is highly heritable. This society is run by an absolute dictator who, like Napoleon, is quite short. Unlike Napoleon, he has seen to it that a necessary condition of success in the society is being under five feet tall. This is a model in which the premises of an analogue of Herrnstein's syllogism are satisfied. But suppose, further, that not *every* short person is chosen to be successful, but that the successful are chosen from the short on the basis of the criterion of being genetically average with regard to stature. That is, the successful are picked from the short on the basis of their being persons such that if they had had average environments, they would have had average heights. One way of doing this would be to pick people whose shortness is due to unusual environmental effects (e.g., double amputation, pituitary damage caused by birth trauma), but whose siblings raised in average circumstances have average heights. Thus in our imaginary society, the successful group is not genetically different in stature genotype from the unsuccessful group and does not satisfy the conclusion to the analogue of Herrnstein's syllogism. The existence of an example, even one as implausible as this, in which the analogues of Herrnstein's premises are true and the analogue of his conclusion false proves that the syllogism is logically invalid.

It is easy to see why the syllogism seemed to be valid. The following conclusion C′ *does* follow from the premises (though we are not claiming the premises are true):

A Mental capacities are (partly) heritable.
B Success requires those mental capacities.
C′ Success requires a characteristic which is (partly) heritable.

But C″ does not follow from A and B.

C″ Successful people differ from unsuccessful people genetically (on the average) with respect to the capacities in question or, what amounts to roughly the same thing, success is (partly) heritable.

Success need not be heritable even if it depends on a heritable characteristic. Herrnstein's conclusion that "success . . . reflects inherited differences" is ambiguous between C′ and C″.

The syllogism is invalid; but is it a good "empirical" argument? That is, if the premises were true, would it in fact be likely that the conclusions would be true too? Or, what is really the same question; is there a set of empirically reasonable statements which, if added to the premises, makes the resulting argument valid? We believe that, with some modification of the premises and with an additional premise, we can construct a version of the argument such that *if the premises are true* then the conclusion is very likely true. More specifically, if

1 IQ is substantially heritable, and if

2 In order to succeed in our society, a person must have IQ at a level which a substantial percentage of the population lacks, and if

2′ There is significant equality of opportunity and significant intergenerational mobility based on IQ or, more generally, if selection for success does not run counter to the genetic contribution, and if

3 Earnings and prestige depend on success, then it is likely that

4 There are some genetic differences among persons with different levels of earning and prestige.

Now let us consider the premises of the argument. What do we know about the truth of premise (1)? Since the success differences in question are differences among adults, the relevant heritability estimate is the heritability of IQ in adults. If H in adults were above .75 and the correlation between IQ and success were .5, a simple mathematical argument shows there would have to be some genetic component to success differences. But what little evidence exists on this point indicates H in adults may be quite low (*see* note 11). In any case, none of the studies used to estimate H in adults have included a representative sample of higher and lower social-class environments.

As to premise (2′), enough is known about the relative unimportance of IQ to the transmission of socioeconomic status between generations to lead one to deny the idea—conveyed explicitly by Herrnstein in his syllogism and implicitly by Jensen—that merely substantial heritability implies the presence of socioeconomic genetic differences. Two recent articles, by Bowles and Gintis and Bowles and Nelson, have argued that the transmission of socioeconomic status from father to son is fairly independent of genotypic IQ.[37] For example, if parents transmitted their socioeconomic status to their children largely by passing on their IQ (through either genetic or environmental mechanisms), then one would expect that the correlation of an individual's socioeconomic status with that of his parents among persons with the *same early IQ* would be much lower than such a correlation among persons in the general population. Taking the correlation of socioeconomic status with socioeconomic background to be .55 in the general population, Bowles and Gintis claim that holding early IQ constant lowers the correlation only to .46. Similarly, Bowles and Nelson show that if childhood IQ were "totally unimportant in the process of intergenerational status transmission, the degree of intergenerational immobility (as measured by intergenerational status correlations) would be lowered by at most 13%."[38] However, if all direct influences of socioeconomic background were eliminated, leaving the genetic inheritance of IQ as "the only mechanism by which a family could pass its level of social and economic privilege on to its offspring, the degree of intergenerational transmission would fall to such an extent that at most one half of one percent of the variance of educational attainments, income or occupational status could be 'explained' by socioeconomic background."[39]

On the other hand, there is recent evidence that indicates IQ transmission plays *some* role. A study by Waller found that the correlation between father-son IQ differences and father-son socioeconomic status differences to be .29 ± .08.[40] That is, there is some tendency for sons who have higher IQs than their fathers to have higher socioeconomic status than their fathers. The Waller study also indicates that, if there are genotypic IQ differences between social classes, they are probably clustered in the middle of the socioeconomic status spectrum. This casts empirical doubt on Herrnstein's prediction of a genetic elite and a genetic residue. We shall return to this issue.

Let us now consider premise (2) of the argument. This premise states that success in our society requires the mental capacities which are indexed by IQ scores. All parties to the debate agree that the effect of IQ on success is mediated mainly through its effect on educational attainment (*Gen & Ed.*, p. 85; *IQM*, p. 126). If one measures occupational success by status, there is no evidence of a substantial independent contribution of IQ; if one measures occupational success by earnings, there is some evidence of an independent contribution.[41] Bowles and Gintis, who have done the calculation for a weighted average of income and status, conclude that most of the IQ-economic-success association is a by-product of social-class background and schooling. Jencks concludes:

> The relationship between test scores and income is thus quite different from the relationship between test scores and occupational status. On the one hand, men with high test scores are more likely to enter high-status occupations than to have high incomes. On the other hand, the effect of test scores on incomes seems to be more genuine than their effect on status, in that more of it persists after we control family background and credentials. In neither case, however, is the direct effect of cognitive skill very large.[42]

If one measures success, using supervisors' ratings or observers' ratings, then again low correlations are found. This point is conceded by Jensen who reports "surprisingly low correlations between a wide variety of intelligence tests and actual proficiency on the job. Such correlations average .20 to .25 and thus predict only four or five percent of the variance in work proficiency."[43] Herrnstein, who agrees that IQ does not have a significant role independent of education in determining success, still thinks that IQ acts as a threshold variable. That is, to succeed one must have a certain minimum IQ level, but once that level is attained rising above it may not make much difference. "As far as IQ alone is concerned, virtually anyone can be, for example, a welder, but half of mankind (the half below IQ 100) is not eligible for auditing, even if the brightest welder may equal the brightest auditor in IQ."[44]

What is the evidence for this assumption? It clearly will not do to examine the distribution of IQs within a given occupation and conclude that nobody with an IQ below the lowest score could succeed in that occupation. The reasoning here will have to be of the form "such-and-such" is the best explanation of the data. Herrnstein's hypothesis is that having a certain level of IQ represents having or being able to acquire a certain level of skill necessary to perform successfully in certain occupations, and that the reason there are few auditors with IQs below 100 is that persons with IQs below 100 are mostly incapable of doing auditing. There are a number of other plausible hypotheses which he does not consider. The main competitor is "credentialism," i.e., the view that high-status occupations have educational requirements that exclude would-be entrants who have not completed certain kinds of formal schooling. Now if it *were* the case that the formal schooling is required to impart or develop necessary skills, this would be consistent with Herrnstein's view, but the credentialist argues this is not the case.[45] After a survey of the available data—which are scanty—Jencks concludes:

> All the evidence we have reviewed points in the same direction. Most jobs require a wide variety of skills. Standardized tests measure only a very limited number of these skills. If an individual with low scores has the necessary noncognitive skills, and if he can get into an occupation, his performance on the job will not usually be appreciably below the norm for the occupation. Entering many high-status occupations does, however, depend on having met formal and informal educational requirements, and meeting these requirements is extremely difficult for people with low test scores.[46]

It would be surprising if a conclusion such as Herrnstein's were correct in light of all we know about abilities to learn and perform in other fields. Knowledge of skills, such as long division, that were once arcane can now be taught to almost everyone. Intellectual achievements such as relativity theory, once claimed to be understood by only three persons, are now accessible to college freshmen. What has made these achievements possible is a better understanding of these subjects and the possession of skills which allow teachers to communicate such subjects effectively. To suppose that any highly motivated individual with the necessary temperamental qualities and an IQ of 90 could not be taught the skills necessary to be an auditor of average ability seems implausible.[47]

What, then, is left of the premise which Herrnstein says "cannot easily be challenged" (*IQM*, p. 215)? The correlation between occupational status and IQ scores is around .50, which means that at least 75 percent of the variance in occupational status is unaccounted for by IQ. Most of that correlation depends on the correlation of IQ with years of education. We argued in part I (sections IV–VI) that more than half of that correlation is artifactual in nature. And we have argued in this section that an alternative hypothesis, that of credentialism, can explain the residual correlation between IQ and status as well as Herrnstein's.

Let us now turn to premise (3)—the dependence of earnings and prestige, and hence, socioeconomic status, as these are conventionally measured, on occupational success. We do not deny that, in our society, this premise holds. The question is whether it will, or must, continue to hold. Herrnstein agrees that the continued force of the syllogism might be challenged: ". . . it may seem more plausible to block the third premise by preventing earnings and prestige from depending upon successful achievement. The socialist slogan 'From each according to his ability, to each according to his needs' can be seen as a bald denial of the third premise." But, he goes on to say, that dictum runs up against human nature. "For what the dictum neglects is that 'ability' is, first of all, widely and innately variable, and second, that it expresses itself in labor only for some sort of gain. . . . Human society has yet to find a working alternative to the carrot and the stick" (*IQM*, p. 216).

The mythology of the carrot and the stick has, as unsupported assertion, been the traditional antiegalitarian argument. But just as Nietzsche observed that it was not the case that all men strive after happiness, just Englishmen, it is not the case that all men "labor for some sort of gain," just men who see little other satisfaction in their work. Is it not at least as plausible that such features of work as flexible hours, the intrinsic interest of the work, the ability to determine what tasks one will perform each day have now, and could have increasingly, a decisive importance in motivating people to work?[48]

We are not arguing that because people could be different than they are, the way they are now cannot be used as a premise. The point is that our present conception of how men are is too uncertain for us to rely heavily on it for policy arguments. It is not the case that our empirical knowledge of human motivation suffices to establish either a definitive case, or in view of the historical evidence even a presumption, that men will only labor for the kind of differential gain that Herrnstein has in mind.

In connection with this premise, it should also be noted that it is not enough to show that differential rewards will accrue or are necessary for different occupations. Despite the fact that Herrnstein lumps together earnings and prestige, it makes all the difference in the social world whether or not a surgeon or executive is rewarded with superior earnings or merely with superior amounts of prestige. As Chomsky has emphasized, the politically significant fact about rewarding in terms of money is that it is cumulative and transmittable.[49] Not only are the children of the successful likely themselves to be more successful than the average child, and hence, to have above average earnings, but they are also going to inherit more than average wealth from their successful parents. With respect to IQ, the phenomenon of regression to the mean acts as an equalizing factor since the children of high-IQ parents will, on the average, have a lower IQ than that of their parents, and the children of low-IQ parents will have, on the average, a higher IQ than that of their parents. It is the concentrations of wealth and, their corollary in a market society, power that produce concern about inequality—not the fact that, say, only a small number of scientists get elected to the National Academy of Science. This brings us to the question of the significance of heritability for the future transmission of intergenerational status.

The Rise of the Meritocracy. From his syllogism Herrnstein derives several corollaries.

> As the environment becomes generally more favorable for the development of intelligence, its heritability will increase. [*IQM*, p. 198]

> The higher the heritability and assortative mating, the closer will human society approach a virtual caste system, with families sustaining their position on the social ladder from generation to generation as parents and children grow more nearly alike in their essential features. [*IQM*, p. 220]

> The syllogism and its corollaries point to a future in which social classes not only continue but become ever more solidly built on inborn differences. As the wealth and complexity of human society grow, there may settle out of the mass of humanity a stratum that is unable to master the common occupations, cannot compete for success and achievement, and is most likely to be born to parents who have similarly failed. [*IQM*, p. 214]

The first point to note is that the relevant development is not that of more *favorable* environments but that of more *equal* ones. Given the assumption that IQ is a linear function of genotype and environment, as more equal environments are provided for all individuals in the population, H will increase because the variance caused by environmental differences will decrease. This means that the variance among individuals will be more exclusively accounted for by genetic factors. Herrnstein claims that this fact alone is important. But clearly it is not. It is only important if three other assumptions are made: that IQ is important, that there will be a large amount of phenotypic variance, and that individual differences, rather than an individual's position above some critical "floor," are important. The crucial role of these assumptions becomes clear if one considers other characteristics. Weight, for example, has a fairly high heritability and there is a considerable amount of phenotypic variance. But who cares? On the other hand, being able to convert phenylalanine into tyrosone is very important and completely heritable, but its variance is such that there are few individuals (something like 1 in 10,000) whose deviation from the norm is great enough to affect their chances for success.

With respect to this premise and its implications, let us consider what Herrnstein says and what he omits to say. He says that as the environment becomes more uniform, the relative contribution of genes to variation in IQ will increase. He also points out in his book what he neglected to mention in the "IQ" article: the possibility of discovering gene-environment interactions of a certain kind, i.e., arrangements of the environment which are more beneficial to those who would normally manifest a low IQ than to those who would normally manifest a high IQ. If we could discover such effects, then making the environments more uniform could reduce variance considerably, in spite of high H.

There is another possibility for reducing differences which does not require making environments more uniform. If we could tailor environments for individuals—on the assumption that different people require different environments to perform best—we might reduce differences greatly. And the heritability of IQ might remain constant or even decrease, since differences in environment might play an even larger role than they do now.

Finally, Herrnstein never mentions the possibility of raising mean IQ significantly by equalizing environments. Nor does he mention the possibility of producing even greater increases by providing new environments. These possibilities are implications of heritability just as much as the ones he mentions. Hence, failure to discuss them presents a one-sided picture which is not likely to further his goal of making possible "a more humane and tolerant grasp of human differences" (*IQM*. p. 222).

Given the link between equalization of environments and increased heritability, Herrnstein's claim about the rise of a caste system still does not follow. In discussing subpopulation differences, it is essential to keep distinct the claims that

(1) Two populations differ on a trait that is substantially heritable.

(2) Two populations differ in genes or gene frequencies affecting the trait in question.

(3) There will develop increasing genetic separation of the two populations.

Herrnstein's premises actually entail only that classes differ in a heritable characteristic. We have seen that one can formulate an argument to show that social classes may differ genetically, but he is now claiming that we are faced with the development of a hereditary caste structure. His argument illustrates in textbook fashion the fallacy of equivocation.

He asserts that as environments become more equal for the development of intelligence, the heritability of intelligence will increase. Then he states that the higher the heritability and assortative mating, the closer will society approach a virtual caste system. As we have pointed out, there are two definitions of heritability, one narrow and one broad (*see* note 8). The degree to which children will resemble their parents in some character, i.e., the extent to which the trait "breeds true," is a function of narrow heritability. In the first claim, Herrnstein is referring to the total genetic variance (broad heritability); in the second, to narrow heritability. Now if these two heritabilities moved up and down together, the equivocation would not be serious. However, they are independent in the sense that large changes in the environment can increase or decrease the broad heritability while doing the opposite to narrow heritability.

Herrnstein's argument is fallacious, and as critics from as different parts of the political spectrum as Chomsky and Eysenck have pointed out,[50] regression to the mean (which, insofar as it is a genetic phenomenon, is determined by narrow heritability) insures that parental status cannot be perfectly transmitted between generations, making the development of a caste system unlikely.

Moreover, there is little empirical evidence that the correlation between education and occupational status has changed in the past eighty years,[51] or that the correlation between IQ and occupational status has changed in the past fifty years, or that the correlation between education and income has changed in the past thirty years, or that the correlation between IQ and income has changed at all.

It might be argued that all of these facts simply indicate that our society fails to adhere to meritocratic standards and that, on moral grounds, we ought to be moving toward a society in which IQ plays an increased role in occupational success. But why should we? If IQ scores did in fact measure a unitary capacity that was in fact essential for success across the whole range of occupations, then perhaps something could be said for the view that increased reliance should be placed upon such scores. In reality, however, as we pointed out above, whatever IQ tests measure seems to have little relationship to successful performance as measured by independent ratings. Perhaps a scale that *really* measured intelligence would show different results.

Those who think that IQ tests are important in predicting success within occupations should ask themselves whether, in thinking of occupations, they are really thinking only of professional occupations and then of quite specific ones such as law and philosophy. It may well be that such occupations call for a very specific range of abilities which people actually like to think of as being intelligence. Even here, though, there is some counterevidence that shows a small correlation between educational performance and professional success.[52] If, on the other hand, one thinks of welders, newspaper reporters, detectives, sales managers, factory foremen, occupational therapists, dentists, novelists, quarterbacks, talk-show hosts, sculptors, accountants, social workers, plumbers, school teachers and locksmiths, it seems less plausible that a society should tend to tie entry into and success within such occupations either to the current battery of tests or to some improved version of such tests. Those who think it is somehow fitting or just that success, assuming it is really dependent on IQ, be rewarded in various ways including increased income, should think about the justice of this practice. Why should high intelligence—which, if Herrnstein is right, is as independent of the individual's efforts as his height or his parent's social class—call for reward when, say, it is clearly considered unjust for one to be rewarded according to the characteristics of his parents?[53] It is certainly true that there are grounds of efficiency for preferring one rationing system over another, but those grounds do not fully account for the strong feelings that somehow ability—even if "innate"—has a just claim for differential reward.

Herrnstein's identification of IQ and intelligence is crucial in arguing for the fairness, justice, and legitimacy of the reward system in our society. For if IQ tests were measures of cultural background, or differential opportunity to master skills, or the economic status of one's parents, or very specific perceptual skills, or some combination of these, then the fact that such tests are fairly good predictors of success would be interesting, but the tests could not be used as a way of legitimizing differential rewards and status. The ideological function of the testing apparatus, the channeling and "cooling out" functions, can only be successful if they are thought of as measuring some cognitive factor required for successful performance in a reasonable society. Further, it has to be a factor that is socially approved. Shrewdness and good memory, for example, would not do—the former having overtones of exploitation of others and the latter being too superficial (sometimes possessed in large degree even by idiots). Intelligence has all the positive connotations needed. Thus, our arguments in part I of this essay against the identification of IQ with intelligence are highly relevant here.

The emphasis placed by Herrnstein and Jensen on individual differences runs through the literature and is aptly reflected in a sentence from Thorndike quoted by Jensen: "In the actual race of life, which is not to get ahead, but to get ahead of somebody, the chief determining factor is heredity" (*Gen. & Ed.*, p. 98). To some extent, the focus on differences is a reflection of the nature of the technique used in these studies, i.e., variance analysis. By definition, this analysis works with differences between individuals. It also reflects the fact that IQ tests are so constructed as to reveal differences—both within age levels and between age levels. That is their function. But in discussions of the sociopolitical significance of IQ, the emphasis on differences must reflect certain assumptions about what is really important in a social context, as the quotation from Thorndike illustrates. It would be unfair to conclude that this emphasis simply reflects a fascination with scarcity and competition characteristic of market societies. For with respect to some things, there is good evidence that it is relative position that matters. The best example of this is economic welfare. Public-opinion surveys over the past thirty years have shown that when Americans are asked what is the minimum amount of money needed by a family to "get by," they give a figure about one half the mean family income at the time.

Is there a similar argument with respect to differences in IQ? The most plausible version is the following. Suppose, it is argued, we could raise mean IQs significantly so that, say, 84 percent of the population would be above the (supposed) minimum IQ required to perform successfully in most occupations. Employers would still choose in order of highest to lowest IQ; the effect would be that those in the bottom half of the new distribution would be no better off than before. Therefore, it is differences that remain crucial.

This argument relies on two false assumptions. First, it assumes that employment decisions are made mainly on the basis of IQ scores—clearly a false assumption. Such decisions are mainly made on the basis of credentials and personality, and while these attributes may be correlated with IQ, we are now assuming that all those above the crucial "floor" are able to acquire them. Secondly, it assumes that if employers could hire directly on the basis of IQ, it would be economically rational for them to do so. At present, there is little data to show that correlation of successful performance with differences above the minimum is significant. Of course, most occupations admit of a range of degrees of competence, and it is possible that for some occupations there is a correlation between competence and the possession of intellectual capacities above those minimally necessary to perform adequately. Whether it is rational to screen people on the basis of such characteristics will depend upon how high that correlation is and whether it is higher than correlations with other characteristics such as personality and motivation.

It might be argued that differences are also important, because they represent differences in learning ability and ease and speed of training. Thus, employers will select in order of IQ to reduce their training costs. The evidence to support this argument is equivocal. L. Tyler comments:

> It seems to be clear, then, that there is no simple way of characterizing the relationships between measured intelligence and learning, but it is possible to get rid of a number of misconceptions about it. In school or out, intelligence tests cannot be expected to tell us how quickly individuals will "catch on" or how much they will improve their performance if the task is one they are all clearly capable of doing. Tests do provide an index of the level of complexity and difficulty in the manipulation of symbols to which any individual may be expected to advance under ordinary circumstances. . . .[54]

Ghiselli, on the other hand, reports an average correlation of .42 of general intelligence tests with speed and ease of training for most common occupations.[55] Suppose there is a modest correlation between IQ and speed and ease of trainability. For persons who are capable of performing satisfactorily in an occupation (but acquire the necessary skills at a slower rate) to be prevented from sharing in the status, income, and intrinsic satisfactions of such occupations is grossly unfair. Considering that the average person spends thirty-five to forty years in his occupation, is there a significant difference between one month and six in training time? At the very least, a decent society should try to find some reasonable trade-off between efficiency and justice.

We are not arguing for either the possibility or the desirability of a society with no differences either in abilities or in rewards. Rather, we envisage both the possibility and desirability of a society in which talent of various kinds is recognized and rewarded in various ways. There will still be competition for the honors and prizes of successful achievement. It is, however, both feasible and desirable to insure that whatever differences in income do accrue are not great enough to result in the vast discrepancies in life chances faced by people at different levels of income. For in a society in which income is the most important measure of merit, there is a tendency for all good things to go together. The amount of income one earns determines, to a large extent, where one lives, who one's neighbors are, what kind of schools are available for one's children, what quality of medical services are available, what class one's spouse will come from, whether one has to deal with state agencies and the associated bureaucracy, how long one's vacation is, how much control one has over the conditions of labor, how one will be treated by the police, how good one's municipal services are, whether one is respected in the community, and how much political power one has.

It is ideology, not psychology, which leads Herrnstein to envisage alternative reward systems in terms borrowed from contemporary reality[56] and to cite E. O. Wilson's *The Insect Societies* as his only reference for exploring alternatives to meritocratic societies (*IQM*, p. 224).

We conclude that Herrnstein is wrong in claiming that "it is likely that the mere fact of heritability in IQ is socially and politically important, and the more so the higher the heritability" ("IQ," p. 58). What is socially and politically important is the ideological message that is conveyed to the readers of the *Atlantic Monthly* and the *New York Times* by statements such as the following:

> The syllogism implies that in times to come, as technology advances, the tendency to be unemployed may run in the genes of a family about as certainly as bad teeth do now.
>
> As the wealth and complexity of human society grow, there will be precipitated out of the mass of humanity a low-capacity (intellectual and otherwise) residue that may be unable to master the common occupations, cannot compete for success and achievement, and are most likely to be born to parents who have similarly failed. ["*IQ*," p. 63]
>
> There are intelligence genes, which are found in populations in different proportions, somewhat like the distribution of blood types. The number of intelligence genes seems to be lower overall in the black population than in the white.[57]

As Marxists are fond of saying, it is no coincidence that such statements are put forward at a time when the society finds itself faced with a no-growth economy (thus sharpening class conflict over distributive issues), and when there is a backlash against the use of the law to foster equal opportunity for women and minorities.[58] The use of scientific credentials and scientific results to justify elitism and racism raises important moral issues for those engaged in such research. It is to these issues that we now turn.

III. Race, Research, and Responsibility

There are several distinct levels at which one can investigate the moral implications and attendant responsibilities involved in the research we have been discussing. One can consider the decision of the individual scientist to pursue certain topics and to make his results available—to which communities and in what form. One can consider the decisions of universities to permit or to promote, in various ways, different kinds of research. One can consider the funding policies of the various governmental agencies that contribute the largest share of research funds. These levels are to some extent independent of each other, and the principles that apply may be quite distinct. Arguments sufficient to show that certain kinds of investigation ought not to be pursued may not be sufficient to show that such research should be forbidden by a university or restricted in other ways.

We propose to discuss the issue mainly in connection with the individual scientist's decision to select a given area of investigation and to publish his findings. It may be that the same considerations which are relevant to these questions also play a role in trying to decide the issue of what research universities ought to allow and support. But that question raises all the difficulties associated with censorship and academic freedom, and deserves separate treatment.

We must first deal with the claim that the probable social consequences of proposed research are irrelevant to the decision whether or not to undertake a given project. According to this view, the only factors that are relevant are those generated by the internal structure of the given discipline: Is the question an important one in the light of the criteria established within the field? Is it likely that answers to certain questions will significantly advance the existing state of knowledge? On this view, the only sin that a researcher can commit is triviality.

This view can be argued for in a number of different ways. One can advance consequentialist arguments to the effect that attempts by individuals to assess the likely social consequences of their work are usually inaccurate and that, on the whole, both the discipline and the welfare of mankind are best served by ignoring such considerations. Or one can argue along nonconsequentialist lines that knowledge and truth are goals that may not be subordinated to other considerations. We shall not argue against the view that the consequences of a given body of research are irrelevant, we shall simply proceed on the assumption that it is false. While we recognize the great value in encouraging the freest possible use of scientific intelligence and the great benefits that have been gained from it, we take it as obvious that this is an area in which there are competing values that must be weighed against one another. We take it as evident that one should condemn the actions of, say, those German scientists who pursued research into atomic phenomena with the aid and encouragement of the Nazi regime, and who knew that the probable consequences of their research, if successful, would be the construction of weapons of immense destructive capacity; the right course of action for the scientists would have been to abandon such research. At some point, the harmful consequences for human welfare of one's research must enter into the decision whether to pursue it—the only significant questions are how, when, and with what weight.

It is interesting to note that Jensen, at least, does not present his research, and hence presumably would not defend it, as being "pure" research or knowledge for its own sake. He presents it in a context which makes it clear that he is interested in finding out certain things because of the practical consequences of such knowledge; consequences he believes will be beneficial for all. He starkly asserts that "compensatory education has been tried and it apparently has failed," and that "if the theories I have briefly outlined here become fully substantiated, the next step will be to develop the techniques by which school learning can be most effectively achieved in accordance with different patterns of ability" (*Gen. & Ed.*, pp. 69, 202). Herrnstein also believes that knowledge about heritability as well as about the specific techniques used to measure various human abilities makes "a more humane and tolerant grasp of human differences . . . possible."

We are going to argue, first, that Jensen does not make a convincing case for there being likely and important beneficial consequences of investigating genetic racial differences in IQ; second, that there are serious and probable harmful consequences which flow from the interpretation likely to be placed on such research; third, that Herrnstein and Jensen were negligent in failing to take precautions in their writing to mitigate such harm; and, finally, that in light of the difficulty of preventing such harms, the proper course would be to avoid undertaking such research altogether.

We shall begin by considering Jensen's arguments in favor of investigating racial genetic differences in IQ. The arguments are basically of two types. The first deals with the educational implications of such knowledge, and the second with the implications for our knowledge of the success or failure of various equal opportunity programs and the avoidance of unrealistic expectations.

In the summary of his book on race, Jensen concludes that "all the major facts would seem to be comprehended quite well by the hypothesis that something between one-half and three-fourths of the average IQ difference between American Negroes and whites is attributable to genetic factors, and the remainder to environmental factors and their interaction with the genetic differences." He then goes on to argue: "If this hypothesis stands up under further appropriate scientific investigation its social implications will be far broader than those that pertain only to education" (*Ed. & Grp. Diff.*, pp. 363–364). But the only implications he mentions are the following educational approaches.

> *Seeking Aptitude x Training Interactions.* This means that some children may learn better by one method than by another and that the best method may be quite different for different children, depending on their particular aptitudes and other personological characteristics. . . .
>
> *Greater Attention to Learning Readiness.* The concept of developmental readiness for various kinds of school learning has been too neglected in recent educational trends, which have been dominated by the unproved notion that the earlier something can be taught to a child, the better (Jensen, 1969d). Forced early learning, prior to some satisfactory level of readiness (which will differ markedly from one child to another), could cause learning blocks which later on practically defy remediation. . . .
>
> *Greater Diversity of Curricula and Goals.* The public schools, in order truly to serve the entire population, must move beyond narrow conceptions of scholastic achievement to find a much greater diversity of ways for children over the entire range of abilities to benefit from schooling—to benefit especially in ways that will be to their advantage after they are out of school. . . . [*Ed. & Grp. Diff.*, p. 364]

It seems to us that no reasonable person could disagree with these as desirable goals—although the notion of greater diversity of curricula and goals could turn out to mean, as it sometimes seems to in Jensen, that some should memorize while others conceptualize. Where we disagree is in supposing that promoting these goals has anything to do with finding out more about racial genetic differences. It is not the case that either race or genes are relevant in these matters. Take race first. Jensen himself points out in the immediately preceding paragraph that "these approaches have nothing to do with race per se, but are concerned with individual differences in those characteristics most relevant to educability" (*Ed. & Grp. Diff.*, p. 364). Quite so. *If* genetic knowledge were important for finding, say, aptitude *x* training interactions—which, as we shall argue shortly, it is not—it would be knowledge of the individual's genetic make-up and not that of any group of which he is a member.

As to promoting the three goals, nowhere does Jensen indicate *how* the knowledge of racial genetic differences he is seeking is supposed to promote their achievement. We already know that the kinds of educational programs pursued so far have not been notably successful. It is therefore reasonable to look for, say aptitude-training interaction effects.[59] Finding these is more or less a matter of trial and error, aided by whatever theory cognitive psychology has developed on these matters; knowledge of genetic matters is not likely to help. The same thing is true for paying greater attention to reading readiness. Surely Jensen is not suggesting that all the controversy and research on this matter should have waited for the kind of knowledge he proposes to acquire or would proceed more fruitfully in the future if we had it. Again, greater diversity in our schools is something that educational reformers have been advocating ever since John Dewey. That this is a good thing does not depend on data about genetic differences, either within or between races. Carl Bereiter, in an article characterized by Jensen as "an exceptionally intelligent and penetrating analysis" (*Gen. & Ed.*, p. 59) argues that with respect to social and racial differences, knowledge of a possible genetic basis is relevant neither to the classroom teacher nor to the educational policy makers.[60] We agree.

Bereiter does argue that knowledge of the heritability of individual differences—and presumably, the analogue for group differences—does permit realistic estimates of what the effects of giving everyone the best currently available environments will be on reducing differences and raising mean performance. While this is true, the reader should refer back to page 488 to see that determining an exact value for E is of relatively little importance, since the reaction range is relatively insensitive to a precise value of E through most of its range. This point applies to race differences also.

On the other hand, it is true that knowledge of a genetic component to racial differences would be relevant to our expectations of the effect on reducing differences of giving everyone, for example, a middle-class environment. No doubt, having realistic expectations about such things is better than not having them, but it seems a small benefit compared with the harm that, we shall argue, might flow from investigating genotypic racial differences. Further, even if racial differences are entirely environmental, it is not clear that the environmental differences are of the sort that can easily be erased. So it is not obvious that finding out that genetic racial differences exist should produce a radical change in our expectations about reduction of differences.

Jensen says

> improving the benefits of education to the majority of Negro children, however, may depend in part upon eventual recognition that racial differences in the distribution of educationally relevant abilities are not mainly the result of discrimination and unequal environmental conditions. None of the approaches that seems to me realistic is based on the expectation of the schools' significantly changing children's basic intelligence. [*Ed. & Grp. Diff.*, p. 364]

But it seems unlikely that the type of educational effort expended in order to change IQ would be very different from the type of effort that would be expended to produce educational (not IQ) gains. If it turns out that there are black-white genetic differences, and this becomes known, one ill effect might be a mistaken cessation of efforts to find new types of intervention aimed at raising IQ.

The second argument is that without research we will not know whether a disproportionately small number of persons from disadvantaged groups in universities or the professions is due to discrimination or to disproportionate abilities. Jensen writes:

> Since much of the current thinking behind civil rights, fair employment, and equality of educational opportunity appeals to the fact that there is a disproportionate representation of different racial groups in the various levels of the educational, occupational, and socio-economic hierarchy, we are forced to examine all the possible reasons for this inequality among racial groups in the attainments

and rewards generally valued by all groups within our society. To what extent can such inequalities be attributed to unfairness in society's multiple selection processes? (Unfair meaning that selection is influenced by intrinsically irrelevant criteria, such as skin color, racial and national origin, etc.) And to what extent are these inequalities attributable to really relevant selection criteria which apply equally to all individuals but at the same time select disproportionately between racial groups because there exist, in fact, real average differences among the groups. . . . This is certainly one of the most important questions facing our nation today. [*Gen. & Ed.*, p. 158]

Herrnstein similarly argues:

> In recent years, efforts have been made to rectify disproportions of racial, ethnic, and sexual groups in schools and occupations, usually on the assumption that, except for prejudice, all groups would be represented with proportionate-to-population frequency in virtually all common occupations and professions. While prejudice doubtless does hold people back unfairly, not to mention illegally, the use of such equalizing quotas may, in time, create unfairness of its own. *Only if it is indeed true that all groups of people have essentially equal talents and capacities can we justly use quotas.* For if there are relatively stable group differences in mental or physical capacity—between the sexes or the races or the nationalities—then the presumption of equality, when translated into quotas, is bound eventually to discriminate against the qualified individual from time to time.[61]

It is essential to bear in mind that both of these are supposed to be arguments for the investigation of genotypic differences among various groups in society. But, as stated by both Jensen and Herrnstein, at most what the argument calls for is an investigation of *phenotypic* differences. For if group A is lacking in certain abilities or does not have the same average level of certain abilities as group B, then *that* knowledge is what Jensen requires to argue that it is not discrimination in selection procedures which completely accounts for the disproportionate representation of group A.[62]

Similarly, Herrnstein's argument relies only on knowledge about phenotypic differences. He is worried about discrimination against qualified people. If by discrimination he means that with quotas a less qualified person may be preferred by a quota system to a more qualified person, then this will happen if, for whatever reasons, there are average group differences. No investigation of the causes of these differences, in particular genetic causes, is required. Therefore, at best, the argument requires the premise, "only if it is indeed true that all groups of people have essentially equal *phenotypic* talents and capacities can we justly use quotas," not the version Herrnstein uses, which must, because of the context in which it occurs (a discussion of black-white genetic differences), be referring to genotypic talents and capacities.

Although this is not the place for a thorough discussion of the arguments for and against quotas, it might be helpful to see that there is a wide variety of such arguments with quite different premises concerning phenotypic or genotypic abilities. Some arguments assume equal phenotypic abilities, claim that any disproportionate representation of the group in question is due to discrimination in selection, and state that there is presently no effective way of preventing such discrimination except by requiring quotas. This argument is used most frequently in connection with assuring proper representation of women in institutions of higher education. Other arguments assume unequal phenotypic abilities, but without any assumption concerning genotypic abilities claim that for pragmatic reasons quotas are justified to achieve important social goals. Thus, quotas for medical schools are argued for as the only practical way of insuring adequate medical care for black communities.[63]

Other arguments do rely on assumptions concerning the causes of phenotypic differences. Compensatory arguments assume that some portion of the phenotypic differences between groups is due to past unjust treatment, and that the society has some responsibility to correct or compensate. Note that past injustice might have resulted in members of a group not being able to express phenotypically a genotype equal to that of the members of the dominant group, or past injustice may have led to differing genotypes.

Other arguments support quotas as a temporary measure necessary to provide suitable role-models and to enhance the self-image of disadvantaged groups. This argument assumes that there are phenotypic differences, and that, in the absence of certain hindrances, one would expect more equal phenotypic performance.

To the extent that arguments are advanced for proportionate-to-population quotas which rely on assumptions about the distribution of genotypic abilities, it becomes relevant to assess the validity of such assumptions.[64]

So far, we have argued that, at present, there is no reason to anticipate that significant benefits will result from the investigation of a genetic basis for racial IQ differences. We now shall argue that, indeed, grave consequences are likely to follow. Serious and harmful consequences arise from the predictable distortion and misunderstanding by the mass media of the technical results of such investigation. Let us look at a sample of that distortion. Jensen says in his *Harvard Educational Review* article that it is "a not unreasonable hypothesis that genetic factors are strongly implicated in the average Negro-white intelligence difference." This becomes translated in a six-page article complete with pictures in *Life* magazine (June 12, 1970) as follows: "More than a year has gone by since Jensen's paper first appeared, reporting that blacks, as a population, score significantly lower on IQ tests than the white population; and attributing their lower IQs primarily to their genetic heritage, not to discrimination, poor diet, bad living conditions or inferior schools." Although that would be an accurate summary of Jensen's position in his most recent book, he nowhere in his article made any quantitative estimate as to the relative importance of genetic versus environmental factors in the black-white difference. Later in the same *Life* article, the following distortion of what Jensen actually said occurs: "Biased or not, the IQ tests have been used by Jensen not only to demonstrate a 15-point black-white intelligence difference, but also to decide that exactly 80 percent of the gap is due to heredity—which would mean the black child's lower IQ is largely invulnerable to any efforts at improvement." Note both the error of transferring the .8 heritability figure from white populations to the black-white mean difference and the incorrect inference about the difficulty of change.

The flavor of the treatment in the mass media can also be seen in the article on Jensen in *Newsweek* (March 31, 1969). Starting off with the headline "Born Dumb?" the article proceeds to develop in the reader's mind the idea that intelligence is fixed and not subject to change.

> Dr. Jensen's view, put simply, is that most blacks are born with less "intelligence" than most whites. . . . Jensen's theoretical views lead him in his article to develop some quite practical policy recommendations. Since intelligence is fixed at birth anyway, he claims, it is senseless to waste vast sums of money and resources on such remedial programs as Head Start which assume that a child's intellect is malleable and can be improved. . . . Instead, programs should concentrate on skills which require a low degree of abstract intelligence.

In fact, Jensen never claims that intelligence is fixed, although he does say that he is skeptical about how much long-range improvement can be brought about by various intervention programs.

Thus, we see how speculations ("a not unreasonable hypothesis") are taken as showing that blacks are genetically inferior. There are grave short-term consequences of such a belief even if it is later shown to be false or to be true only to a very minor extent. Policy makers and social scientists are at present engaged in a very fundamental dispute about what methods to use to alleviate the conditions of the poor in general and blacks in particular. There is a very powerful backlash against the programs of the 1960s—Head Start, War on Poverty, integration of schools, and welfare. There is a growing tendency to "blame the victims" rather than to attack the structural faults of a society which condemns large segments of the population to radical inferiority. Beliefs—even baseless ones—in the genetic inferiority of the poor or the blacks will function as rationalizations to avoid other solutions which will have a profound cost both economically and, perhaps more importantly, psychologically. If, as seems probable, we are at a crucial tipping point with respect to these matters, then it will not be of much help to the victims that in the long run a correct version of the facts may emerge. Educational and social policies which neither follow from the findings nor are even supported by them will be taken to be proved, and other policies which are not contradicted by the results will be taken as refuted.[65] Deepseated stereotypes and prejudices will be confirmed. Attention will be diverted from areas of crucial importance, such as income redistribution, and will be focused on redirecting black aspirations downward and encouraging the adoption of inferior educational programs.

Further, it is likely that results that are significant only statistically will be taken as determinative of the ability of individuals. Indeed, there is an "efficiency" argument to justify this. Making individual determinations is costly; it may thus be more efficient for a large corporation simply to reject all members of a subpopulation known to perform less well on the average—thereby incurring the cost of losing some talented workers—than to design and administer individualized tests. Independent of any of the above political effects, the publicity given to such findings will be an assault on the self-respect and dignity of individual blacks.

Given the very substantial and immediate harms that are likely to occur as a result of pursuing research on the genetic basis for IQ differences, and given that the benefits are minor, what is a reasonable moral decision for individual scientists to make with respect to such investigation? It is clear that, at the very least, investigators have a responsibility to do their utmost to make it less likely that their findings will be misunderstood and misinterpreted. Neither Herrnstein nor Jensen has met this responsibility adequately.

It is certainly true that part of the explanation of the distortion by the mass media lies in the incompetence of reporters and the problem of popularizing difficult material. But if this were the complete explanation, one would expect the errors to be random with respect to their harmfulness. In fact, the errors always go in the same direction. A careful examination of the Herrnstein and Jensen articles will show that at the most critical points where misinterpretation is likely to occur, no adequate steps were taken to guard against it.

Let us consider Jensen's *Harvard Educational Review* article first. Whether or not Jensen made the error of going from within-group heritability to between-group heritability as many of his critics have charged, the reader, after spending many pages on heritability studies and repeatedly encountering the figure of .8 for H, will be likely to extrapolate this figure to the black-white differences.

Consider the following statement by Jensen with respect to the critical question of the significance of high heritability for altering a trait. "The fact that scholastic achievement is considerably less heritable than intelligence . . . means that there is potentially much more we can do to improve school performance through environmental means than we can do to change intelligence" (*Gen. & Ed.*, p. 135). This is simply false.[66] Jensen now concedes this point, but replies that the context of his statement was one relating to reducing group differences and that something like the above statement is true. A careful reading of the original context does not support Jensen's claim; but even if he were right, surely he should have emphasized this point in the very place where he made the statement—a statement obviously subject to misinterpretation.

Perhaps the most culpable error lies in the ambiguous wording of his summary of black-white differences. "So all we are left with are various lines of evidence, no one of which is definitive alone, but which, viewed altogether, make it a not unreasonable hypothesis that genetic factors are strongly implicated in the average Negro-white intelligence difference" (*Gen. & Ed.*, p. 163). Consider the phrase, "strongly implicated." This is ambiguous. On the one hand, it can be read to mean that it is strongly implied (highly probable) that genetic factors play some role; on the other, it may be taken to imply that they play a strong (major) role.[67] In addition to the author's failure to take adequate precautions against possible and likely misinterpretations, the entire structure of the article conveys an implicit message. The article begins by asserting compensatory education has apparently failed to boost the IQs of lower-class children and blacks, goes on to discuss the genetic influence on IQ and the possibility of genetic group differences, and ends by suggesting that educators concentrate on abilities other than IQ for disadvantaged children. One would have to be obtuse indeed not to draw the moral that attempts to raise the IQ of such groups are hopeless, although Jensen never says this explicitly.

This contrast between explicit and implicit message occurs most obviously in Herrnstein's article. Despite all his cautions about intelligence early in the article—e.g., "Even at best, however, data and analysis can take us only so far in saying what intelligence is. At some point it becomes a matter of definition . . . at bottom, subjective judgment must decide what we want the measure of intelligence to measure"—the rest of the article is strewn with references to "not-so-bright accountants," "gifted children," "very dull parents," "natively endowed," and an equation of scores on IQ tests with smartness.

Again, why do we find so much attention being paid to the only three paragraphs in the Herrnstein article devoted to the black-white question? Some of that attention is based on sheer ignorance of what Herrnstein said, but undoubtedly the most crucial fact is the introduction to Herrnstein's article signed by the editors of the *Atlantic Monthly*. As they put it, Herrnstein's article is part of an ongoing discussion that has as its predecessors the Moynihan, Coleman, and Jensen reports, all of which "grappled with the idea that something within the black community itself was holding back its economic and educational advance." With this as the context, it is understandable how *Time* (August 23, 1971) could report that "Herrnstein's study thus becomes the latest in a series of documents that have sparked a continuing dispute over racial differences."

It might be said that, as bad as the introduction was in conveying a certain message to the reader, the responsibility for the introduction does not fall on Herrnstein—unless he approved it. But the latent content is conveyed in subtle ways in the text itself. Consider the interpretation the average reader will place on Herrnstein's quotations of Jensen: "Compensatory education has been tried and it apparently has failed. . . . The chief goal of compensatory education—to remedy the educational lag of disadvantaged children and thereby narrow the achievement gap between 'minority' and 'majority' pupils—has been utterly unrealized." Compensatory education and its failure are now linked in the reader's mind with black-white differences, although most of the children in such programs were white. Then Herrnstein says that, according to Jensen, the reason "why compensatory education has failed is that it has tried to raise IQs, which, he argues, are more a matter of inheritance than environment, and therefore not very amenable to corrective training" ("IQ," p. 54). There then follows a discussion of heritability results. Inevitably, the reader is given the impression that such studies tell us about black-white differences. It is to his credit that Herrnstein explicitly warns against making this error three pages later, but the warning simply illustrates the contrast between the explicit message and the implicit one.

In matters as sensitive as this, when the consequences of misinterpretation and distortion are likely to be so grave, there is a heavy obligation on the part of investigators—particularly those reporting findings in nonscientific journals—to be scrupulously careful to indicate the limitations of their data,[68] to indicate the existing disagreements in the field, to clarify in advance likely misunderstandings of the results, and to caution against their possible misuse in conjunction with mistaken ideological views.[69]

Thus far, we have argued that serious and harmful consequences are likely to occur as a result of the misinterpretation of findings in this area, and that, at the least, researchers must do everything reasonable to guard against such misinterpretation. We now want to go one step further and argue that this may not be sufficient. For if one knows or has reason to believe in advance that one's qualifications and cautions will be ignored, then it is not possible to absolve oneself from responsibility by claiming, quite correctly we can suppose, that one has done all one can.

Furthermore:

> If one believes that the media of communication are biased, either because of political considerations or because of something in the nature of the media (*Time* magazine is not likely to be willing to

> devote the space to the necessary clarifications even if it desired to), and that one's results will be distorted in the direction of producing harmful consequences,
>
> If one believes that the possibilities of countering such distortion are minimal (compare the force of a letter to the *Atlantic Monthly* with that of the original piece one is criticizing),
>
> If one believes that the kind of research in question is burdened with methodological problems, ambiguous in implication, and open to a wide range of interpretations,
>
> If one thinks it unlikely that all sides in the disagreement will get equal access to the mass media or that, in any case, those in power will select from the ambiguous data the findings they need to rationalize their political and social programs,[70]
>
> If one believes that the likely benefits are minimal and the likely harms grave,
>
> Then one has a responsibility to cease such investigations while the above circumstances obtain.

Jensen argues to the contrary:

> In a society that allows freedom of speech and of the press, both to express and to criticize diverse views, it seems to me the social responsibility of the scientist is clear. He must simply do his research as competently and carefully as he can, and report his methods, results, and conclusions as fully and as accurately as possible.
>
> I therefore completely reject the idea that we should cease to discover, to invent, and to know (in the scientific meaning of that term) merely because what we find could be misunderstood, misused, or put to evil and inhumane ends. This can be done with almost any invention, discovery, or addition to knowledge. Would anyone argue that the first caveman who discovered how to make a fire with flint stones should have been prevented from making fire, or from letting others know of his discovery, on the grounds that it could be misused by arsonists? Of course not. [*Gen. & Ed.*, pp. 328, 327]

This argument is irrelevant. What is important is not whether such discoveries could be misused, but whether, on the best available evidence, they are likely to be. Those who argued against the development of the atomic bomb were well aware of the potential beneficial uses of atomic energy; they merely believed, with some justification, that the most likely short-run use would be to incinerate civilians.

The above argument would be considerably weaker if we had reason to suppose that investigation of racial IQ differences was likely to be of fundamental scientific interest, but as Chomsky has argued:

> In the present state of scientific understanding, there would appear to be little scientific interest in the discovery that one partly heritable trait correlates (or not) with another partly heritable trait. Such questions might be interesting if the results had some bearing, say, on hypotheses about the physiological mechanism involved, but this is not the case.[71]

One should, of course, be alert to risks imposed by the policies we suggest. An example is the following. Any environmental explanation for black-white IQ differences, even one which casts blame on a racist society, will have to make use of intervening variables in the explanation—such as family or cultural patterns among blacks. But if it turns out to be the case that genetic factors are the explanation, then one will have been causing many black parents grief which they might have been spared. This is one illustration of the complexity involved in making the kind of cost-benefit calculation we are advocating.[72]

We have not adopted a line of argument that is favored by many participants in the discussion of racial genetic differences. According to this view—held by a number of eminent population geneticists[73]—such questions cannot be settled at this time; thus the moral issue does not arise. We do not choose to adopt this line of reasoning because we are skeptical about its cogency and because we believe it avoids the main issue. We are skeptical because it seems that the standards being imposed on such research are too stringent. In any case, we do not want the moral question to depend on the ingenuity of scientists in designing possible tests of the hypothesis. We prefer to argue the issue on the assumption that some steps toward answering the question might be possible.

The most frequent objection raised to our argument is that we are not likely to convince those investigators, such as Jensen, who are proposing hypotheses about racial differences and designing and carrying out experiments to confirm them. And we should not, in effect, condemn those who, like Scarr-Salapatek, have continued investigation in this area and presented evidence suggesting that environmental differences are the main determinant of group differences in IQ.[74]

We find the suggestion that, in the absence of universal withdrawal, opposing researchers remain in the battlefield to be cogent as a possible "second-best" solution. Our residual reluctance to accept it reflects questions in our minds about the consequences of further investigation of these topics. For in attempting to prevent damage due to one-sided work in the field, one is usually implicitly endorsing both the importance of the questions that are being asked and the methods used to arrive at answers. If the proper object of study, both for theoretical and practical purposes, ought to be different (e.g., norms of reaction as opposed to heritability estimates), then the contribution one makes to legitimizing the current state of the field may outweigh one's efforts to prevent the more obvious forms of error and distortion. In addition, there is the possibility that one may find evidence for a hypothesis that has little scientific or practical interest and whose confirmation can have the kinds of consequences outlined earlier.

In view of the controversial nature of our position, we should like to guard against misinterpretation by emphasizing at this point what we are *not* saying. We are not saying that further research in the heritability of individual differences in IQ scores should stop. We are not saying that investigations that have race as one of the variables—e.g., of phenotypic differences or even of genotypic differences of *some* characteristics—should not be pursued. We are not even saying that, at all times or in all places, investigation of racial genotype differences in IQ scores should stop. What we are saying is that at this time, in this country, in this political climate, individual scientists should voluntarily refrain from the investigation of genotypic racial differences in performance on IQ tests.

Finally, practicing what we preach, we should like to caution against a possible misunderstanding and misuse of our essay. Although we have argued that the best situation would be one in which the scientist avoided research in certain areas and that failure to do so is irresponsible, nothing we have said implies that it is legitimate to interfere with the teaching, research, or speaking activities of researchers who act irresponsibly. The fact that individuals may be acting wrongly does not, by itself, justify the use of coercion against them. We do not approve of those who have threatened violence against Herrnstein and Jensen, nor of the Stanford administration's refusal to allow Shockley to teach a course on these issues.

To emphasize this point is not to absolve ourselves from the possibility that our arguments may be used to support conclusions we reject. We are concerned about this and have tried to guard against misunderstanding. Ultimately we must accept, as our argument affirms, full responsibility for the social consequences of our ideas.

Notes

We would like to thank the following persons for their detailed comments on earlier versions: Gordon Bermant, George Boolos, Richard Boyd, Susan Carey, Susan Chipman, Marshall Cohen, L. J. Cronbach, Norman Daniels, Hartry Field, Jerome A. Fodor, Alan Garfinkel, Clark Glymour, Christopher Jencks, Jerome Kagan, Noretta Koertge, David Layzer, Richard Lewontin, John Loehlin, Richard Miller, Thomas Nagel, Robert Nozick, Hilary Putnam, Thomas Scanlon, Judith Thomson, and Virginia Valian. In particular, we would like to thank the editors of *Philosophy and Public Affairs* for their help and encouragement at every stage of the project. Parts of the article were read at the American Philosophical Association 1974 meetings in St. Louis and San Francisco, at the Boston Radical Philosophers' Group, at the Society for Ethical and Legal Philosophy, and at the University of Washington, Simon Fraser University, and the Massachusetts Institute of Technology. N. J. Block would like to thank the Old Dominion Foundation for a fellowship which facilitated the completion of this essay. Gerald Dworkin would like to express his gratitude to the Battelle Seattle Research Center for a Visiting Fellowship and an ideal environment for intellectual work.

While this essay is a collective effort, N. J. Block had primary responsibility for those portions dealing with theoretical and conceptual aspects of IQ and heritability, G. Dworkin for those portions dealing with social and ethical implications.

Part I

1. This is reprinted in A. R. Jensen, *Genetics and Education* (New York: Harper & Row, 1972), pp. 69–204. Hereafter referred to as *Gen. & Ed.*

2. A. R. Jensen, "How Much Can We Boost IQ and Scholastic Achievement?" *Atlantic Monthly*, September 1971, pp. 43–64.

3. R. J. Herrnstein, *IQ in the Meritocracy* (Boston: Little, Brown & Co., 1973). Hereafter referred to as IQM.

4. A. R. Jensen, *Educability and Group Differences* (New York: Harper & Row, 1973). Hereafter referred to as *Ed. & Grp. Diff.*

5. J. A. Fodor, *Psychological Explanation: An Introduction to the Philosophy of Psychology* (New York: Random House, 1968), pp. xiv–xv.

6. When we ask whether IQ tests measure intelligence, we are asking whether IQ tests measure *mainly* intelligence, not whether IQ tests are *perfect* measures of intelligence.

7. C. Bereiter, "Race and IQ: Blaming the Psychometricians," *Contemporary Psychology* 18, no. 10 (1973): 455–456.

8. While Jensen's explicit message applies to "the disadvantaged" and the lower classes, blacks are often singled out. (*See Gen. & Ed.*, p. 234, for example.) Jensen's argument is based on the claim that the lower classes are deficient in *capacity* for conceptual and cognitive learning and for complex problem-solving (as measured by IQ tests), while they are about equal in associative learning abilities (as indexed by memory tasks). Though we have grave doubts about the scientific merit of Jensen's claim, we do not think his educational suggestion is totally without merit. The lower IQ scores of lower-class children may indicate they are deficient to some degree in certain conceptual *skills*. Thus, they might learn arithmetic more easily if it is taught in the traditional way rather than via the abstract approach of the New Math.

9. W. Shockley, "Dysgenics, Geneticity, Raceology: A Challenge to the Intellectual Responsibility of Educators," *Phi Delta Kappan*, January 1972, pp. 297–307.

10. We use "IQ tests" to refer to those individually administered "intelligence" test best exemplified by the Stanford-Binet and Wechsler scales, and the similar group tests.

11. *See* A. Anastasi, *Psychological Testing*, 3rd ed. (New York: Macmillan Co., 1968) and L. J. Cronbach, *Essentials of Psychological Testing*, 3rd ed. (New York: Harper & Row, 1970).

12. For psychometric viewpoints different in many respects both from those criticized here and also our own, *see* C. Burt, "A Note on the Theory of Intelligence," *British Journal of Educational Psychology* 28 (1958): 281–289; L. J. Cronbach and P. Meehl, "Construct Validity in Psychological Tests," *Minnesota Studies in the Philosophy of Science* 1 (1956): 174–205; and L. J. Cronbach, "Test Validation" in *Educational Measurement*, ed. R. L. Thorndike (Washington, D.C.: American Council on Education, 1970), pp. 443–507, especially pp. 462–484. Roughly speaking, Cronbach tends toward an instrumentalist position, while our position is better categorized as realist. The arguments in this section are directed not against instrumentalism, but against operationalism. For discussion of these "isms" *see* chap. 5 of J. J. C. Smart's *Between Science and Philosophy* (New York: Random House, 1967).

13. *Gen. & Ed.*, pp. 75–76. R. J. Herrnstein makes essentially the same point in *IQM*, p. 107, quoted below (*see* note 31). This sort of operationalism is common in areas of psychometrics other than IQ testing. See R. B. Cattell, "Personality Pinned Down," *Psychology Today*, July 1973, pp. 40–46.

14. A further absurdity they are committed to is that if all thermometers mal-

functioned so as to always read, e.g., 32 degrees F, then all temperatures of measured objects would *be* 32 degrees F.

15. The tests they have in mind when they speak of "intelligence tests" are, of course, IQ tests.

16. For example, the following phrases have the same reference (red) but different meanings: "the color which supposedly excites bulls" and "the color of blood." On the other hand, the English word "red" has both the same meaning and the same reference as the German word "rot."

17. We use "IQ" to refer to whatever quantity or quantities standard IQ tests measure, if anything. We are using this definition not to avoid the question whether IQ so defined is intelligence, but rather to allow us to *ask* this question.

18. *See*, for example, chap. 7 of C. G. Hempel, *Philosophy of Natural Science* (Englewood Cliffs, N.J.: Prentice-Hall, 1966).

19. We will, however, mention one respect in which the application of operationalist doctrine in the social sciences leads to even greater absurdity than in the physical sciences. Percy Bridgman, a physicist, first formulated the modern version of operationalism in detail. *See* his *Logic of Modern Physics* (New York: Macmillan Co., 1927). One of his problems was that for every new way of measuring a quantity, he was required by his doctrine to introduce a new "concept." For example, corresponding to the method of measuring length by measuring rods, he postulated "tactual length"; corresponding to measurement by optical triangulation, he postulated "optical length," and so on. Now all these different ways of measuring length produce the same result, at least in principle. (Accounting for this fact was another serious problem for Bridgman.) But this is not true of the different methods for (supposedly) measuring intelligence. Thus those who accept Boring's definition are stuck not only with as many *concepts* of "intelligence" as there are IQ tests, but almost as many *values* of intelligence as well. Jensen attempts to avoid this difficulty by specifying that by "intelligence" he really means not IQ but *g* or *general intelligence*. (*See* section VIII.)

20. Herrnstein, "IQ," pp. 48–49.

21. We will adopt the somewhat extended use of "measure" common in psychometrics, according to which a test measures *x* insofar as it correlates with *x* (and also satisfies certain formal constraints). This usage has counterintuitive consequences which we shall usually ignore.

22. For a good discussion of this point, *see* A. Heim, *Intelligence and Personality* (Baltimore: Penguin Books, 1971), p. 112.

23. L. M. Terman and Q. McNemar, *Terman-McNemar Test of Mental Ability: Manual of Directions* (Tarrytown-on-Hudson, N.Y., 1941), pp. 3–4.

24. T. Kuhn, "The Function of Measurement in Modern Physical Science," *Isis* 52, no. 8 (1961): 161–193.

25. Historical facts mentioned above are drawn mainly from A. Wolf, *A History of Science, Technology and Philosophy in the Sixteenth and Seven-*

teenth Centuries (London: George Allen & Unwin, 1950); D. Roller, "The Early Development of the Concepts of Temperature and Heat: The Rise and Decline of the Caloric Theory," in *Harvard Case Histories in Experimental Science*, ed. J. B. Conant (Cambridge, Mass.: Harvard University Press, 1948), pp. 117–215; and W. E. K. Middleton, *A History of the Thermometer and Its Use in Meteorology* (Baltimore: Johns Hopkins University Press, 1966).

26. *See* the quote from Herrnstein cited in note 20. On the other hand, Binet did do theoretical work that was good for its time, even if it had little effect on his tests. *See* the article by Binet and Henri in *A Source Book in the History of Psychology*, ed. R. Herrnstein and E. Boring (Cambridge, Mass.: Harvard University Press, 1965).

27. F. S. Freeman, *The Theory and Practice of Psychological Testing*, 3rd ed. (New York: Holt, Rinehart & Winston, 1962), p. 201.

28. H. J. Eysenck, *Know Your Own IQ* (Baltimore: Penguin Books, 1962), p. 8.

29. D. Wechsler, "Intelligence, Definition, Theory and the IQ," in *Intelligence*, ed. R. Cancro (New York: Grune & Stratton, 1971), p. 52. *See also* L. E. Tyler, *The Psychology of Human Differences*, 3rd ed. (New York: Appleton-Century-Crofts, 1965), p. 62.

30. A. R. Jensen, "Can We and Should We Study Race Differences?" in *Compensatory Education: A National Debate*, ed. J. Hellmuth, vol. 3, *The Disadvantaged Child* (New York: Brunner-Mazel, 1970), p. 127, emphasis added. (Factor analysis and *g* are discussed in section VIII.) Of course, even if the measurements constitute proper data, one cannot conclude they are measurements of what they are supposed to measure. See the passage cited in note 13.

31. *IQM*, p. 107. This quotation is less clearly operationalist than the one from Jensen. The sentence about verbal ability and logical power could be read as indicating that Herrnstein sees only a very limited role for stipulation, namely, the role of deciding precisely *which* aptitudes ought to be measured by a test of intelligence. The reader might suppose that there is theoretical work on the basis of which psychometricians have measured verbal ability, logical power, and other aptitudes, and that they have recourse to stipulation only in order to decide which of these aptitudes are the ones an intelligence test ought to measure. But there is no such theoretical work. Indeed, when Herrnstein discusses this matter in his book, he gives the reader the (correct) impression that the use of terms like "verbal ability" and "logical power" to refer to what these tests measure is *itself* dependent on a stipulation (*see IQM*, p. 93). Further, as Herrnstein points out, the correlational analysis he speaks of there depends on arbitrary conventions in other respects: e.g., the choice of which tests to subject to factor analysis and the choice of which of the alternative methods of factor analysis to use. Thus, it does seem that from Herrnstein's position, the claim that IQ tests measure intelligence is to be defended mainly by appeal to convention or stipulation.

32. H. Putnam, "A Philosopher Looks at Quantum Mechanics," in *Beyond the*

Edge of Certainty, ed. R. G. Colodny (Englewood Cliffs, N.J.: Prentice-Hall, 1965), p. 76.

33. This example is from H. J. Eysenck, *The Inequality of Man* (London: Maurice Temple-Smith, 1973), p. 68.

34. L. M. Terman and M. Merrill, *Stanford-Binet Intelligence Scale*, 3rd ed. (Boston: Houghton Mifflin Co., 1960), emphasis added.

35. *See* Eysenck, *Inequality of Man*, pp. 45, 69. He goes on to argue that the popular sense of "intelligence" is the same as the psychometric sense.

36. Quoted (apparently approvingly) in *Gen. & Ed.*, pp. 82–83. The original source is O. D. Duncan, D. L. Featherman, and B. Duncan, *Socioeconomic Background and Occupational Achievement: Extensions of a Basic Model.* Final Report, Project no. 5-0074 (EO-191), U.S. Department of Health, Education and Welfare (May 1968).

37. S. Bowles and V. Nelson in "The 'Inheritance of IQ' and the Intergenerational Reproduction of Economic Inequality," *Review of Economics and Statistics*, February 1974, pp. 39–51, report that an average of parental income, father's schooling, and father's occupational status predicts a child's eventual occupational status substantially better than a 12-year-old IQ does.

38. N. Chomsky, "The Fallacy of Richard Herrnstein's IQ," *Cognition* 1, no. 1 (1972): 38. Reprinted in this volume, pp. 285–294.

39. On the other hand, many "nonintellective" tests are not good predictors. There is enormous disagreement about just how good so-called personality and motivation tests are as predictors, and how much they add to the predictive power of IQ and aptitude tests. Cf. Cronbach, *Essentials of Psychological Testing*, p. 549, and Tyler, *Psychology of Human Differences*, p. 117, for example. In many publications over the last ten years, R. B. Cattell (the foremost exponent of nonintellective prediction) and his associates have claimed that tests of personality and (some tests of) motivation both correlate with measures of achievement about as well as IQ tests do, and further, that the three kinds of tests (IQ, personality, motivation) "overlap" only slightly with one another.

40. "A Non-intellectual Intelligence Test," *Journal of Consulting Psychology* 17 (1953): 242–246.

41. The fact that high-school grades predict college grades better than any IQ or aptitude test has been repeatedly confirmed. Tyler, *Psychology of Human Differences*, p. 114.

42. Eysenck, *Inequality of Man*, p. 72 (emphasis added). *See also* Jensen's "The Heritability of Intelligence," *Saturday Evening Post* 244, no. 4 (1972): 9.

43. Unless otherwise stated, these figures are based on a review by C. Jencks et al., *Inequality: A Reassessment of the Effect of Family and Schooling in America* (New York: Basic Books, 1972), Appendix B. Jencks assumes that measurement error acts independently on the variables whose correlation is being examined, so as to decrease observed correlations by about 10 percent. The figures reported have been increased to offset this phenomenon.

One should distinguish between "predictive" and "concurrent" correlations. For example, consider the correlation of 6-year-old IQ with schooling, and IQ in the last year of school with schooling: the former would be predictive, the latter concurrent. Concurrent correlations are generally higher than predictive correlations. The correlations reported here are based primarily on concurrent data.

44. Tyler, *Psychology of Human Differences,* p. 73.

45. Variance predicted corresponds to the square of the correlation coefficient. We assume here that a linear relation between IQ and intelligence obtains, though we know of no real reason to believe that it does. Without this assumption, inferences to IQ-intelligence correlations of the sort we are examining would be even weaker.

46. Indeed, the correlation of success with intelligence would have to be .87 or higher in order to mathematically guarantee *any* correlation between IQ and intelligence. If a, b, and c are correlations among any three variables, it must be the case that $a^2 + b^2 + c^2 - 2abc$ is less than or equal to 1.

47. Cronbach, *Essentials of Psychological Testing,* p. 281; Anastasi, *Psychological Testing,* p. 392. While we agree with the conclusion, we do not accept one of the arguments often given for it, namely that IQ and achievement tests can be seen to measure the same quantities simply in virtue of their high correlation. (*See* section VIII.)

48. David McClelland, "Testing for Competence Rather than for Intelligence," *American Psychologist* 28 (January 1973): 1–14.

49. *Encyclopedia of Education,* vol. 2 (New York: Macmillan Co., 1971), p. 23. *See also* McClelland, "Testing for Competence," p. 2; Jencks et al., *Inequality,* pp. 186–187; H. J. Butcher, *Human Intelligence: Its Nature and Assesment* (London: Methuen & Co., 1968), p. 290; Tyler, *Psychology of Human Differences,* p. 122; and D. Hoyt, "The Relationship Between College Grades and Adult Achievements: A Review of the Literature," *American College Testing Program, Research Report* no. 7 (Iowa City, 1965).

50. "The Creative and Other Contributions of One Sample of Research Scientists," in *Scientific Creativity: Its Recognition and Development,* ed. C. W. Taylor and F. Barron (New York: John Wiley & Sons, 1963), p. 73. This article is especially impressive because of the wide range of grades in the sample. *See also* McClelland's reply to criticisms, *American Psychologist,* January 1974, p. 59.

51. A. R. Jensen, "Another Look at Culture Fair Testing," in *The Disadvantaged Child,* vol. 3 (cited note 30), p. 63.

52. McClelland, "Testing for Competence," and the study by Wing and Wallach cited there.

53. The only data bearing on this we know of is Yerkes' report that Army Alpha tests given in World War I, before score-causation mechanisms existed, correlated highly with number of years of schooling. But it should be noted that these tests were given to adults who had almost all finished their schooling. Since Army Alpha tests were tests of school learning to an

even greater degree than modern IQ tests, their correlation with years of school should not be too surprising.

54. *Inequality*, pp. 144–145. Jencks's model assumes, of course, that background factors also have a causal effect on IQ.

55. *Gen. & Ed.*, p. 84. *See also* B. Eckland, "Social Class Structure and Genetic Basis of Intelligence," in *Intelligence* (cited note 29); C. J. Bajema, "A Note on the Interrelations among Intellectual Ability, Educational Attainment and Occupational Achievement," *Sociology of Education* 41 (1968): 317–319; Jencks et al., *Inequality*, Appendix B; and Bowles and Nelson, "The 'Inheritance of IQ.' " Both path analysis and linear-regression analysis presuppose that relations between variables are roughly linear. Jencks has investigated the effect of nonlinear relations among the variables IQ, occupational status, education, and income. He found that taking into account the degree of nonlinearity which obtains changes his results only slightly. *See* Jencks et al., *Inequality*, pp. 336–337, and Jencks, "Perspectives on Inequality," *Harvard Educational Review* Reprint no. 8 (1973), p. 111.

56. *See* note 37. Bowles and Nelson also show that the causal effect of socioeconomic background on both occupation and income is two to three times as large as the causal effect of 12-year-old IQ on occupation and income. Calculations in Jencks et al., *Inequality*, Appendix B, are in substantial agreement with these results, though Jencks sometimes ascribes slightly more causal efficacy to IQ. It seems to us that the work by Jencks and Bowles and Nelson represents an advance over earlier work by Duncan which overestimated the causal contribution of IQ to occupational status. Duncan himself now thinks that, roughly speaking, the "truth" lies somewhere between his earlier estimates and those of Bowles and Nelson—as far as he can tell on the basis of current evidence (personal communication, 1974).

57. J. Cattell, "Mental Tests and Measurements," *Mind* 15 (1890): 373: *IQM*, chap. 1.

58. G. D. Stoddard, *The Meaning of Intelligence* (New York: Macmillan Co., 1945), p. 95.

59. D. Wechsler, *The Measurement of Adult Intelligence*, 1st ed. (Baltimore: Williams & Wilkins, 1939), p. 78.

60. L. M. Terman and M. Merrill, *Measuring Intelligence* (Boston: Houghton Mifflin Co., 1937), p. 7.

61. F. Warburton, T. Fitzpatrick, J. Ward, and M. Ritchie, "Some Problems in the Construction of Individual Intelligence Tests," in *Readings in Human Intelligence*, ed. H. J. Butcher and D. E. Lomax (London: Methuen & Co., 1972).

62. L. M. Terman et al., *The Stanford Revision and Extension of the Binet-Simon Scale for Measuring Intelligence* (New York: Warwick & York, 1917), p. 149.

63. Some steps have been taken toward constructing a Binet-type IQ test using tasks Piaget takes to be indicative of basic intellectual development. See Cronbach, *Essentials of Psychological Testing*, p. 244.

64. For example, *see* Wechsler's discussion of his information test in the current WAIS manual, *Measurement and Appraisal of Adult Intelligence*, 4th ed. (Baltimore: Williams & Wilkins, 1958), pp. 65–67.

65. We caution the reader not to suppose that just any old items on which some persons would have an advantage can serve as a basis for the kind of criticism we are making here. Only advantages that tend to add up will do.

66. *See* Cronbach, *Essentials of Psychological Testing*, pp. 60–65, 148, 238–239, 245, 249, 625–630.

67. Both quoted in J. Baker, *Race* (Oxford: Oxford University Press, 1974), p. 447.

68. G. Evans, "Intelligence, Transfer, and Problem Solving," in *On Intelligence: Symposium on Intelligence, Toronto, 1969*, ed. B. W. Dockrell (New York: Barnes & Noble, 1970): p. 217.

69. M. Wallach and N. Kogan, "A New Look at the Creativity-Intelligence Distinction," in *Readings in Human Intelligence* (cited note 61), pp. 131–149.

70. K. W. Eels et al., *Intelligence and Cultural Differences* (Chicago: University of Chicago Press, 1951); p. 29. *See also* Jensen, *Ed. & Grp. Diff.*, p. 296.

71. Wechsler's observations are found in *Measurement and Appraisal of Adult Intelligence*, pp. 81–82, 69, 78. See also Herrnstein's chapter on IQ in Roger Brown and R. J. Herrnstein, *Pscychology* (Boston: Little Brown and Company, 1975), especially p. 500. Herrnstein's emphasis here is quite different from his earlier writings.

72. Anastasi quotations are from "Psychology, Psychologists, and Psychological Testing," *American Psychologist* 22 (1967), pp. 304, 299.

73. *See* Wechsler, *Measurement and Appraisal of Adult Intelligence*, p. 14; and Wechsler, "Cognitive, Conative and Non-intellective Intelligence," in *Intelligence: Some Recurring Issues*, ed. L. E. Tyler (New York: Van Nostrand Reinhold Co., 1969), p. 71.

74. Jencks et al., *Inequality*, Appendix B; L. J. Kamin, *The Science and Politics of IQ* (Potomac, Md.: Erlbaum Associates, 1974). A recent article by Jensen, "Kinship Correlations Reported by Sir Cyril Burt," in *Behavior Genetics* 4, no. 1 (March 1974), agrees with Kamin that much of the data collected by Burt is questionable.

75. S. Vandenburg, "The Nature and Nurture of Intelligence," in *Genetics*, ed. D. C. Glass (New York: Rockefeller University Press, 1968). The data which support these estimates, however, are rather weak and the validity of the ability tests seems at least as doubtful as the validity of IQ tests.

76. Jensen's claims seem to be well known to government officials and may have played some role in the current decline in support of programs such as Head Start.

77. Those who are inclined to disregard this sort of remarks ought to read Tom Cottle's "Look What They Done to My Score," *Science for the People*, March 1974, pp. 26–31, and "What Tracking Did to Ollie Taylor," *Social Policy*, July–August 1974, pp. 21–24.

78. The application of this point to color bias is that blacks do count as a randomly chosen IQ 85 group for our purposes because there is no reason to think their intelligence is lower than the IQ 85 population as a whole. But if we suppose the IQ-intelligence correlation is, say, .5, and that blacks have an average intelligence equal to their expected intelligence of 92.5—they do not count as a randomly chosen group of intelligence 92.5—since they were chosen with the knowledge that they averaged 85 in IQ. That is, there *is* reason to believe their IQ is lower than the 92.5 intelligence population as a whole.

79. Anastasi, *Psychological Testing,* p. 559. For good discussions of the incompatibility between individual and group notions of race or status fairness in terms of the psychometric notion of fairness, *see* R. L. Thorndike, "Concepts of Culture Fairness," *Journal of Educational Measurement* 8 (1971): 63–70; F. L. Schmidt and J. E. Hunter, "Racial and Ethnic Bias in Psychological Tests," *American Psychologist* 29 (January 1974): 1–8.

80. However, IQ tests seem to predict equally well for middle-class blacks and whites, at least as far as college grades are concerned. See *Ed. & Grp. Diff.,* p. 294.

81. Jane Mercer, "IQ: The Lethal Label," *Psychology Today* September 1972, pp. 44–47, 95–97. The differences are sufficiently large that questions about a few indicators are quibbles. A number of other studies have obtained similar results. *See* G. D. Cooper, M. York, P. G. Daston, and H. B. Adams, "The Porteus Test and Various Measures of Intelligence with Southern Negro Adolescents," *American Journal of Mental Deficiency* 71 (1967): 787–792. Arthur Jensen's account of such phenomena would presumably be that associative abilities are more equally distributed among race and status groups than conceptual abilities, and that adaptive tasks, such as Mercer's, tap associative abilities.

82. There are so-called culture-free tests, but these tend to correlate only .4 to .5 with Stanford-Binet IQ for later elementary school grades and above (*see* Freeman, *Theory and Practice of Psychological Testing,* p. 377). What they measure is even less clear than for standard IQ tests. The justification of these tests as measures of intelligence relies heavily on factor-analytic arguments of the sort we criticize in the next section. Whether or not these tests are culture-free, there is a good deal of evidence they are not culture-fair. In *Intelligence and Cultural Environment* (London: Methuen & Co., 1969), p. 25, P. E. Vernon says "the majority of contemporary psychologists have concluded, there is no such thing as a culture-fair test."

83. *See* Eels et al., *Intelligence and Cultural Differences,* p. 21; Cronbach, *Essentials of Psychological Testing,* pp. 60–65, 302–307; D. Johnson and W. Mihal, "Performance of Blacks and Whites in Computerized versus Manual Testing Environments," *American Psychologist* 28 (August 1973): 694–699.

84. Chap. 4 of Terman et al., *Stanford Revision of Binet-Simon Scale.*

85. To avoid the confusing use of different terms, we shall usually use the term "general intelligence." It is thought of by some of its proponents as an

ability which enters into all intellectual tasks. Others now in the majority think of it more narrowly, e.g., as an ability common to all complex problem-solving and reasoning.

86. Wechsler, *Measurement and Appraisal of Adult Intelligence*, pp. 9–11.

87. On the other hand, the usual practice of simultaneously making subtests different from one another by attempting to minimize correlations between subtests cannot hurt, even if general intelligence exists.

88. J. P. Guilford, *The Nature of Human Intelligence* (New York: McGraw-Hill Book Co., 1967), p. 29. We doubt these tests are valid, but this line of defense seems closed to Herrnstein.

89. Anastasi, "Psychology, Psychologists, and Psychological Testing," p. 300.

90. Burt, "The Genetics of Intelligence," in *On Intelligence* (cited in note 68), emphasis added. Guilford's own theory has been attacked as subjective recently by J. L. Horn and J. R. Knapp, "On the Subjective Character of the Empirical Base of Guilford's Structure of Intellect Model," *Psychological Bulletin* 80, no. 1 (1973): 33–43.

91. C. Burt, "The Evidence for the Concept of Intelligence," in *Intelligence and Ability*, ed. S. Wiseman (Baltimore: Penguin Books, 1967), pp. 260–282.

92. Freeman, *Theory and Practice of Psychological Testing*, p. 214; Terman and Merrill, *Stanford-Binet Intelligence Scale*, p. 35.

93. More precisely, the argument need not suppose that the arithmetical items measure arithmetical ability, but only that the various items measure a wide variety of abilities, and that the nonability components can be expected to cancel out over the whole test.

94. J. P. Guilford, "The Structure of Intellect," *Psychological Bulletin* 53 (1956): 267–293.

95. J. P. Guilford, "Three Faces of Intellect," in *Studies in Individual Differences*, ed. J. J. Jenkins and D. G. Patterson (New York: Appleton-Century-Crofts, 1961), pp. 756–772.

96. This sort of view was first put forward by G. H. Thomson in *The Factorial Analysis of Human Ability* (Boston: Houghton Mifflin Co., 1939).

97. Cronbach, "Test Validation," p. 470.

98. Cronbach, *Essentials of Psychological Testing*, pp. 310–311 (emphasis added). We are not saying that Cronbach is a behaviorist, but only that his views overlap with behaviorism insofar as fictionalism with respect to abilities is concerned. The "about" in the middle of Cronbach's second paragraph is included for technical reasons which we will not go into here.

99. *Ibid.*, pp. 334–335. It should be clear that at least as we are using the terms, operationalism is compatible with fictionalism. Indeed, it is hard to see how a proponent of the former could rationally avoid being a proponent of the latter. However, fictionalists often take the two views to be incompatible. What operationalists assert that most fictionalists deny is that theoretical terms can be explicitly defined in terms of observables. We agree with (most) fictionalists on this issue.

100. Fictionalism with respect to abilities may seem more plausible than it is because the word "ability" is used in two quite different ways. The term "arithmetical ability" can mean both *current* arithmetical *skill*, and also arithmetical *aptitude*. Fictionalism may seem more plausible on the former reading, but it is the latter reading on which the social and educational role of "ability" tests is based, and which is needed to justify claims such as Jensen's that general intelligence is something like a capacity for abstract reasoning and problem solving.

101. *See*, for example, J. J. C. Smart, *Philosophy and Scientific Realism* (London, 1963); or Grover Maxwell, "The Ontological Status of Theoretical Entities," in *Scientific Explanation, Space, and Time*, ed. H. Feigl and G. Maxwell, vol. 3, *Minnesota Studies in Philosophy of Science* (Minneapolis: University of Minnesota Press, 1962), pp. 3–28.

102. In our view, a person's having a certain ability is a mental state or condition which—like beliefs and preferences (though perhaps not aches and pains)—can be characterized functionally in terms of its causal relations to sensory inputs; behavioral outputs; and *other mental conditions, states, and events*. It is the causal nature of the view and its characterization of mental phenomena partly in terms of *other* mental phenomena which makes functionalism incompatible with behaviorism. For a criticism of behaviorism and a discusson of one verson of functionalism, *see* N. J. Block and J. A. Fodor, "What Psychological States Are Not," *Philosophical Review* 81, no. 2 (April 1972).

Part II

1. This example is due to R. C. Lewontin. See also R. J. Herrnstein, *IQ in the Meritocracy* (Boston: Little, Brown & Co., 1973), p. 176. Hereafter cited as *IQM*.

2. It makes sense to speak of the heritability of a property because a property can be construed as a numerical quantity whose value is 1 when the property obtains and 0 when it does not.

3. This example is due to A. R. Jensen, *Genetics and Education* (New York: Harper & Row, 1972), p. 120. Hereafter cited as *Gen. & Ed.*

4. A person's phenotypic height is his actual height. A person's genotypic height is the average phenotypic height persons with his genes would have if distributed randomly over environments in the population in question. Genotypic height, like heritability, is a population statistic. What your genotypic height is depends on what population you are considered to be a member of, as well as on your genes.

5. The example is due to R. J. Herrnstein.

6. Jensen uses a claim of high heritability of IQ in the white population as one support of his claim that black-white differences are partly genetic.

See *Gen. & Ed.*, pp. 159–162. The point we sketch here was made in detail in R. C. Lewontin's reply to Jensen: "Race and Intelligence," *Bulletin of the Atomic Scientists* 26 (March 1970): 2–8.

7. Strictly speaking, what we require is that the distribution of the genotypic IQs of the embryos across environments mirror the distribution in the general populaton, and that there is no correlation between the environments of paired embryos.

8. The main assumption, which we make in this paragraph and the next two, but not thereafter, is that the variance in IQ can be divided without remainder into a component due to genetic variation and a component due to environmental variation. Jensen claims this assumption is substantially correct; most of his critics deny it.

9. *See* Cronbach's "Heredity, Environment, and Educational Policy," in *Environment, Heredity, and Intelligence, Harvard Educational Review* Reprint no. 2 (Cambridge, 1969), p. 195.

10. A. Anastasi, *Psychological Testing*, 3rd ed. (New York: Macmillan Co., 1968), pp. 209, 575–576; L. E. Tyler, *The Psychology of Human Differences*, 3rd ed. (New York: Appleton-Century-Crofts, 1965), p. 463. Jensen suggests that this improvement may be due largely to a genetic cause: hybrid vigor ascribable to increases in "outbreeding" since the turn of the century (see *Environment, Heredity, and Intelligence*, p. 230). But his account seems doubtful given that a number of studies have shown that marked changes in educational opportunity in isolated groups produce marked increases in the IQ level of the school population. For example, Tennessee mountain children averaged 11 points higher in 1940 than children largely from the same families in 1930; in the intervening years, educational and economic improvement was considerable. See Tyler and Anastasi, cited above.

11. N. E. Morton, "Analysis of Family Resemblance I," and D. Rao, N. E. Morton, and S. Yee, "Analysis of Family Resemblances II," both in *American Journal of Human Genetics* 26 (1974): 318–330 and 331–359, respectively. Morton estimates "narrow" heritability (*see* note 18) rather than "broad" heritability, the quantity we have been talking about. He does not think broad heritability can be estimated on current data. Authors who estimate both quantities rarely take them to differ by much. For example, Jensen says the best estimates of narrow and broad heritability are .71 and .81, respectively. *See* his *Educability and Group Differences* (New York: Harper & Row, 1973), pp. 172, 178 (Hereafter cited as *Ed. & Grp. Diff.*). While we think that the difference between Morton's estimates of H for adults and children is suggestive, Morton's—like all other estimates of H—depend on a causal model with many unjustified features.

12. L. J. Kamin, *The Science and Politics of IQ* (Potomac, Md.: Erlbaum Associates, 1974). *See* also the articles by Kamin reprinted in this book on pp. 242–264 and 374–382. However, we do not agree with Kamin's conclusion that the data are insufficient to lead a prudent man to reject the "null hypothesis" that H is zero. The hypothesis that H is zero has no more claim to be regarded as the null hypothesis than the hypothesis that H is any

other number. When data are worthless, one should conclude that no estimate can be made.

13. *Ibid.*, chap. 3.

14. The treatment by Christopher Jencks et al., in *Inequality: A Reassessment of the Effect of Family and Schooling in America* (New York: Basic Books, 1972), Appendix A, is rather candid in this regard.

15. *Ibid.*, pp. 67–69. We have phrased these points in terms of abilities and intelligence, not IQ. But the same mechanisms are likely to operate with respect to other characteristics which IQ tests probably also measure to a degree, e.g., personality and motivational characteristics.

16. We are indebted to Jencks et al. (*Inequality*, pp. 66–67) for the idea of this example, as well as for some of the distinctions we will be making in this section.

17. *See* K. J. Hayes, "Genes, Drives and Intellect," *Psychological Reports* 10 (1962): 299–342, for a discussion of one possible mechanism. *See also* Jencks et al., *Inequality*, p. 68.

18. The definition of heritability as the proportion of variance which is genetically caused is often called "broad heritability" by population geneticists, who also speak of "narrow heritability." Narrow heritability is the proportion of the variance which is genetic but not due to various kinds of interactions among genes—interactions which can produce parent-child differences even in characteristics with 100 percent broad heritability, as when two brown-eyed parents have a blue-eyed child. Narrow heritability controls the genetic aspect of parent-offspring resemblance and is the relevant statistic for issues concerning selective breeding, eugenics, and dysgenics.

19. Used here in the usual sense of both direct and indirect genetic variance.

20. R. C. Lewontin, "The Analysis of Variance and the Analysis of Causes," *American Journal of Human Genetics* 26 (1974): 400–411. We are indebted to this article for many of the points in the rest of this section.

21. The upshot of these remarks can be graphically illustrated by thinking of the norm of reaction (the function from genotypes and environments to phenotypes) of a given characteristic as a surface in an *n*-dimensional space where *n*-1 of the dimensions are all the environmental and genetic variables and the *n*th dimension (the "height" of the surface) is the phenotypic value uniquely determined by the genetic and environmental variables. Calculations based on a given population with a particular distribution of environments and genotypes can only give one a local picture—represented by a small area of the surface which represents the norm of reaction. This small area of the norm of reaction will, in general, be a poor guide to other parts of the norm of reaction. Thus, heritability estimates for a particular population will be, in general, a poor guide to the results of changes in either environments or genotypes.

22. This assumption is sometimes expressed as: There is no genotype-environment interaction. The claim that IQ is a linear function of genotype and

environment entails that there is no interaction (though not vice versa). However, the claim that there is no genotype-environment interaction comes to much the same thing as the claim that IQ is a linear function of genotype and environment. For if there is no genotype-environment interaction, the environmental or genetic variables can always be rescaled so as to produce linearity. Indeed, for many environmental variables, the linear scale would be the most natural one.

23. *Gen. & Ed.*, p. 323. It is worth noting that the methods of detecting departures from parallel norms of reaction (interaction effects) that have been deployed are very insensitive (*see* Morton, "Analysis of Family Resemblance," p. 320). Morton (p. 357) and others have argued against the presence of substantial interaction variance on the grounds that if we use a model which *assumes* no interaction, we get a reasonably good "fit" to the data. But this only shows it is *possible* to explain the data without supposing substantial interaction. The question that has not been explored is whether interactionist models can explain the data equally well. Jencks et al. (*Inequality*, p. 266) have argued that estimates of the various components of IQ variance typically add up to *more* than 100 percent, while if there is substantial interaction, they should add up to less than 100 percent. But this argument depends on (mistakenly) taking the heritability data to be sufficiently reliable for quantitative hypothesis testing.

24. Arthur Jensen, "Kinship Correlations Reported by Sir Cyril Burt," *Behavior Genetics* 4, no. 1 (March 1974). Chap. 6 of Kamin's *Science and Politics of IQ* cogently criticizes other aspects of the evidence for parallel norms of reaction. If there is substantial interaction variance and if there is also substantial correlation between genotype and environment, then standard methods of estimating heritability are useless. *See* D. Layzer, "Heritability Analyses of IQ Scores: Science or Numerology?" *Science* 183 (March 29, 1974): 1259–1266.

25. The person associated in the public eye with these issues is Professor Shockley. But *see* Jensen, *Gen. & Ed.*, pp. 331–332.

26. R. Heber, *Rehabilitation of Families at Risk for Mental Retardation*, Regional Rehabilitation Center, University of Wisconsin (Madison, 1969). This study has been criticized on methodological grounds by E. B. Page, "Miracle in Milwaukee: Raising the IQ," *Educational Researcher* 1 (1972): 8–16. We find his criticisms tenuous at best; certainly, they do not provide the "detailed scrutiny" that Jensen claims they do. *See* Jensen, "The IQ Controversy: A Reply to Layzer," *Cognition* 1, no. 4 (1973).

27. Bereiter is instructive on the issue of "failure."

> To treat the eventual vanishing of preschool effects as failure is to imply either that preschool compensatory education is futile or that the effective method has yet to be discovered. Either of these conclusions *could* be true, but those who think they follow from current evidence are applying criteria of success to preschool education that are not applied in any other realm of human effort. They are asking the doctor for a pill they can take when they are ten that will prevent

them from getting fat when they are fifty. C. Bereiter, "Conclusions from Evaluation Studies," *Preschool Programs for the Disadvantaged*, ed. J. C. Stanley (Baltimore: Johns Hopkins University Press, 1971), p.13.

28. For a very detailed and careful analysis of this and other intervention programs for which we have systematic follow-up data on both experimental and control groups, see U. Bronfenbrener, *Is Early Intervention Effective?* (Ithaca, N.Y.: Cornell University Press, 1974). In general, the evidence indicates that failure to involve the family actively in the intervention program produces results that disappear fairly rapidly once the program ends.

29. M. Skodak and H. M. Skeels, "A Final Follow-up Study of One Hundred Adopted Children," *Journal of Genetic Psychology* 75 (1949): 85–125.

30. H. M. Skeels, "Adult Status of Children with Contrasting Early Life Experience," *Child Development Monographs* 31, no. 3 (1969), serial no. 105. It should be noted that IQ tests given before the age of three are very unreliable. In addition, regression effects will account for some portion of the gain.

31. C. Bereiter, "Genetics and Educability: Educational Implications of the Jensen Debate," in *Compensatory Education: A National Debate*, ed. J. Hellmuth, vol. 3, *The Disadvantaged Child* (New York: Brunner-Mazel, 1970), pp. 279–299. (Reprinted in this volume, pp. 383–407.)

Bereiter estimates that, assuming H = .80, compressing the whole distribution of environmental conditions into the top half of the existing distribution would reduce the standard deviation of IQ by about one point. This article is the best available summary on educational implications of heritability.

32. *Ibid.*, p. 288.

33. C. Bajema, "Estimation of the Direction and Intensity of Natural Selection in Relating to Intelligence by Means of Intrinsic Rate of Natural Increase," *Eugenics Quarterly* 10 (1963): 175–187; J. V. Higgins, E. W. Reed, S. C. Reed, "Intelligence and Family Size: A Paradox Resolved," *Eugenics Quarterly* 9 (1962): 84–90.

34. L. L. Cavalli-Sforza and W. F. Bodmer, *The Genetics of Human Populations* (San Francisco: W. H. Freeman & Co., 1971), p. 769.

35. *IQM*, pp. 197–198. A version of this argument was also advanced by Carl Bereiter in his response to Jensen, *Harvard Educational Review* 39, no. 2 (Spring 1969): 162–170.

36. R. J. Herrnstein, "Whatever Happened to Vaudeville?: A Reply to Professor Chomsky," *Cognition* 1, nos. 2–3 (1972): 301, 302. Reprinted in this book, *see* pp. 299–309.

37. S. Bowles and H. Gintis, "IQ in the U.S. Class Structure," *Social Policy*, January 1973, pp. 65–96; S. Bowles and V. I. Nelson, "The Inheritance of IQ and the Intergenerational Reproduction of Economic Inequality," *Review of Economics and Statistics* 56, no. 1 (1974): 39–51.

38. Bowles and Nelson, "Inheritance of IQ," p. 47.

39. *Ibid.*, p. 48.

40. J. H. Waller, "Achievement and Social Mobility: Relationships among IQ Score, Education and Occupation in Two Generations," *Social Biology* 18 (1971): 252–259.

41. O. D. Duncan, D. L. Featherman, and B. Duncan, *Socioeconomic Background and Occupational Achievement: Extensions of a Basic Model.* Final Report, Project No. 5–0074 (EO–191), U.S. Department of Health, Education and Welfare (May 1968).

42. Jencks et al., *Inequality,* p. 221.

43. A. R. Jensen, "Another Look at Culture-Fair Testing," *Western Conference on Testing Problems, Proceedings for 1968* (Berkeley, Calif., 1968), p. 63. Even this "surprisingly low" correlation is probably inflated. Ghiselli, who produced most of the data, was not very precise in describing how job proficiency was measured. Criterion contamination is therefore an obvious problem.

44. R. J. Herrnstein, "IQ," *Atlantic Monthly,* September 1971, p. 51.

45. I. E. Berg, *Education and Jobs: The Great Training Robbery* (New York: Praeger Publishers, 1970).

46. Jencks et al., *Inequality,* p. 187. *See also* P. Taubman and T. Wales, "Higher Education, Mental Ability, and Screening," *Journal of Political Economy,* January–February 1973, pp. 29–55. They present empirical evidence that up to half of the earnings differentials attributed to education is due to its use as a screening device.

47. As an example of the unreliability of standard intelligence tests in predicting trainability, consider the study by M. Jensen, "Low Level Airman in Retesting and Basic Training: A Sociopsychological Study," *Journal of Social Psychology* 55 (1961): 155–190.

48. For an extensive discussion of this point, *see* N. Chomsky, "Psychology and Ideology," *Cognition* 1, no. 1 (1972).

49. N. Chomsky, "Comments on Herrnstein's Response," *Cognition* 1, no. 4 (1973).

50. *Ibid.* and H. J. Eysenck, *The Inequality of Man* (London: Maurice Temple-Smith, 1973), p. 219.

51. Blau and Duncan conclude on the basis of a study of four different age groups of men from nonfarm backgrounds that "there is some indication, although the evidence is by no means conclusive, that the influence of educational attainments on occupational achievements has increased in recent decades." Other of their data indicate "that the influence of social origins on career beginnings has not changed at all in the last 40 years." P. M. Blau and O. D. Duncan, *The American Occupational Structure* (New York: John Wiley & Sons, 1967), pp. 424–425.

52. Jencks et al., *Inequality,* p. 187.

53. For arguments for and against the view that native endowments should not be rewarded, *see* J. Rawls, *A Theory of Justice* (Cambridge, Mass.: Harvard

University Press, 1971) and R. Nozick, "Distributive Justice," *Philosophy and Public Affairs* 3, no 3 (Fall 1973).

54. *Psychology of Human Differences*, pp. 77–79.

55. E. E. Ghiselli, *The Validity of Occupational Aptitude Tests* (New York: John Wiley & Sons, 1966).

56. In discussing a society with a more egalitarian income distribution, Herrnstein characterizes it as "stratified by a mortal competition for prestige." Herrnstein, "Whatever Happened to Vaudeville?" p. 301.

57. A quote attributed to Jensen in the *New York Times*, August 31, 1969, p. 43. This interview is referred to by Jensen as "the most thorough, thoughtful and well-balanced story . . . eminently fair and of meticulous accuracy in summarizing the whole debate up to that time" (*Gen. & Ed.*, pp. 13–14).

58. No conspiracies are implied here, merely a theory, according to which, certain ideologies and theories that are always present find, at certain periods, a favorable environment for their propagation and development because they serve certain needs and interests.

Historically, there is impressive evidence linking the psychologists who brought the Binet Test to America, developed and promoted it (Goddard, Terman, Yerkes, Thorndike, Laughlin, Brigham) to the founding of the eugenics movement, the passing of sterilization legislation in thirty states, and the passage of selective immigration quotas by Congress in 1924 that clearly discriminated against southern and eastern Europeans. *See* C. Karier, "Testing for Order and Control in the Corporate Liberal State," *Educational Theory* 22, no. 2 (Spring 1972): 154–180; and Kamin, *Science and Politics of IQ*.

59. This type of interaction effect should be clearly distinguished from genetic-environmental interaction effects. What is involved here is finding kinds of training that are best for particular kinds of phenotypic aptitude. Genotypic aptitude is irrelevant.

60. Bereiter, "Genetics and Educability."

61. *IQ in the Meritocracy*, p. 187 (emphasis added)

62. Phenotypic differences cannot, of course, account for the fact that a black man with a father in the same occupation and with as much education as the average white winds up in an occupation 19 points below that of the average white, or that blacks who had not only the same amount of education as the average white, but also entered an occupation of the same status as the average white, had an income of 63 percent of the white mean (Jencks et al., *Inequality*, pp. 190, 217–218).

63. T. Nagel, "Equal Treatment and Compensatory Discrimination," *Philosophy and Public Affairs* 2, no. 4 (Summer 1973): 348–364.

64. R. M. O'Neil, "Preferential Admissions: Equalizing the Access of Minority Groups to Higher Education," *Yale Law Journal* 80, no. 4 (1971): 699–767.

65. Partly because of the concern about population growth, partly because of the backlash against welfare programs, there has been a sharp increase in discussions of eugenic sterilization. Professor Shockley, the Nobel-laureate

physicist at Stanford, proposed, at the 1971 convention of the American Psychological Association, to give a bonus to low-IQ parents who submit to voluntary sterilization. A more recent development along these lines is a book published in 1973 by D. J. Ingle entitled *Who Should Have Children? An Environmental and Genetic Approach* (Indianapolis: Bobbs-Merrill Co., 1973). Ingle is a professor of physiology at the University of Chicago and a member of the National Academy of Sciences and the American Academy of Arts and Sciences. In reply to an objection that his program of selective voluntary birth control is a proposal for Negro genocide, Ingle is careful to point out (p. 127): "In my opinion the great majority of Negroes—perhaps 75 percent—are qualified for parenthood."!!

Although this discussion is one of voluntary sterilization, the extension by more coercive means, begins to make its appearance as well. On March 4, 1971, a bill was introduced in the Illinois legislature which would give a welfare mother with two children (born on welfare) the choice of sterilization or loss of welfare. For a recent discussion, see *Eugenic Sterilization,* ed. J. Robitscher (Springfield, Ill.: Charles C. Thomas, 1973). *See also* J. Paul, "The Return of Punitive Sterilization Proposals: Current Attacks on Illegitimacy and the AFDC Program," *Law and Society* 3, no. 1 (August 1968).

66. Hirsch was one of the first to point this out. J. Hirsch, "Behavior-Genetic Analysis and Its Biosocial Consequences," *Seminars in Psychiatry* 2 (1970): 101–102.

67. For further evidence along these lines, *see* J. M. Thoday's review of Jensen's *Educability and Group Differences,* in *Nature,* October 26, 1973, pp. 418–420. (Reprinted in this book, pp. 146–155.)

68. For an example of a study frequently cited in popular discussions of IQ, and used by environmentalists to draw policy conclusions based on questionable data, *see* J. D. Elashoff and R. E. Snow, *"Pygmalion" Reconsidered: A Case Study in Statistical Inference: Reconsideration of the Rosenthal-Jacobson Data on Teacher Expectancy* (Worthington, O.: Charles A. Jones, 1971).

69. We want to make it clear that we are not accusing Herrnstein or Jensen of being racists—either conscious or unconscious. That epithet has been used far too often and far too loosely in the IQ controversy. On the other hand, it would be a mistake to ignore the blatant signs of racism that have been displayed by many of the participants, past and present, in this debate. We present the following as a sample: "Nature has color-coded groups of individuals so that statistically reliable predictions of their adaptability to intellectually rewarding and effective lives can easily be made and profitably be used by the pragmatic man in the street" (W. Shockley, "Dysgenics, Geneticity, Raceology: A Challenge to the Intellectual Responsibility of Educators," *Phi Delta Kappan,* January 1972, p. 307). Karier points out that the work of Terman and others was used to justify the restrictive immigration quota of 1924 which clearly discriminated against southern Europeans. He quotes the following from an address by Terman to the National Educational Association in 1923:

> The racial stocks most prolific of gifted children are those from northern and western Europe and the Jewish. The least prolific are the

> Mediterranean races, the Mexican and the Negroes. . . . It has been figured out that if the present differential birth rate continues, 1000 Harvard graduates will at the end of 200 years have but 50 descendents, while in the same period 1000 South Italians will have multiplied to 100,000. [L. M. Terman, "The Conservation of Talent," *School and Society* 19, no. 483 (March 29, 1924): 363]

Karier also quotes Professor Henry Garrett, the chairman of Columbia University's psychology department for sixteen years and president of the American Psychological Association. In 1966 he was arguing:

> You can no more mix the two races and maintain the standards of White Civilization than you can add 80 (the average IQ of Negroes) and 100 (average IQ of Whites), divide by two and get 100. What you would get would be a race of 90's, and it is that 10 percent differential that spells the difference between a spire and a mud hut; 10 percent or less is the margin of civilization's "profit"; it is the difference between a cultured society and savagery . . . [the state] can and should prohibit miscegenation, just as they ban marriage of the feeble-minded, the insane and the various undesirables." [H. E. Garrett, *Breeding Down* (Richmond, Va.: Patrick Henry Press, n.d.), pp. 10, 17]

70. In Jensen's case the policy maker can select from various positions held by Jensen. *See* G. Dworkin, "Two Views on IQs," *American Psychologist* 29 (1974): 465–467.
71. Chomsky, "Psychology and Ideology," p. 43.
72. In this context, it is interesting to note Jensen's awareness of the fact that he is willing to take certain risks regarding the level of supporting evidence required to advance a speculative hypothesis. Writing in the preface of his book on group differences he says:

> There are always differences among investigators working on the frontiers of a field. They differ in their weighting of items of evidence, in the range of facts in which an underlying consistency is perceived, in the degree of caution with which they will try to avoid possible criticisms of their opinions, and in the thinness of the ice upon which they are willing to skate in hopes of glimpsing seemingly remote phenomena and relationships among lines of evidence which might otherwise go unnoticed as grist for new hypotheses and further investigations. On all of these points we [Jensen and his critics] differ in varying degrees, and my own inclination is perhaps to be somewhat less conservative than would be some other students in dealing with the central topics of this book. My own reading of the history of science, however, leads me to believe that conservatism in generating hypotheses and in seeking means for testing them has not made for progress as often as a more adventurous approach. [*Ed. & Grp. Diff.*, pp. 3–4]

Being adventurous may be admirable when the risks of such adventures are borne by the adventurer or by others who consent to such risks. In this matter, Jensen faces only the risk of being wrong in his speculations; others face the risk of increased political, economic, and educational oppression.

73. W. F. Bodmer and L. L. Cavalli-Sforza, "Intelligence and Race," *Scientific American*, October 1970, p. 28. J. M. Thoday, "Limitations to Genetic Comparison of Populations," *Journal of Biosocial Science, suppl.* 1 (1969), p. 8.

74. S. Scarr-Salapatek, "Race, Social Class, and IQ," *Science* 174 (1971): 1285–1295.

Appendix

Mean. Average.

Variance and Standard Deviation. Both are measures of the "spread" of a distribution, i.e., the extent to which the values of the variables are distant from the mean rather than being clustered around it. The variance is the average of the squares of deviations from the mean. The standard deviation is the square root of the variance, or something like the average deviation from the mean. The variance is a convenient measure of variation for purposes of attributing portions of the variation to different causes.

Correlation. A measure of the covariation of two variables. Suppose we are interested in the covariation of food intake and weight in a large group of people. We make two lists: one lists the people in order of decreasing food intake; the other lists the people in order of decreasing weight. If the people are listed in exactly the same order on both lists, the correlation is 1.0; if the orders are exactly the reverse, the correlation is − 1.0; if there is no resemblance, the correlation is zero. Suppose the distributions of food intake and weight both have the regular "bell curve" shape that so many human characteristics have. Let us also convert the measures of food intake and weight into the kind of scale used in IQ tests: "standard scores." A person whose weight is at the fiftieth percentile i.e., (50 percent of the people are below him) is stipulated to have a standard score weight of 100; a person who is at the eighty-fourth percentile has a standard score weight of 115; and a person at the ninety-eighth percentile has a standard score weight of 130.

The correlation coefficient has two interesting properties. (1) Given a person's food intake, one can use the correlation coefficient to compute his expected weight, i.e., the best estimate of his weight given only his food intake. If the correlation is .5, a person's expected weight is 100 plus .5 times the difference between his food intake and 100. For example, a person with a food intake of

130 would have an expected weight (given only his food intake) of 115. A person who weighs 115 would have an expected food intake of 107.5.

(2) If the correlation coefficient is .5, we can deduce that at most 25 percent (or .5 squared) of the variance in weight is due to the variation in food intake. That is, even if all of the correlation of food intake with weight is due to the causal effect of food intake on weight, still other factors would contribute three times as much as food intake to variation in weight. Put crudely, the potential explanatory power of a correlation is given by the square of the correlation coefficient (always less in absolute value than the correlation coefficient itself), while the power of the correlation coefficient to predict one variable given the other is given by the correlation coefficient itself.

In what we have just said, we have assumed that the relation between weight and food intake is linear; x and y have a linear relation just in case the expected value of x (given only a value of y) is $ay + b$ where a and b are constants, and x and y are expressed in terms of deviations from their means. If this relation obtains, it can be shown that b equals 0 and a equals the correlation coefficient, r_{xy}, whch is defined as

$$\sum_{i=1}^{N} \frac{x_i y_i}{N\sigma^2}$$

where x_i is the deviation from the mean in the quantity x for the ith person, N is the number of people, and σ is the standard deviation (assumed the same for x and y).

Restriction of Range. Since the correlation coefficient measures the covariation of two magnitudes, the correlation of a variable with a constant (e.g., the number 17) is zero (a constant does not vary). The correlation of two variables usually decreases as the variation in one of them is decreased. For example, the correlation of food intake with weight among all adults is probably larger than it is among adults over 300 pounds, since there is less variation in weight in the population over 300 pounds. Further, the correlation might also decrease in the heavy group because perhaps the differences in weight in people above 300 pounds are more a matter of glands than of food intake. In sum, a correlation is only of interest relative to a specification of the range of variation of the variables involved. Many of the technical disagreements mentioned in the text are really disagreements about whether variables have been overly restricted.

Correlation and Causation. Even though neither exerts any causal influence on the other, x and y can correlate 1.0. For example, the high correlation may be due to both x and y being completely causally determined by z. Conversely, x can completely causally determine y even if the correlation between them is zero—if they are related in a complex manner. Differences in age may be the most important causal factor in determining differences in the size of a person's household. Nonetheless, the correlation may be low because children and the middle-aged tend to reside in larger households than young adults and the aged. Correlation coefficients give a false picture in such cases.

Index

Notes on Contributors

Mary Jo Bane is a research associate at the Harvard Center for Educational Policy Research.

Carl Bereiter is professor of education at the Ontario Institute for Studies in Education.

N. J. Block is assistant professor of philosophy at the Massachusetts Institute of Technology.

Noam Chomsky is professor of linguistics at the Massachusetts Institute of Technology.

Gerald Dworkin is associate professor of philosophy at the University of Illinois, Chicago Circle.

Arthur S. Goldberger is professor of economics at the University of Wisconsin.

Richard J. Herrnstein is professor of psychology at Harvard University.

Jerry Hirsch is professor of psychology and zoology at the University of Illinois, Champaign-Urbana.

Christopher Jencks is professor of sociology at Harvard University.

Arthur R. Jensen is professor of educational psychology at the University of California at Berkeley.

Leon J. Kamin is professor of psychology at Princeton University.

Clarence Karier is professor of history of education at the University of Illinois, Champaign-Urbana.

David Layzer is professor of astronomy at Harvard University.

Walter Lippmann was one of the leading columnists and commentators of the twentieth century.

Richard C. Lewontin is professor of biology and zoology at Harvard University.

David C. McClelland is professor of psychology at Harvard University.

Sandra Scarr-Salapatek is professor in the Department of School Psychology and the Institute of Child Development at the University of Minnesota.

Lewis Terman was professor of psychology at Stanford University.

J. M. Thoday is a Fellow of the Royal Society, and professor of genetics at the University of Cambridge.